Einstein and Soviet Ideology

ALEXANDER VUCINICH

Einstein and Soviet Ideology

Stanford University Press, Stanford, California

Stanford University Press
Stanford, California
© 2001 by the Board of Trustees of the
Leland Stanford Junior University

Library of Congress Cataloging-in-Publication Data

Vucinich, Alexander.
    Einstein and Soviet ideology / Alexander Vucinich.
        p.   cm. — (Stanford nuclear age series)
    Includes bibliographical references and index.
    ISBN 0-8047-4209-x (alk. paper)
    1. Communism and science—Soviet Union—History.   2. Communism
and science.   3. Soviet Union—Politics and government.   4. Einstein,
Albert, 1879–1955.   I. Title.   II. Series.

HX514.V83   2001
530'.0947'0904—dc21   2001032266

This book is printed on acid-free, archival-quality paper.

Original printing 2001

Last figure below indicates year of this printing:
10   09   08   07   06   05   04   03   02   01

Typeset in 10/12.5 Minion

To Reginald E. Zelnik

Preface

Einstein and Soviet Ideology is a historical study of one of the most momentous confrontations in the intellectual life of the Soviet Union: the conflict between Einstein's theory of relativity and official Soviet ideology, articulated by dialectical materialism, the Marxist philosophy of nature. It concentrates on the evolution of Soviet Marxist attitudes toward Einstein's alleged physical idealism and mathematical idealism. "Physical idealism" referred to the prominence Einstein attached to the role of such subjective categories as relativity, observer, and pure reason in the construction of physical reality as studied by the scientific community. "Mathematical idealism" was an ideological-philosophical category that referred to Einstein's alleged detachment of complex mathematical operations from their physical content.

The book also examines the evolution of Soviet interest in the humanistic aspects of Einsteinian thought, the philosophical interpretations of the ideological relations of the theory of relativity to conservation laws, the efforts to create a synthesis of Einsteinian and Marxist epistemological outlooks, the diversity of Soviet views on the general theory of relativity, the role of Russian nationalism in interpreting the historical antecedents of the theory of relativity, and the evolution of responses to relativistic cosmology. All topics devote special attention to the conflict between Marxist philosophers as defenders, articulators, and theoretical codifiers of official ideology and a group of leading physicists protecting the interests of science as a system of cultural values and ethical norms.

I am grateful to the Guggenheim Foundation for a fellowship in support of this study and to Professors Nicholas Riasanovsky and Reginald Zelnik for astute criticism, encouragement, and advice. I cannot thank enough Professors Martin Sherwin and James Millar for deep interest in my work. To Ethan Pollock, I am much indebted for help in tracing new sources directly related to archival material. Barbara Mnookin's astute and welcome editorial advice contributed to the improvement in the substance, logic, and writing style of this study. To my wife, Dorothy Vucinich, I am thankful for enthusiastic, devoted, and creative help in every phase of the study.

A.V.

Contents

Einstein and Soviet Ideology

Introduction

Official Soviet ideology, a central theme of this study, was a system of myths, doctrines, and values meant to create a world outlook in the spirit of Marxist theory as interpreted and promoted by the Soviet Communist Party and guarded by the coercive power of the Soviet state. No stretch of the historical imagination or logic is required to recognize four clearly hewn stages of ideological influence in the history of Soviet science from 1917 to 1991: the pre-Stalin age, the era of Stalinism, the post-Stalinist thaw, and Gorbachev's perestroika. Each stage exhibited distinct characteristics and stood out by the dominance of unique clusters of interpretive ideas. At the same time, they were all bound together by an explicit and amply illustrated recognition of the fundamental compatibility of a substantial number of Einstein's ideas with the Marxist philosophy of science. Einstein's belief in causality as the supreme law of scientific explanation was firmly upheld by Soviet philosophers, as was his solid attachment to the idea of continuity both in natural processes and in the growth of scientific thought. As we shall see, every stage of Einsteiniana recognized the revolutionary significance of Einstein's notion of a space-time continuum, the foundations of the relativistic theory of gravitation, the interpretation of Brownian motion, and the theory of photoelectric effect.

The Four Stages in Brief

The first stage in the Soviet stance toward Einstein—the pre-Stalin period—was characterized by a minimum of ideological interference with the theoretical work of the scientific community. The reasons for this restraint are quite obvious. The government was preoccupied, first of all, with repairing the economy devastated by the First World War and the civil war. More important, its philosophical underpinnings were far too shaky to impose ideological controls on the scientific community. The Communist Academy, designed to establish Marxism as a unified system of principles, consisted of individuals with irreconcilable philosophical views. Various official efforts to establish Communist mass organizations showed the earliest signs of a movement toward exerting ideological control over the country. But paralyzed by internal dissension and philosophical uncertainty, these initial moves were largely to little avail. Jehoshua Yakhot was fundamentally correct in asserting that in this stage of Soviet political life, little attention was paid to the ideological struggle as a form of class struggle. The term ideology normally designated the domain of nonmaterial culture—such as art, science, law, and religion—identified as "superstructure" in Marxist social theory.

Ineffective government engagement on the ideological front allowed for a proliferation of different views on Einstein's theory of relativity and epistemological thought. Every possible variant, from full acceptance to total rejection, was in full display. Although no interpretation occupied a clearly marked reigning position, physicists were far more inclined to welcome Einstein's ideas than were Marxist philosophers.

The second stage—the years of Stalinist rule—was dominated by sweeping moves to create and solidify the monolithic unity of Soviet ideology, to eliminate the centrifugal forces in Marxist philosophy, and to build uniform ideological-philosophical foundations for science. In other words, the aim of Stalinist policies, carried out through a continuous reign of terror, was to eliminate the divisive forces in science no less than in philosophy and ideology. As part of a general effort to create the cultural uniqueness of Soviet science based on Marxist philosophy and the distinctive features of the Russian intellectual tradition, Stalinist philosophers waged a bitter war on "idealistic," "bourgeois," and other alien threats to the ideological purity of Soviet science.

In the early 1930's, making ideology a system of quasi-religious beliefs and suppressing ideological enemies became the predominant concerns of the state. In the late 1930's, ideology supplied the justification for sending innumerable citizens to penal colonies and exposing individuals prominent in the arts and the sciences to public vilification. Ideology, in the words of Yakhot, became one of the chief components of the Marxist worldview. Stalin strictly obeyed his own rule: the first step in fighting political enemies abroad was to eliminate ideological enemies at home.

After the Second World War, the campaign to bring science into ideological line was pushed with renewed vigor—to devastating effect. In addition to human sacrifices, it led to the abolition of several sciences and encouraged the creation of a series of pseudosciences. It destroyed the moral foundations and intellectual unity of the scientific community. Although the resistance of the leading scientists prevented the Stalinist ideologues from achieving all their goals, they did not prevent crippling losses in the domains of academic autonomy and professional pride.

In the Stalin era, Marxist philosophers, as interpreters and custodians of official ideology, were granted full rights to participate in the ratification of scientific knowledge. They were guided by the unyielding law that a proposed theory must satisfy not only the rules of scientific procedures but also the spirit of ideological imperatives. One of their basic tasks was to organize attacks on scientists accused of ideological transgressions. Marxist philosophers had sufficient authority to demand the rejection of theories even after they were accepted by the scientific community.

Curiously, not all scientists were held to the same standard of ideological purity. Ivan Pavlov, whose scientific thinking had nothing in common with dialectical materialism and who frequently sent communications to the central government complaining of the harmful effects of certain aspects of its science policy, was never a target of this kind of sustained and organized attack. In 1950, his neuro-physiological theory was officially proclaimed to be fully congruent with dialectical materialism and Soviet ideology, and was made a fundamental component of the entire field of psychology. By contrast, the minor ideological transgressions in physical theory of A. F. Ioffe, L. D. Landau, and Ia. I. Frenkel were regularly blown out of proportion into major crimes against the state.

During this stage, Soviet authorities encouraged Marxist theorists to conduct an open war on the "idealistic" aspects of Einsteinian thought. At the end of the Stalin era, Marxist voices were heard in favor of a total rejection of the general theory of relativity as the grand deception of the twentieth century. Even in the most oppressive years of Stalinist rule, however, a small group of eminent physicists stood firm in their defense of Einstein's contributions to modern science.

The range of thematic interests in Einstein's work in the Stalin era was rather narrow; it barely reached beyond ideological-philosophical comments on the epistemological foundations of the theory of relativity. No effort was made to produce a comprehensive study of Einstein's life and work. Only the most elementary efforts were made to place Einstein's theories within the broader context of modern physics.

Stalinist attacks on Einstein were actually attacks on the scientific community—its right to hold sovereign authority in validating scientific ideas, to maintain a critical stance toward all ideas, secular and sacred, and to protect the ethos of science, the moral foundation of scientific scholarship.

During the third stage—the post-Stalinist thaw—the government inaugurated a cautious effort to eliminate some of the crudest features of Stalinist ideological restrictions on the professional prerogatives of the scientific community. It limited the authority of philosophy as an instrument of ideological control in the realm of science, gave more autonomy to scientific institutions, and terminated the policy of placing excessive emphasis on the national roots of Russian scientific thought. Whereas in the Stalin era, critics dwelled primarily on the incompatibility of Western philosophy with dialectical materialism, they now tried to discover traces of compatibility. Stalinist writers had claimed that a "wrong" philosophy could not produce a "correct" contribution to science; post-Stalinist philosophers were ordered to treat the philosophy and the science of the pioneers of modern physics as separate concerns.

This cautious retrenchment produced formidable results. All sciences hitherto abolished on ideological grounds were fully reestablished. Stalinist pseudosciences—such as the "agrobiologist" Trofim Lysenko's "creative Darwinism"—were brought to an end by the consensus of the scientific community. Abandoning the Stalinist notion of the unity of science, the authorities now openly recognized that Marxism could tolerate more than one theoretical orientation in specific domains. The public chastisement of "errant" scientists and philosophers, a common practice under Stalinism, was terminated.

Many of Einstein's ideas, previously rejected on ideological grounds, were now accepted and made integral parts of Marxist thought. In fact, the most popular Einsteinian orientation in the Soviet Union incorporated all the scientific principles and philosophical ideas built into the theory of relativity. Soviet scientists distinguished themselves particularly in exploratory and integrative work in relativistic cosmology. Einstein's legacy in humanistic thought—in relations of science to art, religion, ethics, and philosophy—attracted much attention, as did his embrace of Dostoevsky's *Brothers Karamazov*. Literature covering every part of Einstein's life and work became a flourishing enterprise.

The fourth stage—perestroika—accelerated these trends. The de-ideologization of science became an endeavor of monumental proportions and unprecedented intensity. Dialectical materialism—the Marxist philosophy of science—lost ground as the dominant branch of philosophical thought and was seriously challenged by non-Marxist orientations. Philosophical pluralism, a popular perestroika expression, emerged as the hallmark of the new age. Some commentators thought that Marxist philosophy should be retained as the ruling orientation but that it should provide channels for expressing different views. Others thought the time had come to do away with the old belief that all philosophical issues could be resolved within the framework of Marxist thought. The rapid disintegration of the Soviet system's longtime philosophical underpinnings destroyed the last bridges linking science with official Soviet ideology. Without the support of philosophy, official ideology ceased to exist as a body of integrated principles.

The time had also come, it was widely acknowledged, for the lingering residues of the Stalinist emphasis on the proletarian and national strength of Soviet culture to give way to a new emphasis on the primacy of world culture. The "new internationalism" was presented as a deep awareness of the preeminence of universal human values over narrow national values. This change was motivated by a strong impulse to transform Soviet culture from a closed to an open system of human aspirations, values, and attitudes. The Stalinist search for Russian priorities and national exclusiveness in science survived only as a haunting memory. The new internationalism represented a massive rejection of Stalinist anti-cosmopolitanism.

Perestroika rejected the Marxist ideologues' traditional concentration on the conflict between materialism and idealism as the basic issue of modern philosophy. It accepted the Einsteinian dictum that a scientist, in his relation to philosophy, must be an opportunist in selecting the views best suited to his specific needs, and that philosophy was helpful to scientists only insofar as it came in many distinct voices. The word went out that the epistemological problems of modern physics, particularly of the kind presented by the theory of relativity, were too complex to be handled by individual philosophies.

The Reception of Einstein's Ideas Before 1917

In 1905, Albert Einstein presented an intricate set of revolutionary ideas in physics that quickly became known as the special theory of relativity, one of the greatest turning points in twentieth-century science. Four years later, the Twelfth Congress of Russian Naturalists and Physicians, held in Moscow in late December 1909 and early January 1910, organized a special and surprisingly well-attended session to discuss the inner fabric and specific implications of the new theory. That session was the first scientific forum in Russia to take up Einstein's revolutionary ideas as a topic of organized public debate.

Three papers were read and discussed at the session. V. S. Ignatovskii offered a method for deducing the constancy of the speed of light from the principle of relativity. P. S. Ehrenfest discussed the applicability of the principle of relativity to rigid bodies, and P. S. Epstein analyzed the relevance of the theory of relativity to understanding the structure of electrons.[1] Although none of the three speakers held a regular academic position, they all subsequently made notable contributions to modern physics. Ignatovskii, after receiving a doctorate in 1909 from Giessen University, stayed on in Germany until 1918, when he returned to Russia to become a research associate at the newly founded State Optical Institute and a professor at Leningrad University. Wolfgang Pauli would take serious note of Ignatovskii's contributions to the mathematical foundations of the special theory of relativity.[2] Ignatovskii's interests drifted from the theory of relativity to other fields in physics and mathematics. The USSR Academy of Sciences elected him a corresponding member in 1932, but soon afterward the secret police accused him of spying for Germany. A Stalinist firing squad ended his life during the Second World War. Paul Ehrenfest, on Einstein's recommendation, moved to Leiden University, where he filled the academic post vacated by the retirement of Hendrik Lorentz, whose ideas in electrodynamics opened one of the major paths to the triumph of the special theory of relativity. P. S. Epstein went first to Switzerland and then to the United States, where he became a professor of theoretical physics at the California Institute of Technology. In 1911, he published a detailed study on relativistic mechanics.[3]

Older and professionally established physicists who presented papers at other sessions of the 1909-10 congress were conspicuous by their unwillingness to take a deeper look into the ideas unveiled by the theory of relativity. I. I. Borgman tackled the problem of "electricity and light" without showing any awareness of Einstein's photoeffect research. Disregarding Michelson's experiments measuring the velocity of light and Einstein's negation of the existence of ether, Borgman claimed "close ties between electricity and ether," even though he was willing to admit that the nature of ether continued to be veiled in mystery.[4] A. A. Eichenwald presented a broad picture of the relations of matter to energy without benefiting from Einstein's principle of the equivalence of mass and energy.[5] In "Time, Space and Ether," delivered at the Second Mendeleevian Congress, held in Moscow in December 1911, D. A. Goldhammer, a Kazan University physicist, showed much more interest in carrying ether to higher levels of structural analysis than in abandoning it altogether, as Einstein had done in 1905. In fact, he specifically declared that "the Einstein-Minkowski formula" did not disprove the existence of ether "as a carrier of electromagnetic energy."[6] Despite their resolute defense of ether, Borgman, Eichenwald, and Goldhammer contributed to an elementary diffusion of Einstein's physical ideas in Russia before 1917.[7]

The general reticence of the older generation of physicists to confront the ideas unleashed by the special theory of relativity was short-lived. Soon after 1910, two eminent physicists became effective proponents of the physical picture of the universe based on the intricate system of relativistic principles: N. A. Umov, a member of the faculty at Moscow University, and O. D. Khvol'son, a St. Petersburg University physicist. The two scholars were united by a firm conviction that the demise of the Newtonian reign in science was imminent and by a keen awareness of the epistemological intricacies of modern physics. Both were active participants in the popularization of scientific knowledge and made extensive comments on the humanistic messages of the philosophy of science.

Nikolai Alekseevich Umov started his scientific activity as a staunch believer in Descartes' mechanistic picture of the universe, which seemed to him more consistent and more complete than Newton's picture.[8] In the late 1890's, he was among the first Russian physicists to acknowledge a rapid decline of the supremacy of the Newtonian worldview in science. "Unexpected" results of experimental studies in radiation convinced him that the time had come to recognize the inner structure and fissionable nature of atoms. Impressed with Darwin's theory of evolution and Mendeleev's periodic table of elements, he reasoned that the time was fast approaching when a study of the evolution of atoms could be seriously undertaken. He saw the future of physics in the reduction of the "chaos" of subatomic structures and motions to the "harmony" of the architectonic principles of the universe. Like Einstein, he hailed the replacement of ponderable matter by weightless electromagnetic energy as the primary building material of the universe. Again like Einstein, he was willing to criticize various aspects of Newtonian science. But he was far from being ready to abandon the mechanistic bent of science altogether; Newtonian science was limited but not erroneous.[9]

In 1910–11, Umov was the first Russian university professor to write about the theory of relativity as the harbinger of a new era in physics, in science in general, and in the relations between science and philosophy.[10] In his philosophical musing, he displayed a strong tendency to include the intellect among the integral parts of physical reality as studied on a scientific level. He offered an explanation of that phenomenon: "The principle of relativity includes the observing intellect as well. . . . The intellect is linked with a complex physical instrument—the nervous system. Therefore, this principle directs moving bodies not only in relation to physical and chemical phenomena but also in relation to the phenomena of life and, consequently, of the human mind. It builds a bridge between two worlds, usually regarded as equal."[11] This statement undoubtedly came under the spell of Einstein's introduction of "observers" and "frames of reference" as essential components of physical methodology. Whereas the idea of relativity was clearly Einsteinian, the explanation of its meaning was strictly Umov's.

Umov's views on ether changed with time. In 1905, obviously unaware of Einstein's dismissal of ether as a useless physical crutch, he claimed that the time had come to replace Newtonian mechanics, centered on the study of matter, with a formulation of new principles centered on the electrodynamic properties of ether, not reducible to the laws of classical physics. By 1914, he was ready to treat ether as an "anachronism."[12] As Umov saw it, the special theory of relativity was a product not of laboratory experiments but of advanced mathematical operations.

O. D. Khvol'son earned international eminence as the author of a monumental textbook in physics—translated into German, French, and Spanish—that inspired the experimental triumphs of many of the pioneers of modern physics. He declared in 1912 that the special theory of relativity was a powerful step toward resolving the mounting challenges to Newtonian science.[13] The Einsteinian revolution, as Khvol'son saw it, was deeper and more extensive than any earlier revolution in scientific thought. Physics based on Newton's mechanics, which had reigned supreme in science for two centuries, was "almost completely" uprooted. "Together with changes in the ideas of space and time came changes in the notions of speed, force, and mass." Already discredited, luminiferous ether, so essential to the Newtonian model of the universe, was eliminated from the realm of science by the special theory of relativity. Khvol'son argued that the Einsteinian revolution was notable also for the abundance of paradoxical thought it presented: no other revolution, in his opinion, had produced so many ideas contrary to common sense.[14]

For Khvol'son, Einstein's ideas signaled the beginning of a thorough theoretical recasting of physics.[15] Whereas Umov conjured up a personal world of relativity in physics, Khvol'son was particularly careful to stay close to Einstein's own formulations. In *The Principle of Relativity* (1914), he offered a popular survey of the main ideas of the theory of relativity, together with a rather sketchy analysis of Einstein's views on the ongoing revolutionary changes in the scientific interpretation of the physical universe.[16] At the time, Khvol'son was an active participant in the growing attacks on materialistic positions in physics and biology.

P. N. Lebedev, the leading Russian physicist of the pre-October era, whose successful measurement of the pressure of light brought him international acclaim, showed no direct interest in the theory of relativity. While admitting that "the new developments in physics," particularly the appearance of the theory of relativity, had made the "old definition" of ether obsolete, he did not go so far as to accept Einstein's idea of its total uselessness as a working concept of modern physics.[17] It may be rightfully claimed, however, that Lebedev's experiments with the pressure of light—which demonstrated that light and electromagnetic energy in general have a certain mass and, consequently, exercise detectable pressure on the objects with which they come in contact—anticipated Einstein's formula equating mass and energy.[18] Lebedev verified a

hint made by James Maxwell without consciously anticipating Einstein's equation.

The diffusion of Einstein's ideas in Russia benefited also from translations
of selected Western studies. In 1912, for example, the series *New Ideas in Physics*
devoted an entire volume to "the principle of relativity." The volume offered
essays by Johannes Classen, G. N. Lewis, and Philipp Frank, as well as Einstein
himself, that showed how vastly the new theory was changing modern science
and philosophy.[19] A reviewer of the book, writing in *Herald of Europe*, was
prompted to evoke Max Planck's statement attributing Copernican proportions to the breadth and the depth of the Einsteinian revolution. A subsequent
volume in the series published an extensive bibliography of pre-1913 Einsteinian studies.[20]

As might have been expected, a flurry of scholarly papers offered ideas
modifying Einstein's conceptualization of relativity. Some writers sought a
new interpretation of the Michelson-Morley experiments with the intent of
defending the existence of ether. In an elaborate array of arguments, relying on
authorities from Johannes Kepler to the French physicist Armand Fizeau, Ia. I.
Grdina "saved" ether by reducing Einstein's ideas on special relativity to the
status of an "incomplete" physical theory.[21] His suggestions precipitated a
lively debate, showing if nothing else the growing interest in and appreciation
of Einstein's formulations. Because of the tantalizing complexity, unexplained
logical leaps, and close allegiance to mechanical explanations of Grdina's suggestions, they were quickly forgotten. I. E. Orlov supported the thesis advanced
by the German physicist W. Ritz that the speed of the rays of light depended on
the speed of the source of light, an idea closer to Newton's than to Einstein's
way of thinking.[22] Many Russian interpreters were inclined to rescue ether
from oblivion and to avoid a sharp break with the Newtonian tradition. Most
critics, however, had no reservations in recognizing the theory of relativity as a
turning point in the history of physics—and of science in general.

Before the appearance of the general theory of relativity in 1916, university
philosophy professors, whose orientation tended generally toward idealistic
metaphysics, paid little attention, if any, to Einstein's ideas. Even the St. Petersburg Academy of Sciences, the mirror of the country's highest achievement
in science, managed to isolate itself from the stormy currents of Einsteinian
thought. Two groups supplied the most serious and enthusiastic initial champions of the theory of relativity: the young generation of scientists, ready to
search for a learned sanctuary outside Newton's thought, and ideologically
attuned writers on philosophical themes close to science. Most members of the
two groups came from middle-class families headed by professionals engaged
in commerce, industrial management, or government administration. Many

had studied in Western universities, where they were exposed to the surging waves of post-Newtonian science.

The growth of popular literature on Einstein's work was rather slow and inauspicious. A small number of articles provided detailed, but oversimplified, accounts of the special theory's basic principles. One example was a seventy-page account by G. A. Gurevich in *Russian Notes* (1916), which not only summed up Einstein's ideas but offered suggestive thoughts on their challenges to the Newtonian tradition as well. Although Gurevich hailed Einstein's creation as a great victory for science, he thought that it would take some time for his ideas to be fully accepted by the scientific community and put into practical application. The experimental testing of the special theory, he thought, was at an initial stage and had produced only a few conclusive judgments. The difference between old and new thinking in physics was so fundamental and intricate that it required critical readjustments in deeply ingrained "habits of thought." But whatever might happen in the future, Einstein's creation would stand as the main force behind "one of the greatest revolutions in the annals of scientific thought."[23]

P. S. Iushkevich, a Menshevik publicist, differed from Gurevich in approaching the special theory not only from a scientific angle but from an elaborate philosophical position as well. In *Materialism and Critical Realism*, an original synthesis of Marxist and Machian epistemologies, he suggested that symbols, with their ever-rising level of abstraction, were a true representation of the external world.[24] As he saw it, the more abstract mathematical symbolism was, the more it truly reflected objective reality. Lenin, who in *Materialism and Empirio-Criticism* reduced all objective knowledge to a mental mirroring of the external world and who was suspicious of "too much" mathematics, predictably made Iushkevich a major target of attack.

In "The Principles of Relativity and the New Theory of Time," published in *Letopis'* in 1916, Iushkevich echoed Einstein's frequent references to Ernst Mach, the founder of a strong current in neopositivist philosophy that accepted the idea of "relativity" in physics without becoming a victim of "relativism" in philosophy.[25] Iushkevich viewed the relativity of time as the cornerstone of Einstein's work in physics. In Einstein's thinking, he said, there was no room for the "monism of time," a conception that allowed for just one absolute and generally uniform categorization of duration. Einstein also replaced the "static" view of time with an evolutionary view. The perception of time varied from one stage of cultural evolution to another, and within individual cultures, from one system of organized experience to another. Hermann Minkowski, according to Iushkevich, carried the mathematization of space and time to an extreme, and as a result, divorced them from physical reality.

Einstein, by contrast, made space and time empirical notions and integral parts of physical reality. Iushkevich preferred Einstein's views but also thought that only the future would tell whether there was a need for corrections and adjustments. Whatever might happen in the future, the idea of absolute time would forever belong to the past.[26]

Iushkevich helped accelerate the diffusion of Einstein's scientific ideas by translating selected Western works interpreting and enriching them. Thanks to his inordinate translation skills and energy, the Russians could read in their own language the relevant works of Felix Klein, Henri Poincaré, Ernst Mach, Max von Laue, Wilhelm Wundt, and many other notable writers on themes intimately related to the genesis and theoretical structure of Einstein's main contributions to modern physics and epistemology.[27] His translations—as well as his own writings—played a major role in advancing the philosophical aspects of the main principles of the theory of relativity.*

*Iushkevich was the first Russian Marxist theorist to undertake a methodical study of relativity. Two major Russian Marxist philosophical works—*Essays in Marxist Philosophy* (1908), written by a group of revisionists seriously concerned with opening dialectical materialism to neopositivist influence, and Lenin's *Materialism and Empirio-Criticism* (1909)—made no reference to Einstein.

The Early Soviet Reception
of Einstein's Theories

Unlike most leading Western physicists, Einstein never visited the Soviet Union. In 1922, on the recommendation of a committee made up of the physicists A. F. Ioffe and P. P. Lazarev and the mathematician V. A. Steklov, the Russian Academy of Sciences elected him an honorary member. The committee recognized him as "the most prominent figure in modern theoretical physics" and singled out four of his contributions for special emphasis: the "principle of relativity," which had to be recognized for its "extraordinary importance," even though it had not received full empirical verification; the inquiry into Brownian motion, which contributed to the development of statistical physics; the introduction of light quanta, which led him to the theory of photoeffects, photochemistry, and the fundamental theory of ray energy; and the application of quantum theory to the study of matter, which created the modern theory of heat capacity, applied to the third (Nernst's) law of thermodynamics.[1]

In the fledgling Soviet state, Einstein's ideas appealed to a wide spectrum of readers in and out of the world of scholarship. *Relativity: The Special and General Theory* appeared in Russian translation in 1921 and went through four printings in two years. Einstein wrote a special preface for the Russian edition, in which he pleaded for strengthened scientific and artistic cooperation among all nations.[2] Among the translated works were Alexander Moszkowski's *Einstein the Searcher*, Ernst Cassirer's philosophical comments on Einstein's physical principles, and Max Born's *Einstein's Theory of Relativity*, offering an astute analysis of the general ideas built into the new theory. During the two-year period from 1920 to 1922, over 100,000 copies of books by or on Einstein were sold in Russian translations.[3] Russian commentators on the revolutionary sweep of the new theory covered a wide array of intellectual backgrounds, ranging from the self-made philosopher of science Nikolai Morozov, to an army of Marxist theorists and a host of experimental and theoretical physicists.

Non-Marxist Views in the Early 1920's

The amorphous and unwieldy group of authors not associated either with physics or with Marxism was dominated by champions of mystical metaphysics, represented by N. A. Berdiaev, who were satisfied merely with noting that the new recognition of the relative value of scientific knowledge underscored the acute need for transcendental contemplation as the only sure path to absolute truth and virtuous life. It also included ethnographers like V. G. Bogoraz-Tan. Excited by the thought that Einstein's theories could be translated into a broad view of cultural relativism, Bogoraz-Tan argued that the special principle of relativity presented a perception of life, and the general principle of relativity presented life itself. Einstein's idea of the space-time continuum, he said, provided superb models and guides for introducing exact measurement into empirical studies of culture.[4] The neurophysiologist V. M. Bekhterev viewed Einstein's "relativity principle" as a law of both the physical universe and human society.[5]

No Russian scholar outside physics went farther afield in drawing general conclusions from the theory of relativity than Pavel Florenskii, a priest, renowned religious philosopher, and mathematician. In *Imaginations in Geometry*, a small volume published in 1922 in Moscow, Florenskii combined mathematical arguments with a sympathetic attitude toward Einstein's strong emphasis on the active and creative role of the observer in relativistic physics. Not a model of precise analysis and logical argumentation, the study advocated a return to the "Ptolemaic-Dantean" model of the universe and to a modified form of the Aristotelian recognition of the lunar (the Sky) and the sublunar (the Earth) universes, each with its own laws of nature. The terrestrial world, reaching as far as the orbits of Uranus and Neptune, obeyed Einstein's limit on the velocity of light; in the sky, the length of every body was equal to zero, mass and time were infinite, and the velocity of light exceeded Einstein's limit.[6] Florenskii did not continue his research in cosmology, however, nor did his contemporaries take his effort to combine Einstein's relativistic thought with medieval astronomy and theological symbolism seriously.

Roman Jakobson, at the threshold of his eminently productive involvement in linguistics, sought to establish a link between Einstein's ideas and modern art. In 1919, he published a programmatic article in which he explored the attributes of futurism, an orientation in art closely related to cubism.[7] In the depth of futurism—and of cubism—he saw a "deformed" presentation of reality, "constructive asymmetry," "color dissonance," and a radical departure from "common sense" that gave added strength to human vision. He traced the links of modern art with three revolutionary developments in science:

Darwin's transformism, Planck's quantum theory and its contribution to the idea of the "disappearance of matter," and Einstein's theory of relativity. As he saw it, Einstein's theory marked two triumphs of modern thought: the end of the static view of natural and social reality—the idea of "rest" had no place in Einstein's thinking—and the victory for a relativistic comprehension of the world. Relying on citations from the Russian physicists O. D. Khvol'son and N. A. Umov, he concluded that the theory of relativity replaced a rigid vision of the architecture of the universe with a fluid vision whose "free lines" helped in producing a much richer picture of the world and in changing the very foundations of our thinking.

A. V. Vasil'ev, a professor of mathematics at Kazan University, a widely recognized expert on Nikolai Lobachevskii's non-Euclidean geometry, and a leading light in the publication of the series *New Ideas in Mathematics*, published an original and inspiring volume on the theory of relativity with a strong emphasis on its philosophical antecedents. The English translation of the book, which appeared shortly after the Russian original in 1923, included an introduction by Bertrand Russell.[8]

Vasil'ev paid special attention to the history of philosophical interpretations of space and time, beginning with Pythagoras and ending with Ernst Mach. His discussion was centered on the growing opposition to Newton's presentation of space and time as objective and absolute realities. In Newton's view, according to Vasil'ev, space and time were objective because they preceded experience and were universally invariant. The first carefully synchronized opposition to that view came from George Berkeley, who saw space and time as subjective phenomena originating in experience, and as relative phenomena that could be identified and fathomed only in reference to objects in motion. David Hume, according to Vasil'ev, expressed essentially the same view as Berkeley. Hume and Berkeley were the first philosophers to become true ancestors of Einstein's theory of relativity.

According to Vasil'ev, Immanuel Kant accepted Newton's view of absolute space and time, but unlike Newton and Hume (and unlike Henri Bergson), he located their origin in "transcendental intuition," totally independent of both experience and the external world. In Kant's view, they were *a priori* categories preceding experience and existing independently of the external world. Despite his obvious reliance on a metaphysical explanation, Kant was recognized, and he recognized himself, as a philosophical codifier of Newtonian notions of space and time, and of the Newtonian picture of the world in general.[9]

In the nineteenth century, according to Vasil'ev, the Newton-Kant conception of absolute space underwent a process of inexorable erosion. The emergence of non-Euclidean geometry led to the idea of the existence of different spaces requiring distinct geometries. Developments in thermodynamics and electrodynamics opened the gates for a gradual drifting from the conception of physical reality as an absolute and objective phenomenon. Ernst Mach, a physicist-philosopher, made the first bold effort to create a new epistemology by recognizing the subjective and relative attributes of physical reality as envisioned by science.

Scrupulously, and not without enthusiasm, Vasil'ev looked closer into Einstein's indebtedness to Mach. From Mach, Einstein had received a solid grounding in the epistemological arguments in favor of a fusion of physics and geometry as a unitary system of scientific operation, and, in general, of the "anthropomorphic"—or subjective—nature of the reality physicists considered their main target of inquiry. Vasil'ev was particularly impressed with Mach's denial of any other reality apart from our sensations, an idea elaborated by a long line of philosophers. That idea, in his view, made a marked impression on Einstein, who linked it with his construction of a relativistic approach to physical reality. Vasil'ev made no effort to conceal his strong affinity with Mach's epistemology. In fact, he was convinced that Mach's ideas had met with complete sympathy by many "learned men and thinkers." "In the works of Stallo, Pearson, and Clifford, and in Poincaré's brilliant books, full of profound and illuminating ideas, there were developed views on the theory of knowledge and, particularly, on the axioms of mechanics which coincided with many of Mach's opinions."[10]

In Vasil'ev's opinion, Mach had announced in the sixth edition of his *Mechanics* (1904) that the ranks of confirmed "relativists" who denied the idea of absolute space and time were growing at a fast pace and would soon obliterate all opposition. The time was fast approaching for mechanics to arrive at a principle uniting "accelerated motion and motion due to inertia." The next step was up to Einstein, who, first, relied on a new concept of space to establish previously overlooked relations between geometry and physics, and, second, united three-dimensional space and one-dimensional time into dissoluble space-time. What Einstein had done, in fact, was to give new depth and concrete expression to Mach's epistemological ideas.

Vasil'ev made no effort to conceal his enthusiasm for Mach's general philosophical stance, which he identified as "relativistic positivism," and for the Machian component of Einstein's thought.[11] Nor did he refrain from placing a metaphysical label on Lenin's protestations against Mach's views.[12] In his opinion, the triumph of the theory of relativity in physics marked the complete collapse of materialism in modern philosophy. It was in opposition to Leninist objectivist epistemology that he accepted Goethe's statement that "we can at will observe nature, measure, calculate and ponder it, but it always remains *our* impression, our world. Man always remains the measure of things."[13]

Small wonder, then, that all Marxist critics considered Vasil'ev an archenemy of dialectical materialism. A. A. Gol'tsman, for example, attacked his identification of "mathematical expressions as a true reality," as well as his rejection of "agnosticism" and the idea of "unfissionable elements of nature."[14] V. I. Nevskii scoffed at his approval of Einstein's favoring of Avenarius's and Mach's "idealistic" road to science and at his harsh words about materialism.[15] Vasil'ev, in turn, argued that Einstein's physical theory, with or without Mach's influence, heralded the downfall of "the antiscientific constructions" of modern neo-Platonists and gnostics, as well as the "narrow fanaticism" of the kind of materialism espoused by Haeckel and Engels.[16] There was no doubt in Vasil'ev's mind that Einstein had wrought a twofold revolution: one in the field of science and the other in the field of philosophy.

Non-Marxist interpretations of Einstein's scientific work were generally marked by a vivid and carefully elaborated philosophical interest. In most cases, they were guided by philosophical views that were neither idealistic nor materialistic in an ontological sense. They were clearly allied to the schools of neopositivism and neo-Kantianism, both of which appeared in many overlapping systems of nuanced philosophical thought. Non-Marxist orientations concentrated primarily on widening the humanistic competence of Einstein's scientific principles. Only in rare cases could they be identified not only as non-Marxist but also as overtly anti-Marxist embodiments of speculative thought.

Marxist Interpretations

Prominent among the Marxist writers who joined the fray was a group that wanted to modernize Marxism by blending it with various currents of neopositivist and neo-Kantian thought. Considered ideological traitors by those who aspired to become official articulators of Marxist thought, these men had a strong interest in the philosophical foundations of modern physics. Since at the time there was still much ideological uncertainty and philosophical confusion, some of them were members of the Socialist (after 1923, Communist) Academy. The most influential was A. A. Bogdanov, the main target of Lenin's *Materialism and Empirio-Criticism*. Bogdanov was the principal contributor to the symposium *Einstein's Theory of Relativity and Its Philosophical Interpretation* (1923), which also included a Russian translation of an abridged version of Moritz Schlick's *Space and Time in Modern Physics* and special articles by such synthesizers of Marxism and neopositivism as V. A. Bazarov and P. S. Iush-kevich.[17]

Bogdanov discussed the main contributions of the theory of relativity to the emancipation of modern thought from the fetters of mechanistic science and philosophy, that is, from epistemological "subjectivism" and materialistic re-ductionism. He viewed the theory of relativity, which he considered fully veri-fied, as a powerful new method for translating the subjective knowledge of dis-crete individual experiences into the objective and generalized knowledge of scientific significance—a new method for the "socialization" of knowledge.[18] The theory of relativity, he reasoned, did not destroy the objective (that is, so-cially significant) forms of space and time but simply gave them a more precise interpretation and wider flexibility in the study of the dynamics of social and cultural phenomena.

Bogdanov thought that Einstein's theory made a powerful contribution to philosophical monism, which reduced all natural and cultural phenomena to universally shared structural principles.[19] It was, in his view, one of the major expressions of the twentieth-century striving for a full understanding of cos-mic and cultural unity. Some other crowning results of this striving were, in his judgment, the growing interest of mathematicians in structural analogies ex-pressing the unity of nature and the triumph of energeticism, thereby provid-ing unifying links for all forces of nature and human society. At this time, Bog-danov was deeply involved in building the logical and sociological edifice of "tektology," which he envisaged as a new scientific study of the key attributes of cosmic organization. He claimed that the "structural symmetry" of the uni-verse, the basic idea built into the general theory of relativity, expressed the central postulate of his own world outlook.[20] He worked at bringing Marxist sociology and Machian "subjective" theory of knowledge into a unified system of thought.[21] Predictably, Marxist critics were consistent in identifying Bog-danov's views with philosophical relativism, a subjectivist position in the the-ory of knowledge and comparative sociology.[22]

Marxist philosophers were the most active—and the most inconsistent—interpreters of Einstein's theory. Much of their uncertainty stemmed from Lenin's warning that the modern revolution in physics was not only a great leap in man's incessant effort to unlock the mysteries of nature, but also an en-ticing invitation to physical idealism. As defined by Lenin, physical idealism included all philosophical orientations in modern physics that refused to view matter as the primary substratum of physical reality, emphasized the subjective origins of scientific knowledge, and challenged the effectiveness of causality as the basic explanatory principle in science.[23] But despite the heavy burden of the Leninist legacy, it was not unusual for individual Marxists in the 1920's to view the October Revolution and the revolution in science wrought by Einstein's theories as historically related, mutually supporting developments. In their opinion, just as Marx and Engels redefined the structural principles of human society, so Einstein redefined the structural principles of the universe. At this time, a solid group of Marxist philosophers exalted the grand scope and depth of Einstein's physical thought; but they also lamented Einstein's "personal" inclination toward philosophical idealism combining a total irreverence for ontological questions with epistemological relativism.

Speaking for a typical group of Marxist theorists, V. I. Nevskii wrote in 1922: "No one denies the great value of the theory of relativity in physics and mathematics; at the same time, however, every unbiased person sees clearly that the philosophical foundations of Einstein's theory are pure idealism."[24] Nevskii represented the group of writers who thought that a fuller and more definitive evaluation of the theory of relativity should wait for additional empirical testing of the basic principles built into Einstein's contribution to physics.

The Orthodox Marxists

The orthodox Marxist interpreters of Einstein's theory formed three clearly demarcated groups. One group rejected the idea of physical relativity on both scientific and philosophical grounds. It considered Einstein's views a backward step in the evolution of modern physics and the scientific world outlook. The second—and largest—group avoided both a full rejection and a full acceptance of the new theory. It produced a plethora of suggestions for placing the theory of relativity within the philosophical framework of Marxist materialism. The third group accepted Einstein's basic scientific ideas and emphasized the full compatibility of the theory of relativity with the scientific and philosophical outlook of dialectical materialism.

The leader of the first group was A. K. Timiriazev, a man with academic training in classical physics, a strong flair for philosophical debate, and an unlimited dedication to Marxist socialism. As a professor first at Sverdlov Communist Academy and then at Moscow University, he offered courses in the kinetic theory of matter and theoretical physics. The most noted of his works was *Natural Science and Dialectical Materialism*, a collection of essays he published in *Under the Banner of Marxism* and related journals from 1920 to 1925.[25] Initially, in the early 1920's, Timiriazev hailed quantum theory as a revolutionary step in the progress of modern science based on what he saw as three strengths in particular: it blended the bold new steps in physical research with the classical tradition that triumphed in the work of Rudolf Clausius, James Maxwell, and Ludwig Boltzmann; it expanded the base of the practical potential of theoretical research; and it presented new evidence showing the work of dialectics in natural processes.[26] But by 1925 he had soured on the matter, charging the pioneers of quantum theory with an excessive emphasis on mathematical formalism and an unwarranted tendency to downgrade the scientific merit of the law of causality.[27]

As the most influential scientist in the Communist Academy of the Social Sciences and the most active member of the editorial board of the theoretical journal *Under the Banner of Marxism* (*Pod znamenem marksizma*), Timiriazev worked tirelessly at organizing a broadly conceived attack on the basic scientific postulates and philosophical underpinnings of the theory of relativity.[28] From a strictly philosophical vantage point, Einstein's theory, as Timiriazev saw it, opposed ontological materialism and epistemological objectivism; it built its scientific theories on the subjective principles of Mach's theory of knowledge, rather than on objective generalizations grounded in experimental data.[29] It could boast only three scientific predictions, the perihelion of Mercury, the bending of light rays, and the displacement of lines in the solar spectrum—a scientific output that, "in comparison with other theories," fell "far below the norm." Even "these 'brilliant' prophesies, deduced from the general as well as from the special theory of relativity, and at times hailed by the world press, [were] a long way from an adequate empirical verification."[30] Einstein's heavy reliance on non-Euclidean geometry, in Timiriazev's view, represented a flagrant betrayal of one of the most truly scientific branches of knowledge—Euclid's geometry.

Nor was Timiriazev inclined to accept Einstein's (and Sir James Jeans's) assertion that geometry was a product of pure reason.[31] Generally, most Marxist theorists found it exceedingly difficult to bring their thinking in tune with Einstein's claim, made at a session of the Prussian Academy of Sciences in 1921, that mathematics, "a product of human thought which is independent of experience," rose above all other sciences because its propositions were "absolutely certain and indisputable."[32] To Timiriazev, a Marxist close to Lenin's epistemological views, mathematics was neither independent of experience nor the only sanctuary of absolute truth. Timiriazev rejected categorically the usefulness of "thought experiments" as sources of new scientific theories; "thought experiments," he said, "are excellent for illustrating the course of thought in the study of a complex problem, but they are useless as substitutes for real experiments."[33] Undue emphasis on thought experiments could only open the door to philosophical idealism.

Particularly after the publication of Lenin's essay "The Importance of a Militant Materialism" in 1923, Timiriazev argued against hasty and sweeping attacks on modern theories in physics. In fact, he advocated a close reexamination of the methodology and philosophical premises of both the natural sciences and dialectical materialism.[34] But underneath all his knotty reasoning remained a firm conviction that no modern theory supplied arguments convincing enough to challenge Newtonian science. Although he insisted on a careful examination of modern physics, he quickly arrived at the conclusion that the theory of relativity represented an intellectual hoax of vast proportions. Denigrating the classic experiments of Michelson and Morley, he claimed that no scientist had produced satisfactory proofs for the constancy of the speed of light, the backbone of the special theory of relativity.[35] All major theories advanced by Einstein must be rejected even as promising hypotheses. After all, they made absolute truth permanently inaccessible to man. Relativism, he said, stood in categorical opposition to dialectical materialism, which taught that the relative and imperfect picture of the world was becoming ever more absolute and complete.[36] He was the first Soviet physicist to claim that Einstein's general principle of relativity and theory of gravitation were in no way related to each other.[37]

Inflexible and dogmatic to the extreme, Timiriazev hurled at his enemies a mixture of scientific anachronisms and personal ridicule. The fury of his self-styled anti-Einsteinian crusade reached a peak at a time when most leading physicists in the Soviet Union and the West had accepted the theory of relativity as the mainspring of the twentieth-century scientific revolution.[38] His *Natural Science and Dialectical Materialism*, overflowing with anti-Einsteinian statements, served for several years as a textbook in Marxist schools and as a propaganda weapon in the war against the new physics. A contemporary Marxist critic praised his *Physics*, a popular and elementary survey of the main branches of physical theory, as the first systematic effort to apply the ideas of Engels' *Dialectics of Nature* to modern physics.

At first engaged primarily in bitter feuds with Soviet scholars who defended the theory of relativity, even on a limited basis, Timiriazev gradually transferred his major concern to "the waves of idealism" that originated in the West and drew support from Einstein's scientific and epistemological ideas.[39] He severely attacked all Einstein's theoretical claims as contrary to the postulates of materialism in scientific and philosophical thought. Combining sarcasm and acrimony, he scolded the British mathematician and astronomer Sir Arthur Eddington for his effort, in *Relativity and Quanta*, translated into Russian under the editorship of Boris Hessen, to substitute a "world of symbols" created by mathematics for the physical universe in its real existence.[40]

Timiriazev considered Einstein's theory of relativity a major threat to the health of modern science; on a much smaller scale, as we saw, he quickly turned against quantum mechanics as well. He expressed several misgivings about the latter theory, particularly about its shift of emphasis from "physical content" to "mathematical formalism," its introduction of a teleological element into physics, its disregard of continuity as an essential attribute of physical processes, and its reliance on probabilistic approaches to the physical micro-universe.[41]

After 1926, the year in which Werner Heisenberg offered the uncertainty principle and Niels Bohr formulated the principle of complementarity, Marxist philosophers intensified and consolidated their attack on the epistemological foundations of the Copenhagen school. The defenders of Marxist orthodoxy gradually became accustomed to considering quantum mechanics as ominous an enemy as the theory of relativity. In addition to their specific challenges, the two revolutionary branches of modern physics were seen as having two common characteristics dissonant with Marxist ideology: excessive reliance on mathematical symbolism and excessive reliance on the basic elements of subjective epistemology.

Timiriazev's attacks on the theory of relativity received consistent support from I. E. Orlov, a regular contributor to *Under the Banner of Marxism* and an energetic crusader against physical idealism. In his view, Einstein made several glaring errors. He did not link theoretical considerations to an empirical substratum, and he did not have much use for intermediary generalizations, without which no experimental study could produce useful results.[42] The theory of relativity, he argued, presented a graphic example of "mathematical fetishism," a methodological posture that substituted mathematical abstractions for physical reality. In Einstein's thinking, according to Orlov, mathematical methods assumed the power of the laws of nature. The substance of physics gave way to the formalism of mathematics. Einstein's thinking sacrificed physical relativity to philosophical relativism, a bottomless fountain of fictitious conclusions, and the arbitrary uses of coordinate systems. Orlov preferred Newton's physical picture of the world, which "excluded all arbitrariness in the choice of coordinates," did not substitute mathematical functions for physical causality, and did not consider the motion of heavenly bodies a function of man-made clocks.

The second group of Marxist interpreters of Einstein's ideas was unusually fluid, generally leaderless, and much less militant than the Timiriazev-Orlov group. It included a wide range of people who recognized the revolutionary impact of those ideas on modern physics, but who were especially eager to point out the critical need for accommodating them to Marxist philosophy. On the whole, this group tended to see an energeticist bias in the principle of the equivalence of mass and energy and were suspicious of the overpowering influence of Mach's "subjective" theory of knowledge. They resented Einstein's repeated statements detaching mathematical operations from empirical reality.[43] Some thought that Einstein did not show sufficient respect for causality as the true key for scientific explanations of the work of nature. Others did not appreciate Einstein's favorable reference to a finite or spherically closed universe. Many joined Orlov in complaining about Einstein's alleged transformation of physics into a branch of geometry.[44]

A. A. Gol'tsman, a very active member of the group, was an outspoken partisan of the theory of relativity, which he considered the most impressive turning point in the history of scientific thought. But this ebullience did not prevent him from recognizing widespread "philosophical confusion" as a major stumbling block in understanding the theory's full impact on modern science.[45] He too thought that Einstein's scientific ideas, like those of Newton and Planck, were too deeply rooted in theoretical abstractions and mathematical symbolism to allow for an effective analysis of their social and cultural foundations.[46] The sociology of knowledge, he contended, needed much methodological improvement before it could undertake a successful analysis of the theory of relativity as a creation of modern culture.

Gol'tsman directed the attention of Marxist theorists to what he labeled a major inconsistency in Einstein's use of the word "relativity": when it referred to physical phenomena, the word was irreproachable as an indicator of the idea of cosmic dialectics; but when it referred to human knowledge, it was on shaky grounds because of its close alliance with idealistic philosophy.[47] Gol'tsman advised Marxist scholars to forgo every attempt to examine Einstein's "philosophical physiognomy" as part of a general analysis of the scientific foundations of the theory of relativity. He clearly implied that Einstein's science, separated from Einstein's philosophy, was the most potent weapon in the war on philosophical idealism. He was the first Soviet Marxist to advocate keeping separate account of current developments in science and philosophy.

Gol'tsman rejected Timiriazev's claim that Einstein's principle of the equivalence of mass and energy represented a direct attack on the ontological primacy of matter, the philosophical essence of Marxist philosophy. Lenin, he said, made it clear that different physical theories stressed the primacy of different states of matter—some theories, for example, emphasized the primacy of electromagnetism—and that all these theories were materialistic insofar as they recognized matter as an objective reality. Einstein, according to Gol'tsman, recognized energy both as a measure of the mass of a substance and as an objective reality existing independently and outside human consciousness. New developments in physics appeared to Lenin, according to Gol'tsman, as discoveries of new ramifications of matter; they marked new conquests for physical theory and new victories for the philosophy of materialism.

Gol'tsman also rejected Timiriazev's view that non-Euclidean geometry involved a new look at the physical role of space in both the micro-universe and the macro-universe. For him, non-Euclidean geometry marked a triumph of materialism in mathematics, strengthening its empirical base and expanding its vision far beyond the narrow limits of the earth's "flat" surface. The reliance of modern physics on more than one geometry gave materialism a broader and deeper view of the architecture of the universe.

A. A. Maksimov, the most aggressive member of the group, concentrated above all on epistemological aspects of the theory of relativity. He criticized the idealistic undertones of Einstein's "Geometry and Experience" and was one of the most persistent and determined critics of "neopositivists" and "neo-Kantians" like Moritz Schlick, Hans Reichenbach, A. V. Vasil'ev, and Ernst Cassirer who relied on Einstein in building the ominous edifice of "physical idealism." Whereas Timiriazev was ready to dismiss Einstein's theory as unfounded and totally erroneous, Maksimov wanted to save it by separating it from the superstructure built by idealistic philosophers with direct help from Einstein. The strength of dialectical materialism, according to Maksimov, lay in its rejection of the idea of closed systems in philosophy and its exclusive reliance on concrete facts offered by the natural sciences.[48] Despite its idealistic transgressions, the theory of relativity, Maksimov argued, must be considered a contributor to the modern erosion of philosophical idealism.

Maksimov laid the groundwork for a long Marxist tradition in the Soviet Union that concentrated much more on the philosophical flaws than on the scientific strengths of Einstein's ideas. Maksimov resented the current popular writing on Einstein, mainly translated works, 75 percent of which, in his estimation, consisted of "the propaganda of idealism, metaphysics, and . . . scientific quackery."[49] Although his criticism reached far afield, his heaviest blows were aimed at what he identified as an undue emphasis on the subjective factor and "free imagination" in the definition and analysis of physical reality, on logical constructions unrelated to empirical evidence, on the "relativization" of scientific methodology, and on "metaphysical assumptions" in general.[50] He viewed these and similar departures from the Marxist interpretations as special adaptations to "bourgeois ideology."

Like Maksimov, Z. A. Tseitlin, another prominent member of the group, was interested in both Einstein's science and Einstein's philosophy, primarily in the area where they met. He, too, showed a lively interest not only in Einstein's explicit philosophical pronouncements, but also in the philosophical implications of the theory of relativity. In Tseitlin's opinion, Einstein was indebted to a stream of modern philosophical thought that had evolved from Kant's broad generalizations to the philosophies of science advanced by Mach, Richard Avenarius, Karl Pearson, William James, Henri Poincaré, Hermann Cohen, and Josef Petzoldt.[51] Unlike Timiriazev, he anticipated a great future for the theory of relativity, both special and general. To triumph, though, Einstein's physics must be recast so as to make the premises more consistent with dialectical materialism. Physics must go back to ether in a more wholesome fashion, and it must produce a clearer picture of physical space, untrammeled by the metaphysical fictions of the idealistic philosophers.

Tseitlin believed that Newton showed more "dialectical perspicacity" than some of the Soviet Marxist supporters of Einstein's theory.[52] He envisioned the future of physics as a dialectical synthesis of Newtonian and Einsteinian thought, but only after the latter discarded the unwelcome elements of idealistic thought. Dialectical materialism, according to Tseitlin, viewed the history of science as a development from relative to absolute knowledge. He resented Einstein's implied claim, as well as the views of his typical Marxist supporters, that the history of modern physics was marked by a shift from Newton's postulates of absolute truth to Einstein's affirmation of relativity as a guide to the previously unexplored depths and mysteries of the universe. He ignored the claim of those supporters that relativity was only a heuristic device opening new paths to absolute truth—or, in Marxist vernacular—to closer approximations to absolute truth. While defending the vital role of "dialectical synthesis" in the history of science, Tseitlin argued that Newtonian thought offered strong support to various principles counted among the pivotal premises of the theory of relativity. Newton, for example, showed a clear awareness of exceptions to his treatment of space and time as absolute categories, independent of physical content. Tseitlin went so far as to claim that Einstein's analysis of space differed from Newton's mainly in the use of a more elaborate mathematical apparatus. In his opinion, many Marxist historians of science erred in not placing sufficient emphasis on the continuity in the growth of modern physics that made the emergence of the Einsteinian revolution possible in the first place.

Tseitlin disagreed with the total negativism of Timiriazev's critique of Einstein's theory of relativity. Although he found much to criticize in Einstein's physical theories, he professed a fundamental agreement with "the principle of the relativity of motion" and "the conception of space as a physical or material body" built into the theory of relativity.[53] Still, on most questions concerning Einstein's theory, he said, he valued Timiriazev's views.[54] For example, he agreed with Timiriazev's categorical assertion that Einstein tried "to erase from physics all revolutionary developments started by Faraday and Maxwell and to replace physics by mathematical-Machian propositions."[55] Tseitlin admitted that it was exceedingly difficult to draw a picture of Einstein's revolutionary role in modern physics that would be acceptable to physicists without offending the custodians of Marxist thought. Unlike Timiriazev, however, he showed unambiguous flickers of hope that the theory of relativity, corrected and elaborated, would become a dominant force in the future progress of physics and the philosophy of science.

The third group of Marxist interpreters, led by S. Iu. Semkovskii and B. M. Hessen, considered Einstein's scientific ideas a powerful confirmation of the basic postulates of the Marxist philosophy of science. Semkovskii, a member of the Ukrainian Academy of Sciences and a leading authority in the Marxist philosophy of science, is credited with making the first comprehensive analysis of the philosophical problems of relativistic physics in the light of dialectical materialism.[56] In his extensive writing, he treated the theory of relativity as a sound combination of materialism and dialectics and as a "brilliant confirmation" of the Marxist philosophy of science.[57] He praised Engels' *Dialectics of Nature* as a philosophical codification of dialectical materialism, which, in his view, provided the points of departure for developments in science that triumphed in the emergence of the theory of relativity.[58] As David Joravsky has pointed out, "Semkovskii felt that great revolutions in natural science coincide with periods of great social revolutions, and he recalled Engels' insistence that materialism must take on a new form with every advance in the scientific understanding of matter. Dialectical materialism must accordingly absorb the new insights provided by the theory of relativity."[59]

Semkovskii concentrated on Einstein's use of motion as both an objective and a subjective category. Motion was objective in the sense that it was meaningful to a physicist only when treated as a reality independent of human consciousness. It was subjective in the sense that it could be understood only when viewed in relation to a physical point of reference that was also in motion and therefore eliminated the need for the Newtonian conceptualization of the state of rest as a privileged frame of reference and as a standard of absolute measurement.[60] In Einstein's conceptualization of space and time, Semkovskii saw a bold attack on Kantian apriorism and Machian subjectivism, the archenemies of materialistic epistemology. "Einstein's theory," he wrote, "shows that space and time are inextricably connected with matter, and designate not unchanging uniformities but changing structures determined by material masses in motion and by their gravitational fields."[61] Space and time were not "pure" or "*a priori*" elements of thought but "empirical" and "physical" forms of reality.[62]

Semkovskii argued that an analysis of the interrelations of dialectical materialism and the theory of relativity must concentrate "not on Einstein's individual philosophical views and statements . . . but on the ideas that are inextricably woven into the fabric of the theory of relativity"—the ideas built not on Mach's subjective "elements-sensations" but on objective reflections of "matter."[63] It was true, Semkovskii noted, that Einstein expressed great admiration for Mach, and that occasionally—as, for example, in the opening paragraphs of "The Mathematical Foundations of the Theory of Relativity"—he expressed himself in Machian terms. But it was equally true that he respected Mach primarily as a "historian of science" and that he showed a full awareness of the historical and personal conditions that prevented Mach from developing a sense of urgency for a study of "the simultaneity of spatially separated events."[64]

In Semkovskii's view, there was no ground for allying Einstein's physical thought with physical idealism of any kind. For him, the dual linkage of space and time and of space-time and matter-in-action stood out as Einstein's key contribution to modern physical theory; and dialectical materialism and the theory of relativity agreed on the fundamental principle that there was no absolute rest and no absolutely privileged systems of coordinates. Semkovskii lamented the "scholasticism" of many Marxist philosophers who, overcommitted to speculative generalities, made little effort to keep up with the ongoing revolution in science and, as a result, failed to appreciate the full scope of the ties between dialectical materialism and the theory of relativity.

Semkovskii dealt extensively with the relation of the theory of relativity to Kantian, Hegelian, and neopositivist (primarily Machian) philosophy. He was especially eager to dissociate himself from Vasil'ev's discourses on the philosophical antecedents of Einstein's ideas. His basic aim was to refute Petzoldt's claim that none of Einstein's basic principles contradicted Mach's philosophy.[65] On the epistemological level, Einstein, according to Semkovskii, supported two principles of dialectical materialism: the objective nature of both physical nature and scientific knowledge; and the law of causality as the chief mechanism for explaining natural processes. In the victory of causality, Semkovskii recognized irrefutable signs of the demise of the last residues of mysticism in science.[66]

The reaction of the world of scholarship to Semkovskii's forceful and precisely argued defense of Einstein's ideas was far from uniform. V. Iurinets, writing in *Under the Banner of Marxism*, credited Semkovskii with "convincing proofs" that Einstein's theory of relativity not only did not challenge the theoretical foundations of dialectical materialism, but, on the contrary, gave Marxist thought another powerful tool in building a new world of ideas.[67] Semkovskii deserved much credit, he noted, for showing the bankruptcy of mechanistic materialism and for unveiling the "absurdity" of Kantian and Ma-. chian interpretations of Einstein's relativistic principles. Less charitable was a small but vociferous faction of Marxist theorists who resented Semkovskii's manifest inclination to follow a "centrist" line, that is, to avoid full identification with either the mechanists or the dialecticians, the two feuding factions of Soviet Marxists.[68]

An unusually large number of Marxist critics indicated that the label "relativity" clashed with the sensitivities of the more orthodox defenders of dialectical materialism, who, conditioned by Lenin's *Materialism and Empirio-Criticism*, preferred to characterize scientific knowledge in absolute terms. Other critics challenged Einstein's mass-and-energy equation as promising a revival of the much-detested Ostwaldian energeticist ontology and giving weight to the widely circulated notion of a "de-materialization of atoms" as the central idea of modern physics. The relativistic conceptions of space and time, some critics alleged, reinforced Poincaré's vigorous defense of the conventional foundations of modern science. Still others were put off by Einstein's enthusiasm for the mathematizing of all sciences as the main vehicle for future advances of knowledge; the claim provoked much skepticism among Marxist scholars swayed by Lenin's provocative statement that so much emphasis on mathematics opened the doors for the debilitating interference of idealistic philosophy into scientific thought. These and similar questions not only plagued Semkovskii but continued to be permanent fixtures on the philosophical agenda of Marxist scholars.

In a caustic review of Semkovskii's book on dialectical materialism, Tseitlin accused him of flirting with idealistic metaphysics—to the point, he thought, that Einstein himself came closer to materialistic positions in philosophy than Semkovskii.[69] Semkovskii was particularly guilty of tolerating the idealistic leanings of Ostwald's energeticism and of rejecting the existence of ether, which led him to the "idealistic blunder" of believing in empty space. Tseitlin argued that the general theory of relativity, by Einstein's own admission, assumed the reality of ether. For all that, he found Semkovskii's book valuable in one important respect: it provided a classical example of how *not* to apply dialectical materialism to the study of the theory of relativity.[70] Other Marxist critics found Semkovskii guilty of treating Lenin as one of Plekhanov's understudies.[71]

Semkovskii received scattered plaudits from physicists who at that time showed little inclination for philosophical discourse. One was O. D. Khvol'son, who commended Semkovskii for his "philosophical erudition and intimate familiarity with the salient developments in modern physics."[72] (Khvol'son, it will be recalled, was a representative of the "old school" of Russian physics and an early Einstein enthusiast. As he stated in his 1914 book on the theory of relativity, Einstein's studies presented "a detailed, integrated, and well-rounded picture of the world."[73]) Another Semkovskii supporter was the influential A. F. Ioffe, who pronounced himself in full agreement with Semkovskii's argument in favor of the total compatibility of dialectical materialism with the basic principles of the theory of relativity.

Much like Semkovskii, Boris Hessen gave Marx credit for anticipating the methodological strategy and epistemological premises of the theory of relativity.[74] Hessen saw the real source of the human understanding of the world—and the cardinal feature of both dialectical materialism and the theory of relativity—in the interaction of the subject and the object. True materialism, Marx wrote, should not ignore the role of the subject in the construction of scientific knowledge. Hessen thought that he followed both Marx and Einstein when he asserted that the subject was the prism refracting the streams of sense perceptions on their way to socially articulated and structured knowledge and channels of thought. The refracting power of the subject received full recognition in Marx's *Theses on Feuerbach* and Einstein's theory of relativity.[75]

Hessen was the first Soviet Marxist philosopher of science to broach the question of the subjective aspect of scientific knowledge, a rather unpopular topic in Marxist circles until the 1960's.[76] Leaning on Marx, he observed that it was not that idealistic philosophers were guilty of failing to recognize the role of the subject in mirroring the external reality, but that they attributed absolute authority to it. He made a partial concession to subjectivist epistemology for the sole purpose of accommodating dialectical materialism to Einstein's "idealistic" orientation in the theory of knowledge.

Hessen's views of the relation of dialectical materialism to fundamental changes in modern physics were unorthodox. Instead of waging a determined war on the physical ideas that contradicted Marxist thought, as was customary at the time, he argued in favor of making dialectical materialism a living philosophy capable of absorbing the philosophical challenges of quantum mechanics and the theory of relativity. Dialectical materialism, he thought, could grow and prosper only by encouraging a positive attitude toward the theoretical insights and challenges of the new philosophy of science—only by a readiness to absorb ideas responsible for revolutionary changes in physics. He felt that the preoccupation with negative criticism threatened to make dialectical materialism a narrow system of petrified principles, a negative intellectual force.

Hessen, for example, fought against his Marxist colleagues who saw the new emphasis on "statistical laws" in physics as an attack on determinism—on the reign of causality as the chief tool of scientific explanation.[77] The study of statistical regularities in the physical world, he argued, must be viewed as a positive effort to reach far beyond the limits of classical determinism. He recommended that Marxist philosophers penetrate the depths of modern physics for the purpose of adding new substance, insights, and vibrancy to dialectical materialism. Quietly and unobtrusively, he fought the pronounced tendency in the Soviet Union to isolate Marxist thought from the swelling currents of Western philosophy.

In Hessen's view, the theory of relativity explained many riddles of the physical universe that were totally beyond the competence of Newtonian science, and, unlike classical mechanics, it needed no ancillary propositions to extend the power of the laws of physics. In direct opposition to Newton's ideas, Einstein showed that space and time did not exist outside matter—that matter was their true reality and their synthesis.[78] In Hessen's view, no idea built into the theory of relativity was in any way contrary to the general theory of dialectical materialism. Einstein's theory was in fact a dialectical union of classical and modern physics, denying the universality of Newtonian mechanics but not its role as an essential part of the larger picture of the universe. The revolutionary strides in contemporary physics, he argued, made it imperative to advance dialectical materialism as a flexible and open philosophy.

Hessen came under savage attack, the target of both ardent defenders of Marxist orthodoxy and young talented physicists like Lev Landau and George Gamow, who resented philosophical interference with the pursuit of science. Later Soviet critics claimed that Hessen "overestimated his capacity to offer an adequate view of [modern physics] and especially of the general theory of relativity and quantum mechanics."[79] He also erred in placing too much emphasis on philosophical considerations. His unorthodox thought brought him great distress in the years to come. In the mid-1930's, he became one of the countless victims of Stalinist terror.[80]

But Hessen in fact consistently cautioned Marxist scholars not to commit the grave error of accepting Einstein's philosophical ideas whole cloth, of assuming that they formed a complete system of well-rounded propositions and were fully harmonious with the scientific principles of the theory of relativity. He kept reminding Marxist philosophers that the problem lay not in the accelerated mathematization of science, but in the arguments of idealistic metaphysicians like Florenskii, who "abused" Einstein's mathematical symbolism by submerging it into a world of mystical ideas. In his view, the health and future growth of dialectical materialism depended on its capacity to keep up with advances in scientific thought. Einstein's rejection of the idea of absolute motion, he thought, constituted the key dialectical pillar of the theory of relativity; it made Einstein's theoretical principles fully congruent with the Marxist philosophy of science.

In accepting the revolutionary triumphs of the theory of relativity and the idea of a fundamental congruence of Einstein's ideas and Marxist philosophy, Semkovskii and Hessen depended more on the results of laboratory experiments and astronomical observations than on logical constructions built on selected premises of Marxist theory, always with clear ideological implications. They contributed to an ongoing tradition in Soviet thought that made Marxist philosophy a supreme judge of scientific ideas with ideological overtones. Despite these limitations, they were the most astute Marxist defenders of the theory of relativity and quantum mechanics as turning points in the growth of scientific thought.

In their argumentation in favor of the full compatibility of the theory of relativity and dialectical materialism, Semkovskii and Hessen relied little on Marx's writings. Instead, they referred specifically to Lenin's *Materialism and Empirio-Criticism* and Engels' *Dialectics of Nature*, the two Marxist works directly involved in a discussion of the philosophical intricacies of natural science. Although their concerns were different, Engels' preoccupation being with the last phase in the reign of the mechanistic view of the universe, and Lenin's with the philosophical signs of the emergence of a post-mechanistic or post-Newtonian orientation in science, the two works helped the more persevering Marxist theorists to become aware of the critical indicators of a major crisis in the epistemological and methodological foundations of modern physics.

In *Materialism and Empirio-Criticism*, published in 1909, Lenin had not displayed any awareness of Einstein's scientific work. That did not prevent him, however, from waging a bitter war on the new epistemological premises that found their way to the theoretical setting of the principle of relativity. The nearest Lenin came to a confrontation with Einstein was in his heavy attack on Ernst Mach, the physicist–philosopher who, in Einstein's own view, had been the leading influence on his theoretical ideas. Semkovskii and Hessen depended on Lenin's essay on "militant communism" only insofar as it appealed to Soviet intellectuals to be receptive to all Western ideas that could be used profitably in laying the foundations for a socialist society.

Engels' *Dialectics*, published in 1925 in Russian and German, was a never-completed manuscript that he had stopped working on in 1883. The work was noted primarily for its sustained critique of "mechanical" or metaphysical materialism. The manuscript had languished in the archives of the German Social-Democratic Party until 1924, when Eduard Bernstein, a tempered Marxist and the author of *Evolutionary Socialism*, showed it (or part of it) to Einstein for advice on its publication value. Einstein responded that its only possible value was in throwing light on Engels as a "historical person"; it would not in his view be of "particular interest either to modern physics or to the history of science."[81] Semkovskii and Hessen could rely on Engels' newly published work, written long before the revolutionary changes they were wrestling with, only insofar as it offered statements in favor of making dialectical materialism a flexible philosophy, dedicated to keeping up with and absorbing new developments in science. In defending the theory of relativity, Soviet Marxists of the 1920's quickly learned that they must depend more on their own intellectual ingenuity and ideological sensitivities in interpreting the spirit of the Marxist legacy than on quotable authorities.

The Eclipse of Mechanist Views

The Communist Academy served not only as a center of inquiries into the philosophical foundations of the new physics but also as the main battlefield for the feuding factions of Marxist philosophers. The two main factions—the mechanists and the dialecticians (also known as Deborinites)—found themselves irreconcilably split on the issue of the theory of relativity as the theoretical base of modern physics. A typical mechanist—such as A. K. Timiriazev—demonstrated a distinct predisposition to deny any scientific value to Einstein's work. A typical dialectician, such as A. M. Deborin, did not bother much with the details of the theory of relativity or of modern physics in general. Indeed, Deborin argued that dialectical materialism, as a philosophical method, should be concerned not with the questions of physics but with the theory of scientific knowledge, its own realm of inquiry.[82] That argument did not deter him from hailing relativistic physics as a magnificent corroboration of the strategic principles of Marxist philosophy. According to Deborin, the new physics came on the wings of Engels' prediction of the forthcoming downfall of the "mechanistic worldview" that dominated classical physics.[83] In *Anti-Dühring* and *Dialectics of Nature*, Engels had unleashed a savage attack on popular efforts to subsume all natural processes under the laws of mechanics. The theory of relativity and quantum mechanics, in turn, adduced strong arguments supporting Engels' attack on the reign of mechanism in natural science. Deborin exhibited much more interest in fighting the general philosophical outlook of the mechanists than in providing a systematic, coherent, and comprehensive elucidation of the dialectical qualities of modern physics. The mechanists criticized him and his followers for their view of Engels as an errorless philosopher and a true prophet of the twentieth-century revolution in physics.

As the theoretical complexity and research intricacy of the new physics grew at an accelerated pace and the fundamental unity of the theory of relativity and quantum mechanics developed deep and secure roots, the philosophical discussion became increasingly more vague and removed from theoretical issues of primary importance. This was most clearly manifested in the interminable debates sponsored by the Communist Academy. S. F. Vasil'ev, a relatively flexible Marxist philosopher and historian of science, recorded that the Deborinites went far afield in distorting the theoretical positions of the mechanist leaders. He observed that the long succession of verbal skirmishes between the two groups were sordid displays of mutual misunderstanding, usually caused by deliberate or unintentional misrepresentations of the opponents' views.[84] The situation was made all the worse by the mechanists' wont to depend heavily on what they considered scientific arguments, and the Deborinites' wont to depend almost exclusively on philosophical metaphors. Where the one felt secure in the tradition of Faraday and Maxwell, the other looked forward to the unity of the new physics and Marxist thought.

By the end of the 1920's, it had become clear to higher-ups that the philosophical and scientific feud had taken on strong ideological and political overtones, and that there could be no reconciliation without outside help—or coercion. As a first step, that help could come only from someone not actively involved in the conflict and acceptable to the highest echelons of the Communist Party. The question of the Marxist position on the compatibility of the theory of relativity with dialectical materialism had emerged as one of the major issues requiring official clarification.

O. Iu. Shmidt, a member of the Communist Party and, in the words of Joravsky, "the most important Bolshevik on the scientific front," was decidedly an ideal person to act as mediator.[85] By the time Shmidt graduated from Kiev University (1913), he had already published three works in mathematics. He went on to earn a master's degree at his alma mater by publishing a rather sophisticated survey of the basic propositions of group theory, one of the most abstruse and challenging branches of mathematics, and then, in 1917, took up a teaching position there. Soon after the October Revolution, he identified himself with the cause of Bolshevism, and in December 1918, he became a full member of the Communist Party, for which he received government rewards of impressive magnitude. He occupied many influential positions in institutions close to the central government and played a major role in formulating the guidelines for the establishment of a national system of state-supported publication activities, an undertaking of enormous proportions. The Marxist authorities expressed full confidence in Shmidt's loyalty by appointing him editor-in-chief of the *Great Soviet Encyclopedia*. His grand plan to publish a general history of science from a Marxist standpoint collapsed under the burden of over-ambitious expectations and the unsettled state of Marxist philosophy.

As it turned out, Shmidt did something more than mediate. Leading the delegation of the Communist Academy to the Second All-Union Conference of Marxist-Leninist Scientific Institutions in 1929, he unleashed a biting attack on Marxist theorists who, looking for idealistic impurities in Einstein's philosophical views, overlooked the "materialistic core" of the theory of relativity. In endorsing Semkovskii's line of reasoning, he explained that "the elements of dialectics found a clearer expression in the theory of relativity than in any other modern theory." "The theory of relativity," Shmidt concluded, was "thoroughly dialectical."[86] In addition to Semkovskii, Shmidt relied on the authority of Hessen, who a year earlier had published a lengthy essay on the relationship of the theory of relativity to dialectical materialism in which he boldly asserted that Einstein's theory gave "concrete expression to the dialectical conceptions of space and time."[87] Shmidt asserted that Hessen's study offered a more thorough application of dialectical materialism to the theory of relativity than any other work on the subject. Indirectly, he accepted the idea of complete harmony between the two, and sided with a more liberal and open-minded interpretation of Marxist philosophy. In an article published in 1937, Ioffe credited the 1929 conference with providing the "sound" basis for the continued discussion that led to a general acceptance of the theory of relativity, a law of nature "as firmly established [now] as the law of the conservation of energy."[88]

Shmidt pointed out the essence of the Einsteinian revolution. Newtonian or classical physics, he said, separated space from time and both from matter; Einstein united space and time and made space-time and matter inseparable. In all this, he also eliminated ether from the legitimate domain of physics. Shmidt was firmly convinced that the positive curvature of cosmic space, as Bernhard Riemann saw it in his version of non-Euclidean geometry, provided the true key for the explanation of universal gravitation. These ideas formed what he identified as "the materialistic kernel of the theory of relativity." He advised Marxist scientists and philosophers to concentrate on strengthening and broadening the materialistic base of Einstein's theories by eliminating the last residues of Machian philosophical influence.

Shmidt's address did not end the anti-Einstein campaign, but it did contribute to a lessening of its intensity. Although Timiriazev did not withdraw from the battlefield, his arguments became less logical and scientific and more quarrelsome and ideological. In 1930, for example, he wrote that the excitement about the originality of the special theory of relativity had no basis in fact; Einstein himself had admitted that all he did was to give a mathematical formulation to the ideas advanced by Ernst Mach. And as Lenin had declared in *Materialism and Empirio-Criticism*, a mathematical formulation that failed to draw a line between "functional" and "causal" analysis did not make Mach's ideas any sounder on philosophical grounds.[89]

Shmidt's proclamation contributed to a partial and transitory retreat of the Timiriazev forces, but it did not contribute much to a consolidation of antimechanist factions. Nor did it prevent Einstein's philosophical theory from continuing to cause consternation, anguish, and much confusion among the articulators of Marxist theory, and from keeping alive the deep and incongruous rift between physics and philosophy. For years to come, in fact, the oscillating intensity of attacks on the "idealism" of Einstein's theory of knowledge served as a barometer of ideological pressure.

Until the end of the 1920's, and perhaps a few years later, Soviet philosophers were the most active participants in discussions on the theoretical foundations of the theory of relativity. In shaping their view and conclusions, they acted independently of the scientific community, which had made perceptible progress in absorbing, diffusing, and advancing the avalanche of new ideas triggered by modern physics and had shown little interest in questions of philosophical import. The debates between various Marxist groups, Joravsky has noted, had no relevance "to the professional problems and disputed issues of physicists who took relativity and quantum mechanics for granted, or worked in specialties untouched by them."[90] In 1929, an editorial in *Under the Banner of Marxism* announced that the time had come to make the philosophy of science the exclusive domain of people with proven professional competence in "individual disciplines and their dialectics."[91] The appeal marked the first serious effort to attract active scientists to the crucial issues confronting the philosophy of science and to build more versatile ties between science and philosophy.

Until close to the end of the 1920's, theoretical controversies in the community of Marxist philosophers paralyzed all government efforts, diffuse as they were, to forge ideological unity in the assessment of Einstein's legacy. But that situation was soon to end as the central authorities set out to give official ideology a monolithic structure and to saturate every phase of social and cultural activity with its norms, guarded by the coercive power of the state. Efforts to make Marxist philosophy a unified system of thought were now in full force.

The Scientific Community and the Theory of Relativity

In the pre-Stalin era, the scientific community faced the task of learning to live with the sweeping effects of two revolutions, each profoundly affecting a distinct domain of human experience and reality. The Bolshevik Revolution demanded accommodations to radical changes in the fabric of society, to rising pressures of a new and alien political ideology, and to a philosophical outlook dominated by elaborate systems of materialistic principles, often in open conflict with the spirit of modern science. Increasingly involved in a search for help from various dynamic philosophical schools, mainly with neopositivist and neo-Kantian leanings, scientists were suddenly confronted with a rigid, ideologically saturated, and officially articulated system of philosophical thought.

The other revolution, totally unrelated to the first, was the one that had taken place in physics, the most dynamic and advanced science. It undermined the traditional faith in the golden laws and unlimited intellectual power of Newtonian science, which viewed the universe as a rigid system of mechanical principles. The new physics challenged the invincible authority of both Aristotelian logic and Euclidean geometry. It gave a powerful advantage to a generation of physicists unburdened by a heavy load of traditional habits and methods in pursuing scientific thought.

In the prerevolutionary years, some physicists were reluctant to accept post-Newtonian ideas. This was generally true of older scholars who held tenured university positions, wrote textbooks based on traditional sources, and considered themselves founders of scientific schools. Nevertheless, during the 1920's, the acceptance of new ideas was widespread and enthusiastic. By then, the theory of relativity and quantum mechanics were so firmly established in the West they could no longer be considered passing fads. A large influx of Jewish scholars into the Soviet community of physicists also contributed much to the growing acceptance of new ideas and theories. Previously limited to only a fringe representation on university faculties, those scholars had no vested interests in the tattered theoretical or paradigmatic commitments of "old science" and were ready to meet, and to add to, the challenges of the most recent developments in both scientific theory and experimental methods. They also brought a distinct culture that underscored disciplined participation in the burning intellectual problems, past and present, and were inspired by Einstein as a model of daring spirit, pioneering zeal, and superior achievement in the pursuit of science. Their models included also Georg Cantor and Hermann Minkowski, the founders of new branches in mathematics that became important tools of the twentieth-century revolution in science. The families of both Cantor and Minkowski had historical roots in Russia.

The Confrontation Between the Old and New Physics

The first representative of the Soviet scientific community to discuss the theory of relativity was a defender of the Newtonian legacy. In an article published in the bulletin of the Russian Academy of Sciences in 1919, N. P. Kasterin, a physicist from Odessa University, could find neither experimental nor mathematical support for Einstein's ideas and tried to create the impression of a full detachment of the theory of relativity from physical reality.[1] The paper impressed most contemporary Soviet physicists as an expression of nostalgia for the rapidly declining reign of mechanistic philosophy in physical thought. Kasterin was least willing to part with the key notion of ether.

Kasterin recognized two principles of relativity, one connected with the name of Galileo and built into the grand edifice of Newtonian science, and the other associated with the name of Einstein and built on the Lorentz transformation and the Maxwell-Lorentz electrodynamics.[2] He thought that Einstein erred in identifying the Galilean—or "mechanical"—principle of relativity as "approximative" and his own ("electrodynamic") principle as "absolutely exact." Obviously favoring the Galilean principle, he criticized Einstein's special theory of relativity as lacking empirical support. His verdict was based on the physical experiments of Alfred Bucherer and Walter Kaufmann, which he interpreted as proof that Einstein was incorrect in claiming that the speed of light represented the maximum speed in the universe. But in his eagerness to refute the theory of relativity, Kasterin had failed to take note of the fact that the Bucherer-Kaufmann views were now generally recognized as being founded on inadequately conducted experiments. The publication brought Kasterin the reputation as the chief critic of Einstein's theory of relativity within the Soviet scientific community.

A. K. Timiriazev, whose activities were discussed in Chapter One, was a trained physicist who, in the Soviet era, undertook no physical research of consequence, but instead concentrated on organizing a philosophical campaign against the theory of relativity. Once he became firmly entrenched as a Moscow University professor of physics, he helped bring Kasterin to the capital for the purpose of strengthening the "scientific" sources of anti-Einsteinian ideas. Holding temporary positions first in the Communist Academy and then at Moscow University, Kasterin was given a chance to conduct independent research in the hope of strengthening Timiriazev's crusade against the "idealistic" underpinnings of Einstein's theoretical thought.[3] Although Kasterin was a lethargic and undisciplined researcher with little standing among his peers, Timiriazev continued to give him his whole-hearted support. In the meantime, Timiriazev continued to fuel his own anti-Einsteinian philosophical attacks with "scientific" ideas produced by the rapidly shrinking number of Western physicists, such as Werner Lenard in Germany and Dayton Miller in the United States, deeply involved in a crusade against the theory of relativity.

Unlike the philosophers, the physicists presented a strong front in defense of the new physics. As a rule, their writing aimed not at producing a popular version of the theory of relativity but at presenting an accurate and technically competent interpretation of the great physicist's major scientific ideas. Most felt that a philosophical discussion would distract them from the real substance of Einstein's theoretical thought. They were ready to acknowledge the revolutionary significance of relativistic principles in physics. In the words of O. D. Khvol'son, there was an honored link between classical and modern physics: "Along with the collapse of the old scientific edifice, there emerged a new theory whose beauty and inner harmony were unsurpassed. It laid the foundations for a modern world outlook, deeper and broader than the one connected with the name of Copernicus."[4] Although the general theory of relativity owed its existence only to Einstein, it represented, according to Khvol'son, a unity of three historical tributaries: Mach's idea of the relativity of motion; the non-Euclidean geometrical ideas advanced by Nikolai Lobachevskii, Karl Gauss, and Bernhard Riemann; and the idea of the curvature of space-time. In another study, quickly translated into German, he stated that "in the entire history of man, no scientific advance had made a deeper and more lasting impression on the broad and varied strata of civilized nations than Einstein's theory of relativity."[5]

In 1922, Khvol'son published an updated version of *Einstein's Theory of Relativity and the New Worldview*, a work he wrote before the October Revolution in an attempt to popularize Einstein's principal ideas and to depict the theory of relativity's revolutionary impact on scientific thought.[6] What Khvol'son found particularly impressive in the new direction in physics, which he accepted without major reservations, were Einstein's fearless departures from conventional wisdom and his emphasis on the active role of the observer as a key factor in the scientific construction of physical reality. Maksimov praised the style and substance of the Khvol'son book, but was put off by its animated excursions into the world of idealistic philosophy.[7] Khvol'son was a founder of the strong Soviet tradition that treated the theory of relativity not only as a revolutionary step in the advancement of modern science but also as an appealing aesthetic creation.

Khvol'son's was one of a flurry of works in the early 1920's on the changes taking place in science. In 1921, A. F. Ioffe, a newly elected member of the Russian Academy of Sciences, a noted expert on the electrical qualities of quartz, and a section head in the recently founded State Roentgenological and Radiological Institute in Petrograd, discussed "the new paths in physics," giving due attention to the contributions of the special and general theories of relativity to the most modern interpretation of universal gravitation based on non-Euclidean geometry.[8]

In the same year, V. K. Frederiks became the first Soviet physicist to present a systematic exposition of the general theory of relativity.[9] In 1924, V. A. Arnold published *Einstein's Theory of Relativity*, an expanded version of a lecture he had delivered in 1923 at a joint session of the faculties of the Institute of Transportation Engineers and the Institute of Civil Engineers.[10] He fully endorsed Jean Becquerel's claim that Einstein's ideas made it impossible to return to the old views in physics.

Among this spate of publications were two penned by Iakov Frenkel, one of the most talented and productive Soviet physicists. Born into a cultivated middle-income professional family, he had even as an adolescent displayed a strong inclination toward mathematics and established himself as an accomplished violinist and portrait painter. His involvement in physics was characterized by a diversity of interests, particularly in the general fields of atomic physics and electrodynamics. Although the theory of relativity was not his major concern, his contribution to the diffusion of Einstein's ideas in the Soviet Union was outstanding. In 1923, he published *The Theory of Relativity*, a careful effort to present the intricate conceptual structure of Einstein's theories shorn of mathematical complexities.[11] That was followed two years later by a popular article tracing the history of ether from its early setting in Christiaan Huygens's physical theory to its full rejection by the special theory of relativity in 1905.[12] Useful "scaffolding" for the physical theory of the Newtonian mechanical orientation, ether could not find a functional place in post-Newtonian physics. It was defended by conservative physicists like Lenard in Germany and Oliver Lodge in England, whose undying loyalty to Newton made them skeptical about the gushing streams of new scientific ideas. Frenkel told his readers that the death of ether marked the emancipation of science from its last links with mysticism.

Awarded a Rockefeller Foundation grant, Frenkel spent two years (1925–26) in Germany, France, and England working on a comprehensive review of electrodynamics, with special emphasis on its microphysical aspects. Volume One of that study, presenting a "general mechanics of electricity," dwelt exclusively on the links between the special theory of relativity and electrodynamics.[13] The book appeared simultaneously in German and Russian editions. In Frenkel's opinion, the origin of the theory of relativity was to be sought in "the fiasco of numerous efforts to place electromagnetic processes within the framework of a mechanical orientation in physics."[14]

Frenkel's major interest was in adducing convincing evidence that the safest path to an understanding of the absolute aspects of the laws of nature lay in the relative attributes of methodological operations in physics. In his opinion, the "theory of the absolute" was the most appropriate name for the theory of relativity.[15] In making that claim, which was identical with a suggestion Minkowski had made in 1908, Frenkel undoubtedly wanted to allay the apprehension of Marxist critics—with whom he generally had very little contact—about making relativity a quintessential feature of scientific knowledge. In most of his writings, he manifested an impressive awareness of the epistemological complexities of the new physics.

Western tours—including a one-year (1930–31) visiting professorship at the University of Minnesota—brought Frenkel in close touch with many leading physicists working on the new frontiers of science. He considered himself particularly fortunate to have had rewarding interchanges with Max Born and Albert Einstein. More than anything else, personal contacts with the leaders of world science helped bring the Soviet scientific community closer to the spirit, style of work, and inexorable philosophical involvement of the architects and masters of post-Newtonian science.

Around the mid-1920's, a certain caution crept into the physicists' writings, demonstrating that the times were changing. In 1927, when Ioffe returned to the world of Einstein's ideas by informing *Pravda* readers that the theory of relativity was both the leading body of knowledge in contemporary physics and "the best foundation for a materialistic world outlook," he took care to note that some of the claims of the general theory still awaited experimental verification.[16] He obviously made that concession to the Marxist critics of the theory of relativity to avoid publicly aired and generally unproductive controversies.

In 1928, S. I. Vavilov, an expert in luminescence, showed somewhat less discretion. In a long and perceptive article on the experimental foundations of the theory of relativity, he gave a public airing to the view, quietly shared by many physicists, that "the philosophical controversy over relativity, space, and time" had no relation to "the question of the correctness of the theory."[17] Vavilov argued that the special theory of relativity rested on such a solid foundation that all its theoretical claims could be considered verified even where additional mathematical operations had not yet been completed. To be sure, the general theory of relativity needed additional verification[18]—it required a more precise explanation of the physical attributes of the electromagnetic field, the nature of the electron, "certain cosmic phenomena," and links with quantum actions—but its theoretical core rested on a solid experimental foundation. It belonged among theories that grew and were perfected. Vavilov concluded that "while the number of experiments supporting the general theory of relativity was small, none conducted under rigorous experimental conditions contradicted it."[19] His critical statements, like those of Ioffe, did little to disguise his great admiration for Einstein's achievement in physics, and in science in general, but they helped soften the intensity of ideologically motivated Marxist attacks on the scientific community.

Every chapter in Vavilov's study was introduced by an epigraph from Newton's works, a choice that, in the estimation of Vladimir Vizgin, the leading Soviet expert on the genesis and early evolution of the general theory of relativity, went beyond simply showing there was little in Einstein's system of theoretical principles that would not be acceptable to Newton and that was not anticipated in his writings.[20] After all, he might have selected epigraphs from the works of Galileo or Descartes—or, for that matter, from those of such modern "classics" as Hermann von Helmholtz and James Maxwell. Vizgin offered several explanations for Vavilov's decision. At the time of his writing, the scientific community was engaged in preparations to commemorate the bicentennial of the death of the great English scientist. Moreover, at that point Vavilov had just produced the first Russian translation of Newton's *Opticks* and published an article on the Newtonian theory of light. Newton was also a good choice because he represented the purest expression of the classical orientation in physics, just as Einstein represented the purest expression of the relativist orientation.

Vizgin added another very important reason for Vavilov's decision. Newton figured prominently in the feud between the mechanists and the dialecticians: the mechanists tried to strengthen their anti-relativist arguments by invoking the authority of Newton, and the dialecticians cited Newton for the purpose of showing the deep historical roots of Einstein's ideas. As Vavilov's earlier writings indicated, he sided with the dialecticians and used every opportunity to bolster their effort to attune dialectical materialism to the theory of relativity. He pursued his task without becoming involved in either the raging controversies over the merits of the new physics or the tedious philosophical debates over Einstein's ideological leanings.

Vavilov knew that all groups, despite major differences in attitudes toward Einstein's achievement, continued to hold the highest respect for the colossal proportions of Newton's contributions to science. He also knew that the supreme architects of official Soviet ideology acknowledged the Newtonian foundations of three cardinal principles of Marxist thought: the principle of causality as the pivotal mechanism of scientific explanation, the principle of continuity in natural and cultural processes, and the principle of the objective nature of physical reality as studied by science.

In 1929, the twenty-two-year-old Matvei Petrovich Bronshtein pushed the debate a step further toward confrontation. In a short article in the journal Man and Nature, he elaborated on Einstein's current search for a unified field theory. In his opinion, Einstein worked on a general theory that promised to cover both universal gravitation and electromagnetism. Einstein, he thought, built the geometrical foundations of gravitation and the inner logic of covariance in the study of electromagnetism, but he bypassed the physical micro-universe unveiled by quantum mechanics. That omission notwithstanding, Bronshtein was under the impression that the unified field theory was well on its way to successful completion.[21] Although he did not touch on the matter of philosophy, the idea of the geometrical foundations of the unified field theory, as he presented it, clearly contradicted Marxist claims of the primacy of physical empiricism over geometrical formalism—of physics over geometry. Like Vavilov and Ioffe, Bronshtein made little use of Marxist metaphors and allusions.

At roughly this time (the end of the 1920's), Igor Tamm, in response to popular demand, offered a course at Moscow University on the general theory of relativity, presenting an earnest effort to examine the intricate texture of theoretical structures involved in Einstein's search for a unified field theory. This was also the time when two other prominent physicists, V. A. Fock and D. D. Ivanenko, came up with original suggestions for building a bridge between the theory of relativity and quantum mechanics.[22] Their arguments, insufficiently elaborated, did not make a strong impression on contemporary physicists. However, several Russian physicists, led by Tamm, showed some interest in exploring the English physicist Paul Dirac's successful steps toward creating a relativistic orientation in quantum mechanics.

A. A. Friedmann and the Idea of the Expanding Universe

As the work of Fock and Ivanenko showed clearly, Soviet physicists did not limit their work to a mere interpretation of Einstein's scientific work. They also made bold efforts to add original ideas to the new physics. Among them the place of greatest distinction belonged to Aleksandr Aleksandrovich Friedmann, a scholar of diversified interests, a keen mathematical mind, and unbounded intellectual resourcefulness. The son of a musician with broad cultural interests, Friedmann was born in St. Petersburg in 1888. As a high school student, he teamed up with Ia. D. Tamarkin—one of his youthful peers and, much later, a mathematics professor at Brown University—in investigating the ever-challenging Bernoulli theorem. He was just eighteen years old when a paper he coauthored with Tamarkin was published in *Mathematische Annalen*, the most prestigious German mathematical journal, and one of the best, if not the best, in the world. In 1910, Friedmann graduated from St. Petersburg University and was retained by his department as a "professorial candidate"; at that point, he was a pure mathematician with a strong emphasis on differential equations.

During the First World War, Friedmann served in a research branch of the imperial air force, working chiefly on problems related to weather forecasting and air navigation. His effort at that time to formulate a theory of atmospheric turbulence was more an exhibit of daring and skillful mathematical constructions than a model of impeccable physical analysis. In 1918, after working at the Pavlovsk Meteorological Observatory (soon to be renamed the Main Physical Observatory) for a while, he was transferred to Perm to introduce courses in higher mathematics at the newly founded local university and to help in carrying out the educational policies of the new government.[23]

In 1920, Friedmann returned to Petrograd, where he immediately plunged into many professional activities. As a member of the Atomic Commission of the State Optical Institute, he delivered two papers on the mathematical aspects of quantum theory. At different times, he offered courses on various branches of mechanics at Petrograd University, the Institute of Communication Engineers, the Petrograd Polytechnical Institute, the Naval Academy, and the Mathematical Bureau of the Main Physical Observatory.[24] He was the initiator of and the main contributor to the revitalization of the Petrograd Physical and Mathematical Society, an important organizational link between scientists scattered throughout the many academic institutions of the great northern city and its region. His success in teaching, research, and key administrative assignments brought him many privileges, including an opportunity to visit Western centers of science. His selection to serve as editor of the geophysical section of the *Great Soviet Encyclopedia* came as no surprise to experts in the field. In 1925, shortly before his untimely death from typhoid fever, he was appointed director of the Main Geophysical Observatory, whose mathematical section was much stronger than its experimental and observational base.

In 1917, the Berlin Academy of Sciences had published Einstein's study "Cosmological Considerations on the General Theory of Relativity," which laid the foundations for relativistic cosmology, one of the fastest growing and most exciting branches of scientific knowledge in the twentieth century.[25] The universe, as seen by the new cosmology, was homogeneous, isotropic, closed (spherically bounded), and static (unchanging in time), a picture of harmony, symmetry, simplicity, and unity that, in Einstein's view, made the scientific study of the structure of the universe possible in the first place. Five years later, in 1922, Friedmann surprised the scientific world by challenging the cosmological implications of the general theory of relativity in *Zeitschrift für Physik*, the German journal that, in publishing the fundamental papers of Born, Heisenberg, Pauli, and many other leaders of modern science, played a major role in transmitting the international advances in the new physics. In a daring and unexpected mathematical interpretation of the general theory, Friedmann contended that its mathematical formulas permitted the assumption that the space curvature of the world was "nonstatic" and even "expanding."[26] Friedmann's was the first major deduction from the grand logic and mathematics of the general theory of relativity. In his treatment of that theory, no less than in his work in theoretical meteorology, Friedmann relied on his own adaptation of differential equations with partial derivatives. Oddly enough, the same issue of the *Zeitschrift* carried an appeal to Western scientists to help in collecting recent journals and books for the benefit of Soviet scholars, isolated as they were from the Western centers of learning by the years of war and revolution.

Although Friedmann stood by the mathematical rigor of his grand deduction (even though he was rather skeptical about the feasibility of empirical support), Einstein, in a brief note published in the same journal a few months later, dismissed the new idea on both mathematical and physical grounds. But within a matters of months, he reversed himself. In another note to the journal in mid-1923, he said that, after further study, he was now convinced that Friedmann's contribution was both "correct" and "innovative."[27] The Soviet historian of science George Gamow, who readily admitted that he learned his relativity from Friedmann, wrote many years later that Einstein told him that his initial uncertainty about the idea of the expanding universe was "the biggest blunder" he had made in his entire life.[28]

What made Einstein change his mind on the scientific merit of Friedmann's daring proposal? A letter he received from Friedmann immediately after his published criticism, offering additional methodological explanations of his claim, may have had something to do with it. In it, Friedmann also stressed his keen interest in deducing a picture of constant negative curvature from Einstein's equations.[29] But what probably most persuaded Einstein to take a fresh look was the defense of the notion of the expanding universe by Iu. A. Krutkov, a Soviet theoretical physicist, during a discussion at the home of Paul Ehrenfest, a well-known Leiden University physicist, in the spring of 1923.

At any rate, by 1931, in a short paper published by the Berlin Academy of Sciences, Einstein was ready to state that Friedmann's proposition had convinced him that "the static nature of space" could no longer be defended, and that the time had come to integrate the new idea into the general theory of relativity.[30] In 1945, in an appendix to the second edition of *The Meaning of Relativity*, he noted that Friedmann's theory "found surprising confirmation in Edwin Hubble's discovery of the expansion of the stellar system."[31] In the words of Steven Weinberg, Friedmann's analyses based on Einstein's original field equations provided "the mathematical background for most modern cosmological theories."[32]

In 1924, Friedmann published *The World as Space and Time*, a general study of Einstein's principles designed to find a happy middle-ground between excessive popularization and burdensome technical detail. In perhaps the most attractive and challenging part of the book, he concentrated on Hermann Weyl's bold effort to combine an experimental study of matter with a picture of the universe drawn from Bernhard Riemann's non-Euclidean geometry, David Hilbert's work on axiomatics, and Einstein's theory of relativity. Friedmann presented Einstein's theory as the prime mover in the modern search for a synthesis of the most advanced branches of physics, cosmology, and mathematics. He based his main arguments on Einstein's idea that all attributes of matter, which in his view consisted of electromagnetic processes, came from the geometrical characteristic of the world. In the final analysis, he wrote, the world was made of geometrical features.[33]

Friedmann was firmly convinced that Einstein's theory had experimental support, and that its real strength was not only in answering previously asked questions, but in advancing new perspectives for unheralded and fundamental developments in physics. He told his readers that Einstein's theory applied to the entire universe and opened the gates for new uses of astronomical data. Above everything else, it suggested new lines of development in some of the most advanced branches of mathematics. Although the book was written in a lucid style, it proved to be too intricate to attract many readers and to evoke much discussion.

In the same year (1924), Friedmann teamed up with Frederiks in writing a mathematical treatise on the theory of relativity intended for university students specializing in physics. His sudden death in 1925 cut short that ambitious enterprise, but the one volume they managed to produce showed convincingly that Friedmann held an atypical position in his judgment on the future development of the theory of relativity. Contrary to prevalent opinion in Marxist circles, he believed that the tools of mathematics provided the only reliable path to the elaboration and enrichment of Einstein's physical theory. As might have been expected, the published volume of *Foundations of the Theory of Relativity* dealt exclusively with tensor calculus. The next two volumes were expected to treat other relevant branches of mathematics, in addition to the fundamental principles of electrodynamics, the conceptual gateway to the world of relativity. The Friedmann-Frederiks collaboration was a most fortunate one, at least according to Fock, who heard both physicists present reports and participate in discussions: whereas Frederiks had an excellent command of modern physics, but did not show much interest in mathematics, Friedmann was a master mathematician with only a secondary interest in physical theory.[34]

Marxist critics were quick to attack Friedmann and Frederiks for their acceptance of Einstein's ideas of the "arbitrary" and "hypothetical" nature of mathematical formulations in physics.[35] They were also quick to condemn all favorable comments on the growing trend toward a full axiomatization or geometrization of physics and on current proposals to make logic the chief guide and source of principles in building mathematical structures. Friedmann reasoned that the theory of relativity, as the key to a new and more comprehensive interpretation of cosmological problems, was slow in developing because its guiding physical principles were far ahead of the available mathematical tools necessary for theoretical constructions.[36]

Soviet astronomers and most Marxist commentators on theoretical developments in science referred to Friedmann's achievement as little as possible. The general Marxist attitude was ambivalent and adrift. On the one hand, Marxist writers were quick to point out the "idealistic" subtext of the theory of the expanding universe; on the other hand, they hailed but did not elaborate on Friedmann's "rejection" of Einstein's "static" approach to the universe. After all, the Marxist philosophy of science favored a dialectical approach to nature, a unique form of historical orientation.

Marxist writers were strongly inclined to interpret Friedmann's cosmological theory as an attack on the idea of the infinity of the universe, an exalted pillar of official Soviet ideology. Ideological considerations like this played an important part in making Soviet astronomers reluctant to discuss the scientific merits of the idea of the expanding universe. On the ideological level, Friedmann's cosmological idea opened the doors to creationist ideas, which Marxists considered the most pernicious adversary of dialectical materialism. What made the situation all the more alarming was the fact that the head of the Vatican Academy, Georges Lemaître, who was also a prominent cosmologist, was generally considered a major contributor to the idea of the expanding universe.[37] It came as no surprise to alert contemporaries that Marxist writers chose "fideism" as the most appropriate label for the philosophical message of the notion of the expanding universe. Both Marxist writers and professional astronomers felt uncomfortable about the chain of paradoxical events that saw a triumph of Soviet science darkened by the suffocating effects of official ideology. Ironically, the Soviet Union was one of the last countries to accord full recognition to Friedmann's cosmological ideas.

A. V. Vasil'ev: Historical Roots of Einstein's Ideas

Friedmann involved himself in carrying the theory of relativity to new heights and to new truly revolutionary ideas in science. The mathematician Aleksandr V. Vasil'ev worked in the opposite direction: he delved deeply into history to demonstrate that behind the theory of relativity there were long and fertile sequences of scientific and philosophical thought. That fact, he reasoned with admirable clarity, in no way belittled the grandeur of Einstein's achievement; on the contrary, those historical roots helped put it into sharper focus and give it a more definite place in the evolution of scientific-philosophical thought.

Vasil'ev's historical analysis of the ideas and currents that provided valuable hints for the philosophical foundations of the theory of relativity were discussed in Chapter One. Here we shall concentrate on his views and comments on the streams of thought that initiated the twentieth-century revolution in science. As a scientist with a deep interest in philosophy and fully formed epistemological views, Vasil'ev was inspired by, and attracted to, scientists who showed a sophisticated level of philosophical involvement.

Vasil'ev's *Space, Time, Motion* appeared in 1923.[38] Thanks to a flaw in the rapidly expanding machinery of government censorship, the book was published despite its unmitigated attack on the "metaphysical" foundations of dialectical materialism, a dismissal of the ideas of Marx, Engels, and Lenin that was particularly pronounced in the English translation published the same year. Vasil'ev here provided the lone Soviet voice in defense of Ernst Mach as a major contributor not only to the philosophical side of Einstein's theory but also to its scientific grounding. At the time, all Marxist philosophers were united in considering Mach's theoretical views an unfailing sign of the decadence of Western thought and society—the only exceptions, the Marxist "revisionists" represented most ardently by A. A. Bogdanov, having been almost completely silenced by now. Vasil'ev introduced Soviet readers to the view of Mach's influence on Einstein that was predominant in the West; it was a picture of the theory of relativity that was substantially different from the one presented by Marxist writers.

Vasil'ev deserves credit as the first Soviet scholar to make a comprehensive effort to trace the scientific antecedents of Einstein's theory. For him, two major currents of thought in the nineteenth century had laid the foundations on which the grand edifice of Einstein's scholarship was built: the emergence, evolution, and recognition of non-Euclidean geometries and the rapid rise of the scientific awareness of physical problems that classical—Newtonian—physics could not resolve. As a result, by the beginning of the twentieth century, the time had come for bold efforts to establish broader theoretical perspectives for higher levels of integration of the existing principles of scientific knowledge, particularly in physics. Vasil'ev saw both the theory of relativity and quantum mechanics as manifestations of accelerated moves or leaps that did not violate the law of continuity in the growth of scientific knowledge. He credited Poincaré with merging non-Euclidean geometry with the theoretical perspectives in physics advanced by brilliant scientists from Maxwell to Lorentz.

No Russian scholar concerned himself more thoroughly and understandingly with Lobachevskii's non-Euclidean geometry, its inner structure and history, than Vasil'ev. He traced the tangled story of non-Euclidean geometry from the day in 1826, when Lobachevskii presented his first model, to 1916, when Einstein made Riemann's model an integral part of the general theory of relativity. In a magnificent example of international cooperation in science, Karl Gauss and Felix Klein in Germany, Eugenio Beltrami in Italy,

William Kingdon Clifford in Great Britain, and Henri Poincaré in France paved the way that led from Lobachevskii's original idea to Einstein's explanation of the nature of gravitation. Newton claimed that gravitation worked at a distance, but he could not explain what made it work in the first place. The general theory of relativity, relying on modern field theory, rejected the Newtonian idea that gravitation worked at a distance and explained what made it work.

Although Vasil'ev recognized Lobachevskii as the man who started the series of developments that made non-Euclidean thought a major component of modern science, he considered Riemann's contribution the decisive link between the general theory of relativity and non-Euclidean geometry. In a paper presented at Kazan University in 1926, on the occasion of the centennial of the first non-Euclidean geometry, Vasil'ev also acknowledged Lobachevskii's contribution to the foundations of axiomatics—a reduction of mathematical or physical principles to a logically arranged and integrated system of axioms, first undertaken by David Hilbert in *The Foundations of Geometry* and strongly supported by Einstein.[39]

Vasil'ev's brave, though tortuous, tour through post-Maxwellian developments in physics demonstrated the fading power of Newtonian science to explain some key results of both experimental research and theoretical thinking. His main task here was to show how the development of electrodynamics from Faraday to Poincaré had laid the groundwork for Einstein's special theory of relativity by producing a fusion of physics and geometry. By his account, Armand Fizeau's "ether-drag" experiments, Albert Michelson's "ether-wind" experiments, the FitzGerald-Lorentz contraction hypothesis, the Lorentz transformation equations, and the Eötvös experiments had all provided concrete ideas that found their way into the theory of relativity.

The mathematician William Clifford, the first English supporter of non-Euclidean geometry, was in Vasil'ev's eyes the true link between Lobachevskii and Einstein. Not least was Clifford's role in making Lobachevskii's views known to his countrymen in a lecture delivered at the Royal Institution in London in 1873, where he compared Lobachevskii with Copernicus: "Each of them has brought about a revolution in scientific ideas so great that it can only be compared with that wrought by the other. And the reason of the transcendent importance of these two changes is that they are changes in the conception of the Cosmos."[40]

Clifford, as Vasil'ev portrayed him, not only appreciated the vast proportions of Lobachevskii's contribution to the scientific view of the universe, but also anticipated the appearance of the general theory of relativity. In his meditations on Riemann's creation of a distinct non-Euclidean geometry,

Clifford considered the notion of curved cosmic space a powerful spring-board for future developments in geometry and cosmology. Vasil'ev was im-pressed with Clifford's repeated statement that physical changes were closely connected with changes in the curvature of space.[41] With that insight, he be-lieved, the idea of the curvature of space was well on the way to Einstein's space-time curvature as the key factor explaining gravitation.

Of all the ideas that provided a background for the theory of relativity, Einstein, according to Vasil'ev, assigned the most important place to those that came from Mach as both a scientist and a philosopher. As a scientist, Einstein deeply appreciated Mach's persevering criticism of the mechanical exclusiveness of Newtonian science. From Mach he received the challenging and fertile claim that the law of inertia must consider "the masses of the uni-verse," and that "only relative motions exist."[42] (Long after, Einstein himself, in his "Autobiographical Notes," published on the occasion of his seventieth birthday, acknowledged Mach's strong influence on the development of his ideas, though in his later life his thinking had gone in other directions.[43])

The principle of relativity, in Vasil'ev's view, expressed the previously overlooked unity of mechanical and electromagnetic phenomena.[44] It re-solved the Newtonian impasse by submerging the mechanical view of the physical world into a larger, more flexible, and more fertile outlook: frag-mentary but cumulative antecedents did not prevent Vasil'ev from viewing the theory of relativity, both special and general, as the creation of one man—Albert Einstein. All leading physicists agreed with him.

The Spirit of the Age

During the 1920's, the intellectual climate in the Soviet Union encouraged diversity in the interpretation of the main theoretical advances in modern physics. Stimulated particularly by the excitement created by the unortho-dox ideas of the theory of relativity, Soviet intellectuals often sought to go beyond the limits of Einstein's claims. In a 1929 letter to Einstein, Max Born told him he had met a young Russian who formulated intelligent arguments in favor of a six-dimensional theory of relativity.[45] Persuaded by Born to read the young man's paper on the subject, Einstein concluded that it presented "original and formally well-developed" ideas and that steps should be un-dertaken to make it possible for the young man—Iu. B. Rumer—to continue his work. Born and Rumer wrote a joint article on nuclear physics, and many years later, after his release from a Stalinist forced labor camp, Rumer joined Landau in writing a popular account of the basic principles of the theory of relativity.

At the end of the decade, most leading Marxist philosophers recognized the fundamental compatibility of the scientific foundations of the theory of relativity with the philosophical principles of dialectical materialism. But they did not acknowledge a similar compatibility of Einstein's philosophical views with dialectical materialism. From now on, Soviet philosophers took on the arduous task of creating a Marxist theory of relativity as a blend of Einstein's scientific principles, modified or not, and the philosophical foundations of dialectical materialism. This task was made all the more difficult by the enormous difficulty of establishing a clear line between Einstein's science and Einstein's philosophy. Generally, however, Marxist theorists manifested a strong predisposition to consider most of Einsteinian science a specific mode of philosophical expression.

A. M. Deborin, the head of the Deborinite or dialectical faction, was the architect of a strong, but paradoxical, Marxist approach based on two principles working at cross-purposes. On the one hand, he was eager and ready to recognize the theory of relativity—as well as quantum mechanics—as a revolutionary step in the advancement of science and to argue in favor of a basic integration of Einstein's scientific principles into the intellectual framework of dialectical materialism. On the other hand, he viewed Marxist philosophy as a superior science and as such, the ultimate judge of scientific theories, an idea widely used and abused in the era of Stalinism. He approved of the theory of relativity not because it met the standards of scientific verification, but because, in his opinion, it measured up to the prescribed norms of dialectical materialism .

At this stage, the leading Soviet physicists showed little affinity for or contact with Marxist philosophers. They avoided dialectical materialism—the official Soviet philosophy of science—as a government-managed instrument of ideological control over the professional activities of the academic world. In the pre-Stalin era, no well-known physicist contributed to *Under the Banner of Marxism* or other leading Marxist journals. No leading physicist of the 1920's allowed himself to take part in a public debate on the relations of dialectical materialism to science. Timiriazev represented a minuscule number of physicists who chose to switch their full attention from exploratory physical research to the defense of dialectical materialism. Although he made extensive use of elaborate scientific arguments, his choice of topics was always guided by philosophical considerations, primarily by insisting on the full consistency of the pivotal postulates of Faraday-Maxwell-Hertz physics with the basic epistemological principles of dialectical materialism. He made extensive and thoroughly uncritical use of the offerings of current experiments aimed at discrediting the theory of relativity. He, for example, was one of the most enthusiastic supporters of Dayton Miller's claim to have fully discredited the Michelson-Morley experiments that led Einstein to dispose of ether as a physical object.[46]

In short, where Marxist philosophers were wholly preoccupied with the question of whether the theory of relativity bolstered or challenged their ideology and world outlook, the matter was of no concern to most physicists. For them, Einstein's theory and dialectical materialism were noncomparable systems of thought, each belonging to a clearly demarcated and separate realm of ideas. Most of the leading physicists concentrated more on understanding than on advancing the intricate web of mathematical and physical principles built into the theory of relativity. Several, including Ioffe, Frenkel, and Vavilov, tried to reach the general public with moderately popular expositions of Einstein's scientific—but not philosophical—ideas.

V. I. Vernadskii, a pioneer in the scientific study of the biosphere, came close to expressing the prevalent attitude among members of the scientific community toward Einstein's place in modern science. In a paper titled "Thoughts on the Importance of the History of Knowledge," presented at the 1926 inaugural meeting of the Commission on the History of Knowledge, organized in 1925 by the USSR Academy of Sciences, Vernadskii declared that Einstein's revolutionary changes in the conceptualization of space-time and gravitation marked a major break with the Newtonian tradition and had helped usher in a qualitatively different scientific picture of the world. He warned that it would take a long time for the scientific community to carry Einstein's theoretical principles to their logical conclusions. He also asserted that the Einsteinian revolution in science was much broader in compass and deeper in meaning than what usually went under the name of "great scientific discoveries in the nineteenth century."[47]

For all the solid backing within the community of physicists for Einstein's two theories, there was considerable disagreement over some of the individual principles embodied in them. Einstein's views on the role, if any, of ether in the theory of relativity appeared as one of the more notable sources of discord. Einstein did not help much in resolving the controversy. On April 2, 1920, he presented a paper at Leiden University in which, in an apparent reversal or modification of his earlier position, he asserted that "according to the general theory of relativity, space without ether [was] unthinkable; for in such space, not only would there be no propagation of light; there would also be no possibility of existence for standards of space and time (measuring-rods and clocks), and therefore no space-time intervals in the physical sense."[48] Einstein's unexpected move caused some confusion and uncertainty among Soviet physicists known for their enthusiastic acceptance of the theory of relativity. Most physicists were neither disturbed nor confused: they were firmly convinced that the "ether" Einstein made essential in that 1920 paper was something other than the "ether" he discarded in the special theory of relativity in 1905. But isolated physicists responded to the Leiden address by claiming that the time had come to reopen the question of ether for further discussion.[49] Bronshtein, for example, had no argument with Einstein's statement that "the ether of the general theory of relativity is a medium which is itself devoid of *all* mechanical and kinematical qualities, but helps to determine mechanical (and electromechanic) events."[50] He approved of Arthur Eddington's somber warning that "we need an ether" because without it the physical world could not be divided into "isolated particles of matter and electricity."[51]

In an essay titled "The Mystique of World Ether," published in 1925, Iakov Frenkel expressed the dominant view of the scientific community: the idea of ether was dead and buried, and no argument was forceful enough to justify a call for its revival.[52] He traced the history of ether from Huygens's original formulation in the seventeenth century to Lorentz's view at the end of the nineteenth century and Einstein's rejection in 1905. He took no note of Einstein's Leiden address. For him, though the notion of ether changed meaning in the course of more than two centuries, it always represented a phenomenon closer to religious mysticism than to the real world of science. The fall of ether, in Frenkel's view, marked a major intellectual conquest for modern physics. Nevertheless, even among some strong supporters of Einstein's theories, the question of the existence or nonexistence of ether was not finally resolved until the mid-1930's.

Under the Banner of Marxism, the chief Marxist theoretical journal from its founding in 1922, and the bitter debates between mechanists and dialecticians inaugurated in 1925 were the most notable vehicles for expressing the temper of the time and the dilemmas of unsettled conditions in the years between the October Revolution and Stalin's rise to political power. On the one hand, they demonstrated the presence of the full range of disagreements in Marxist views on the relations of dialectical materialism and official ideology to the key developments in contemporary science. *Under the Banner of Marxism*, for instance, offered articles on Cantor's set theory in mathematics, the place of biophysics in the family of science, Freud's psychoanalysis, Darwin's theory of organic evolution, modern genetics, Pavlov's neurophysiological approach to psychology, and quantum and relativity theories, to name a few, expressing as wide a diversity of opinion as existed in the West. On the other hand, Marxist philosophers, who faced diminished opposition with the gradually unfolding oppressive measures applied by the Bolshevik

government, worked steadfastly on saturating science with ideological premises and directives. They were the prime movers in making philosophy both a
servant of the official version of Soviet ideology and an ideological clearing-
house for scientific ideas. The rampant and tortuous wavering in Marxist
interpretations of modern thought, however, prevented them from making
the ideological control of science a firmly rooted, comprehensive, and consistent effort, at least during the 1920's. It helped give the Einsteinian debate
the semblance of an exchange of ideas uninhibited by political tutelage and
control.[53]

Despite chronic confusion and particularistic tendencies, a political
house-cleaning in the Academy of Sciences at the end of the 1920's signaled
the incipient mobilization of the scholarly community for active and direct
participation in philosophical discussions aimed at contributing to the cen-
trally guided and controlled unity of official Soviet ideology, Marxist philosophy, and the world of scholarship. It signaled, in other words, the onset
of Stalinism and the concerted attempt to mold the Soviet Union into a
monolithic state.

History, however, presents no true examples of fully formed monolithic
political systems, and the Soviet Union was no exception. The 1920's produced inner contradictions in the political-ideological structure of the Soviet
state that no totalitarian methods, even of the Stalinist kind, could eliminate.
Nowhere was that phenomenon more clearly evident than in the insistence
of leading physicists that scientists, in the pursuit of their professional duties,
should be above ideological interference by political authorities and their
spokesmen.

The deep and irreconcilable differences between the two groups broke out
in the open in 1928, when the young and fiery Ivanenko told the Sixth Congress of Soviet Physicists that the time had come for theoretical physics to
replace philosophy.[54] To be sure, this was a time of profound epistemological
inquiries into the theories of quantum mechanics and relativity. But behind
Ivanenko's statement was an open rebellion against the ambitious efforts of
Marxist philosophers to hold the upper hand in validating new physical
thought. Ivanenko paid a heavy price for his critical comments: though one
of the most cited Soviet physicists in Western scientific literature,[55] he never
became a member of the USSR Academy of Sciences.

By the beginning of the 1930's, those physicists who had taken a position
on the relations of physics—particularly of the theory of relativity and
quantum mechanics—to dialectical materialism formed two distinct groups.
One group, represented by Ioffe and Vavilov, made occasional, mainly
nominal, concessions to the rhetoric of dialectical materialism. Their obvi-

ous aim was to placate the guardians of official Soviet ideology as part of a sustained effort to protect physics and physicists from the vituperative attacks of Marxist zealots. The second group, identified by Marxist critics as the "Leningrad school," was dominated by Frenkel, Landau, Gamow, Ivanenko, and Bronshtein. It produced many direct and indirect signals indicating its firm belief that there was no way to establish sound cooperative relations between modern physics and dialectical materialism. Marxist attacks on that group continued until the end of the Stalin age. But most physicists did not in fact belong to either group. Although some joined the interminable Marxist campaigns against unwanted intrusions of Western thought, the majority lived in amorphous isolation from ideological issues and activities.

Early Stalinism and
Einstein's Theory

In 1929, the Second All-Union Conference of Marxist and Leninist Scientific Institutions declared the mechanist wing of Communist theorists a party outside of and in opposition to true Marxist thought. The dialecticians had won the furious and generally inflammatory battle that had raged since the mid-1920's. But their victory was short-lived. At the end of the year, a new group of Marxist philosophers consolidated its ranks and immediately began a scornful campaign against the dialecticians, especially against Deborin.[1] The Deborinites became targets of slanderous attacks for their alleged tendency to downgrade the place Lenin occupied among the pioneers of Marxism, to inject an overdose of Hegelian "idealism" into dialectical materialism, to exhibit a favorable attitude toward Menshevik philosophy and ideology, and to isolate Marxist theory from the practical problems of the five-year plan and socialist construction. They were also accused of disregarding moral obligations called for by *partiinost'*, a Stalinist innovation that made it mandatory for philosophers and scientists to follow a strict Party line in interpreting the policies of the Soviet government and the edicts of high functionaries in the Communist Party.[2] A resolution issued by the Communist organization of the Red Professoriat that December identified the views of the Deborin group as an eclectic fusion of materialism, Hegelian idealism, and "errors of the Feuerbachian and Kantian variety." Labeled "Menshevizing idealists," the dialecticians were considered less dangerous than the mechanists and were exposed to milder ideological attacks. Deborin in fact was elected to the Academy of Sciences the following year. But he never recaptured his position as one of the most erudite and influential authorities in the Marxist philosophy of science.[3]

The Rise of Soviet Orthodoxy

In the fierce discussion preceding the passage of the Red Professoriat resolution, the Stalinist faction attacked Deborin for supporting Boris Hessen's interpretation of the theory of relativity. Pressed to the wall, Deborin conceded that Hessen had incorrectly interpreted "matter, space, and time." He nevertheless continued to defend his original claim that Hessen's views on the relations of Einstein's theory to dialectical materialism were on the whole accurate even though they led to a few "wrong formulations." In 1934, M. B. Mitin, a dominant figure in that debate, accused the Deborinites of treating every scientist as a bona fide dialectical materialist.[4]

Mitin, then busy constructing a Stalinist version of Marxism-Leninism, gave Stalin full credit for laying the groundwork for the downfall of both mechanist and Deborinite orientations. "The shattering of the theoretical base of right opportunism signified a simultaneous attack on the entire mechanist revision of Marxist methodology. In the course of several years, Deborin's school waged a 'war' against mechanism, but it could not produce beneficial results because that war was too abstract, too speculative, and too isolated from the links between theoretical-philosophical and actual political problems. The works of Comrade Stalin have led to a full uprooting of mechanism."[5] Those works, according to Mitin, fashioned the unity of "philosophy and politics, theory and practice."

The "new" philosophers worked not only on reinterpreting Marxism-Leninism, which was now the one and only legal philosophy in the country, but also on organizing themselves to prevent the dual "enemy" from regrouping and reappearing in overt or covert counteroffensive moves.[6] Their continued attack on the mechanists and dialecticians was a way to clarify their own philosophical positions and to identify the most dangerous residues of opposition to Stalin's political philosophy. In making the mechanists and dialecticians easy targets in the total war on aberrant views in the Marxist camp, the architects of philosophical Stalinism did not hesitate to misrepresent and oversimplify their respective theoretical positions. The attackers also resorted to the indiscriminate use of political labels to discredit the two deviant philosophical orientations. The mechanists were said to support the strategic theoretical positions of "vulgar mechanicism" or "mechanical materialism," identified with N. I. Bukharin's sociological theory, the ideological backbone of so-called right deviationism in Soviet politics, which considered "social equilibrium" rather than "dialectics" the basic law of social development. The dialecticians were said to exhibit an affinity with revived Menshevism of a Trotskyite type.[7] The Stalinist philosophers also recognized another, equally dangerous enemy: the large number of scientists who ignored dialectical materialism altogether. Their attacks were thus aimed in part at non-Party philosophers, in part at Marxist theorists who apparently espoused one of the two deviant positions, and in part at the scientists who thought science an activity sufficient unto itself.[8]

In January 1931, the presidium of the Communist Academy issued a special resolution specifying the current tasks of Marxist-Leninist philosophy. The resolution encouraged Marxist philosophers to keep a watchful eye on the ideological implications of the theory of relativity and to exercise special care when comparing Einstein's scientific principles with the spirit of dialectical materialism. It, for example, criticized Hessen's "uncritical" claim that Einstein's views "on the interrelations of space, time, and matter were in principle congruent with the views of dialectical materialism."[9] The resolution also stated that a critical attitude toward the theory of relativity should not be part of an indiscriminate rejection of its physical principles, as was done by Timiriazev and several other mechanists and their allies. In general, the members of the academy were advised not to extend their attack on Einstein's philosophy to the realm of Einstein's science. Ernest Kol'man summed up the situation when he called Einstein "a great scientist and a poor philosopher."[10] Despite the relatively moderate tone of the resolution, it soon became eminently clear that Marxist philosophers were not disposed to draw a firm line between philosophy and science. They read philosophical implications into the full spectrum of Einstein's scientific ideas.

In their search for an ideologically pure and philosophically acceptable view of science, the guardians of Soviet ideology concentrated on physics, the discipline presenting the strongest challenge to Leninist epistemology. As they saw it, they had to go no further than Lenin's *Materialism and Empirio-Criticism* for a full supply of ready-made answers to all fundamental questions. Modern physics, they held, was the product of the kind of dialectical synthesis Lenin proposed, a blend of contending orientations initiated by the inner logic of the growth of scientific knowledge. They considered matter an epistemological rather than an ontological category: in Lenin's words, "the sole 'property' of matter with whose recognition philosophical materialism is bound up is the property of being an objective reality, of existing outside the mind."[11] In mass and energy they saw properties of matter, the former denoting inertia, the latter motion. In space-time they recognized a distinct form of "matter in motion." Resonating declarations of "the disappearance of matter" or "the de-materialization of atoms" were the prime targets of their attacks. In Lenin's claim that "the disappearance of matter" meant only "the disappearance of the limit within which we have hitherto known matter," they saw a recognition of historical relativity in the evolution of scientific knowledge—a recognition that scientists were penetrating deeper into the innermost secrets of matter, and that certain attributes of matter (impenetrability, inertia, and mass), previously regarded as "absolute, immutable, and primary," had been discovered to be relative.[12] The new philosophers attributed more flexibility to the law of causality than Lenin, but they continued to treat it as the key element in the scientific explanation of physical reality. The philosophers conceded that rapid advances in mathematics added immensely to the efficiency of modern scientific methodology, but they feared that the same advances threatened to invite a massive revival of idealistic philosophy.

Soviet physicists with a flair for philosophy and deep loyalty to Stalinism worked primarily on discovering appropriate means of applying the limited number of Lenin's epistemological principles to the expanding and unsettled physical thought of the day. Their most important duty was to make Lenin's ideas sufficiently flexible to harmonize current developments in physical theory with Soviet ideology as translated into the language of philosophy. To be on safe ground, they avoided a broader treatment of Lenin's ideas and handled individual propositions with the utmost economy of words. That approach produced an abundance of philosophical activity but little effort to advance a systematic and comprehensive Marxist philosophy of science in general and of physics in particular. Not until the late 1940's was an earnest effort made to consolidate the philosophy of natural science as an "independent discipline," that is, to codify the fundamental principles of dialectical materialism as a philosophy of nature.

Although the philosophical debate was fragmented, narrowly conceived, and pushed more or less militantly, depending on the individual's strength of feeling, one issue loomed particularly significant: the growing threat of "physical idealism," a casting of the epistemological foundations of modern physics in the light of various neopositivist and neo-Kantian orientations. Like other science-related issues directly linked to ideology, the specter of physical idealism generated much bitterness and passionate disavowal. To the defenders of Marxist orthodoxy, the term covered all epistemological orientations that violated Lenin's view of knowledge as a reflection of the external world and challenged causality as the basic explanatory principle in science.

Confronted with the task of defending Leninist epistemological objectivism as a pillar of Soviet ideology, Marxist philosophers aimed their deadliest blows at the Copenhagen school.[13] In its "idealistic" leanings, they saw a devious effort to bestow "the freedom of the will" on atoms and to picture the world as "total chaos." Nor did Einstein's general theoretical views escape bitter attacks on several fronts. Soviet philosophers took particular exception to Einstein's repeated emphasis on the primary role of pure reason, intuition, and mathematical axiomatics in the development of the mainstream of modern scientific thought. Einstein's philosophy, in their view, was flawed on two counts: it denied the treatment of scientific knowledge as a copy of the objectively existing universe, and it refused to view the sciences as systems of theoretical elaborations distilled from and tested by experience.

Einstein and Mathematical Idealism

Marxist philosophers emphasized the empirical roots and practical orientation of mathematics. They criticized Einstein's categorical claim that mathematical abstractions were products of "mere imagination," and that mathematics was "a product of human thought which is independent of experience."[14] In their relentless war against "mathematical idealism," they criticized the leaders of the Moscow school of set theory for their "solipsistic" and "immanentist" leanings. In Lobachevskii's non-Euclidean geometry, by contrast, they saw a triumph of the empirical and practical orientation in mathematics. In the St. Petersburg school of mathematics, founded by P. L. Chebyshev in the nineteenth century and given extraordinary depth by A. A. Markov and A. M. Liapunov, they saw a perfect expression of an orientation firmly rooted in concrete reality.

Marxist critics viewed mathematical idealism as a product of the adverse influence of idealistic philosophy, such as Machism, on the conceptualization and methodology of physical research. That philosophy, they said, made "a strong imprint on the leading physical theories and reduced their credibility as parts of science."[15] The leading neopositivist schools in philosophy were roundly attacked also for their excessive emphasis on mathematization as the safest guaranty for the steady progress of science. Marxist philosophers attributed much weight to Lenin's warning that too much mathematics left science open to the corruptive influence of idealistic philosophy. One author cited Einstein's "The Method of Theoretical Physics" to show the negative effects of the dependence of physical idealism on the tools of mathematics. Totally overlooking the qualitative features of physical reality, he wrote, Einstein made the mathematical method the only motor of creative thought in physics. He wanted no part of Einstein's statements that "nature is the realization of the simplest conceivable mathematical ideas," and that "experience may suggest the appropriate mathematical concepts, but these most certainly cannot be deduced from it."[16]

The general run of Marxist philosophers believed that the mathematization of physics "created an illusion of the omnipotence of equations and of the possibility of replacing physical reality by mathematical symbols."[17] It contributed to the rebirth of Kantian philosophy, which claimed that "reason prescribes the laws of nature." Like physical idealism, the mathematization of physics was a strong representative of philosophical relativism, "a negation of positive and absolute truth," and a treatment of all knowledge as "subjective" and "conditional."

Marxist philosophers and scientists acknowledged and welcomed the primary role of mathematics in modern science. They were consistent, however, in claiming that mathematics could not answer all the methodological needs of science and that excessive mathematical formalism divorced from the realities of nature was nothing else but an invitation to the corruptive work of philosophical idealism. Einstein's complex mathematical expressions of the equivalence of inertia and gravitation and of the relations of the covariance of differential equations to the universality of the laws of nature were more often than not attacked as exercises in mathematical formalism unrelated to concrete reality.

Variations in Views on Einstein

In essence, the defenders and articulators of Stalinist orthodoxy all aspired to create a Soviet or Marxist theory of relativity by modifying Einstein's scientific principles in such a way as to make them consonant with dialectical materialism. But Maksimov, Kol'man, and Mitin, the three leading and most aggressive members of the group, though firmly united in their war on physical idealism, were not of the same mold. They manifested marked personal differences in temperament, educational background, style of philosophical argumentation, communication techniques, and tactics.

Maksimov, one of the most belligerent critics of Einstein's philosophical ideas, felt no constraint about making scientific judgments. He stuck closely to Lenin's dictum that the difference between idealistic and dialectical-materialistic epistemologies inevitably produces irreconcilable differences in the views on the logic and methodology of scientific inquiry. He argued that Einstein, Erwin Schrödinger, Werner Heisenberg, Niels Bohr, and Pascual Jordan were "all idealists of the Machian variety," and that their idealism made them enemies of dialectical materialism on epistemological, logical, and methodological grounds. For that reason, he said, "the struggle for Bolshevism in science [was] the struggle for a fundamental reconstitution of science."[18] The principle of relativity, the key concept of Einstein's scientific thought, he argued, led—and could only lead—to a denial of the objective nature of motion. Only occasionally did Einstein's theory of relativity reveal materialistic nuclei—or "rational kernels"—of objectivity characteristic of true science.[19] Maksimov saw in the theory of relativity an offshoot of Machian philosophy, the main source of a subjective-idealistic view of space, time, and matter. He did not hesitate to lambaste even so prestigious an academician as O. Iu. Shmidt for accepting the theory of relativity without any criticism of its idealistic foundations at the Second All-Union Conference of Marxist-Leninist Scientific Institutions. Philosophical bombast, rhetorical hyperbole, and acrimonious attacks were his normal tools in polemical engagements.*

Kol'man, with an interest in the history of mathematics, differed from Maksimov by a deeper involvement in the intricacies of physical theory and a livelier interest in formulating a Marxist theory of relativity. In addition to philosophical criticism, his published papers offered informative surveys of the growth of modern physical thought and demonstrated an awareness of the rising complexity and thoroughness of the mathematical foundations of quantum mechanics and the theory of relativity. Although he avoided the excesses of ideological rhetoric, his allegiance to Lenin's epistemology was rigid and narrowly fashioned. He tried to bring modern physical theory in tune with Lenin's ideas on "the physical and philosophical attributes of matter and the empirical nature of space and time," "the inexhaustibility of the electron" (for explaining the historical relativity of scientific knowledge), the incontrovertibility of conservation laws, and the need for a broader and more flexible conceptualization of causality.[20]

Kol'man made a concerted effort to please both the community of physicists and the community of Marxist philosophers. To conciliate the first group, he spared no words in making ebullient statements on the epochal significance of Einstein's scientific ideas, from the unity of space and time and the mathematical correlation of mass and energy to the law of gravitation and the unity of mechanics, electrodynamics, and thermodynamics. To satisfy the second group—actually his main task—he produced long lists of objections to Einstein's views, both philosophical and scientific. His analysis was fluid, always reflecting the direction of the current ideological winds.†

*Maksimov graduated from Kazan University's Department of Physics and Mathematics in 1913 and joined the Communist Party in 1919. He specialized in the Marxist philosophy of science, a subject he taught at the Institute of the Red Professoriate, Moscow University, and the Communist Academy. In 1934, Moscow University granted him a doctorate in philosophy, and in 1943 the Academy of Sciences elected him a corresponding member. His publications dealt mainly with the philosophical aspects of quantum mechanics and the theory of relativity. During the Stalinist war on cosmopolitanism, he published a book on the materialistic tradition of Russian science. For more details, see Sonin, *"Fizicheskii idealizm,"* pp. 27–32

†Kol'man was born in Prague, Czechoslovakia, where he graduated from Charles University's Department of Mathematics in 1913. A year later, he completed an officers' training

If Soviet physicists found Kol'man's philosophical discourses somewhat more tolerable than those of Maksimov, it was because his attacks were milder and his logic more rigorous.[21] By making public statements on the desirability of philosophical discussion of the type engaged in by Kol'man, the physicists helped weaken the unity of the philosophical front and the intensity of external interference with scientific activity. In comparison with Maksimov, Kol'man was more careful in keeping separate accounts of scientific and philosophical aspects of Einstein's theoretical principles; he was also more inclined to take issue with Einstein's scientific ideas that did not require philosophical consideration. In retrospect, some students of the Stalinist attacks on science consider him the most dangerous critic of both the philosophical foundations of modern science and the theoretical contributions of the theory of relativity and quantum mechanics.[22]

Of the three, Mitin showed the most restraint. Obviously uninformed about developments in modern physics, he limited his comments to matters of a general nature. He rejected Maksimov's claim that Leninist philosophy demanded a "total reconstitution" of the logical, methodological, and epistemological foundations of modern science. He charted a middle course between the total rejection of the theory of relativity by most mechanists and its total acceptance by most dialecticians. Although his effort was neither systematic nor comprehensive, he undertook the difficult task of keeping separate account of the incontrovertible achievements of modern physics and the unwelcome suppositions and deductions of physical idealism. In the early 1930's, Mitin thought that Einstein, despite his admiration for Machian philosophy, stood essentially on a materialistic position, and that Heisenberg, despite his profuse idealistic pronouncements, came close to recognizing the dialectical nature of the unity of the observed object and the observing subject (and his research equipment).[23] It would not be unrealistic to assume that by following a course of relative moderation, Mitin tried to prevent a total alienation of Soviet physicists from questions of a philosophical nature.

During the 1930's, Mitin was instrumental in tying the basic principles of

course and was sent to the Russian front. He was captured by a Russian army unit in 1915 and then imprisoned, in 1917, on charges of spreading Bolshevik propaganda. He joined the Bolshevik Party after the October Revolution and performed various duties connected with political propaganda, including assignments as a political commissar in the Red Army and undercover work in Germany. At the end of the 1920's, he concentrated on the Marxist philosophy of science, with particular interest in modern physics and mathematics. In 1934, Moscow University granted him a doctorate in Marxist philosophy. After the Second World War, he was made a member of the Czechoslovak Academy of Sciences (Kedrov et al., "Ernest Kol'man"). He emigrated to Sweden in the 1970's. For more details, see Sonin, "*Fizicheskii idealizm*," pp. 32–42.

Leninist philosophy to the guiding precepts of Stalinist ideology. Indeed, he is considered the chief architect of Stalin's position in that phase of his thinking Although Mitin avoided any personal abuse of Soviet physicists and refrained from political threats, he played a major role in reducing the Leninist philosophy of science to a short list of epistemological slogans with forthright ideological messages. Mitin thought that Einstein favored two philosophical views in direct opposition to the claims of the Copenhagen school: he recognized causality as the prime tool of scientific explanation and he accepted sense perception as a true reflection of the objectively existing world of nature.[24]

Alongside these conscientious and deeply engaged champions of Stalinist orthodoxy stood two groups of unorthodox philosophical interpreters of Einstein's scientific and philosophical principles. The first, led by Timiriazev, Mitkevich, and Kasterin, was homogeneous mainly in the sense that its members uniformly rejected Einstein's science, as well as his philosophy. The most vocal of the three was Timiriazev, who though badly bruised by the acrimonious attacks on him as the leading mechanist, continued to be unwavering in his claim that the one was as unacceptable as the other. He pointed to Einstein's own admission that he was indebted to Mach for all his scientific principles, and that all he had done was to give those principles mathematical expression. In that process, Timiriazev contended, Einstein substituted "function" for "causality," a mode that Lenin criticized severely. [24] Timiriazev was an outspoken supporter of Stalin's war against the scientific community in 1937-39. By his lights, to be in favor of the theory of relativity was to be criminally disloyal to the Soviet system.

Mitkevich, a member of the academy and a physicist of the old school, essentially walked in lockstep with Timirazev. He too depended heavily on philosophical arguments and ideological pronouncements; saw the theory of relativity and quantum physics as imbued with idealism, the last gasp of "dying capitalism"; and relied on personal insult and political threat as offensive weapons. He was no less extravagant than Timirazev in expressing his loyalty to the Stalinist regime.[25] Kasterin, from whose scientific work Timiriazev and Mitkevich sought help, was of a far different stripe. He avoided philosophical arguments and ideological belligerence. Concerned only with refuting the body of Einstein's scientific work, he went so far as to propose a synthesis of aerodynamics and electrodynamics as a substitute for the combined theory of relativity and quantum mechanics.[26]

For the philosophers in the other group, the theory of relativity, far from contradicting Leninist epistemology, offered indisputable confirmation of its basic principles. The most outspoken champions of that position—Semkovskii and Hessen—perished in Stalinist penal camps. In one of his last articles, Semkovskii compared Engels' dialectical view of nature with the theory of relativity as an eminently successful effort to present a "trans-relative" and "invariant" picture of the physical world.[27] In the special theory of relativity, he added, Einstein depended on the Lorentz transformation of different reference frames to move from "relative perspectives" of individual observers to "invariant" and "objective" reality, fully independent of both coordinate systems and observers. In Semkovskii's opinion, Einstein's refusal to recognize "privileged" coordinate systems was part of his search for objective and universal truth, and was fully consonant with dialectical materialism.

Semkovskii honored Engels by considering him an ancestor of the theory of relativity. He reasoned that all efforts to mark the physical theory of relativity as incompatible with the philosophical theory of dialectical materialism were based on a complete misunderstanding either of Einstein's ideas or of Engels' legacy—or both. Only in his appraisal of Einstein's work on a unified field theory did Semkovskii express a critical view. Through the theory of relativity, Einstein made geometry a branch of physics; but in his work on the unified field theory, he was engaged in making physics a branch of geometry, an effort that Semkovskii did not accept.

Both groups were pariahs in the eyes of the Stalinist orthodox. In 1931, as we saw, the presidium of the Communist Academy condemned both Timiriazev's full rejection of the theory of relativity and Hessen's full acceptance of it. But it went further in Hessen's case by censuring him for his statement in 1928 that the theory of relativity generally agreed with the views of dialectical materialism.[28] Haunted by senseless and intimidating police surveillance and violently criticized and ridiculed by Marxist writers, Hessen became increasingly demoralized and effectively withdrew from professional activities.[29] He wrote little about Einstein thereafter, and when he did, it was with due caution. For example, he was obviously swayed by his loss of authority as a Marxist philosopher when, in preparing the philosophical part of Einstein's biography for the *Great Soviet Encyclopedia*, he chose to criticize Einstein's total rejection of discontinuity (or discreteness) in physical processes. "The one-sided emphasis on continuity," he wrote, "gives the theory of relativity a nondialectical character and makes it a theory pervaded with the spirit of continuum."[30] That omission, he argued, prevented Einstein from establishing working contact with quantum mechanics. In another article he wrote for the encyclopedia, he argued that the question of the existence of ether had not yet been resolved.[31] Despite isolated disagreements like these, his respect for Einstein's science and theory of knowledge continued unabated. He is remembered primarily for his book on Einstein, published in 1928, which made him one of the most astute and forthright Soviet champions of the full integration of Einstein's ideas into Marxist thought.

Digging deep into Soviet archives, G. E. Gorelik has been able to unravel the tangled threads of Hessen's tragic life.[32] On the surface, as Gorelik shows, Hessen emerges as a well-placed expert in the area where physics met Marxist philosophy and Soviet ideology. He was the first dean of Moscow University's faculty of physics and then director of its Institute of Physics; subsequently, he became deputy director of the Institute of Physics of the Academy of Sciences, which represented a major promotion. He served on the editorial boards of the *Great Soviet Encyclopedia* and *Advances in the Physical Sciences*, one of the best Soviet scientific journals. In 1931, he presented a widely acclaimed paper on the socioeconomic foundations of Newtonian mechanics at the Second International Congress of Historians of Science in London. That long paper, commonly conceded to be the first Marxist study in the history of science, made Hessen one of the better-known leaders in the field. In 1933, the Academy of Sciences elected him a corresponding member.

But Hessen's professional success was more apparent than real; in fact, it was a mirage. During the last eight years of his life, he was the target of devastating attacks by a succession of groups. In the beginning, they came from Marxist zealots speaking for themselves. Then they came from young physicists eager to protect their science from what they considered adverse ideological infringements. Finally, they came from Communist authorities, dressed in different verbiage and carrying more serious signals of doom.

Marxist zealots, ready to join the Stalinist camp and presenting themselves as philosophers concerned with the questions of science, found much to criticize in Hessen's published work. They accused him of intolerant schemes to modify dialectical materialism in order to make it consonant with the theory of relativity, instead of recommending modifications in the theory to bring it closer to dialectical materialism. They chastised him for overlooking idealistic flaws in Einstein's thinking, for not showing sufficient flexibility in applying Marxist thought to the Newtonian mechanistic legacy, and for placing undue emphasis on the Hegelian side of the dialectical equation.

Hessen irritated a group of young physicists with his assertion in the *Great Soviet Encyclopedia* that the existence of ether remained an open question. Upon reading the article, Lev Landau, George Gamow, and a few others sent him a sarcastic telegram suggesting that, after having "rescued" ether from oblivion, he should now concentrate on doing the same for phlogiston, caloric, and electrical fluid, the notorious crutches of bygone science.

The telegram may have been an innocent and humorous prank by youthful physicists. But there was nothing innocent or humorous about the letter Gamow sent to Stalin identifying Hessen's view on ether as an idealistic onslaught on Soviet physics. He reminded Stalin that to revive ether was to ally Marxism with a long-rejected physical theory supported by the "reactionary" Phillipp Lenard, the "fideistic conservative" John Joseph Thomson, and the "spiritualist" Oliver Lodge—that is, by men who had become "a laughingstock even among progressive bourgeois physicists."[33] Gamow expressed the hope that Stalin would undertake the steps necessary to free physics from anti-Marxist thought.

Gorelik located Gamow's letter in the archive of Ernest Kol'man's papers.[34] He was unable to determine whether it ever reached Stalin or whether Gamow received an official reply. He is certain, though, that Gamow's interest in protecting dialectical materialism was nothing more than a ploy to marshal Stalin's help in what he considered an unwarranted effort to discredit Einstein. In 1932, when he wrote the letter, Gamow was viewed by Marxist critics close to Stalinist orthodoxy as a prominent member of "the Leningrad school of physicists," headed by the "long-time Machist" Ia. I. Frenkel. At the time, Marxist ideologues viewed Machism as the most pernicious manifestation of physical idealism.

In 1936, when Stalin's Great Terror shifted into high gear, the police authorities accused Hessen of having participated with Trotsky and Zinov'iev in their conspiracy to overthrow the Soviet government. He was arrested on August 31, 1936, and was executed by a firing squad within months, on December 20. Informed about his imprisonment, but not about his death, the Marxist activists in the academy's Institute of Physics summoned a meeting in April 1937 to discuss—among other things—Hessen's arrest. The senior members of the institute—Vavilov, Tamm, and G. S. Landsberg—admitted that they were close to Hessen but avowed that they had no evidence whatever to link him with terrorist activities of any kind. Several junior members of the staff spoke up to criticize Hessen's departures from philosophical orthodoxy and his neglect of administrative duties. One, M. A. Divil'kovskii, a research associate, went so far as to charge him with instituting a hiring policy based on "nepotism," not on merit—but none of them had seen any evidence of terrorist activity either.[35]

That the consolidating Stalinist regime caused Einstein's admirers to watch their tongues is not in doubt. Men like Hessen and Semkovskii expressed their approval of Einstein's views during the 1930's less forcefully, in less detail, and with more reservations than in the 1920's. To compound the difficulty of the state of Einsteinian studies, the 1930's did not produce a single new Marxist philosopher ready psychologically and morally to resist the marauding anti-

Einstein forces encouraged and shielded by the highest echelons of the Communist authorities. At the same time, it is fair to note that Marxist critics did not declare total war on Einstein's ideas at that point.[36] Nor did they try to explain the sharp contrast between their fleeting expressions of respect for the greatness of Einstein's overall scientific achievement and their general disrespect for his philosophy and selected parts of his science. Paradoxes of this kind contributed to the accelerated alienation of the scientific community from Marxist philosophy.

Physicists and Philosophers: Simmering Controversies

In the 1930's, the philosophers involved in attacks on physical idealism—a base of "bourgeois ideology" in physics—found it exceedingly difficult to secure the cooperation of the leading physicists. The situation became particularly complicated in 1933, when the distinguished physicist I. E. Tamm, in an article published in the Marxist theoretical journal *Under the Banner of Marxism*, accused the philosophers of an appalling ignorance of current developments in physics. "The real evil," he wrote, was "the fact that an overwhelming majority of the representatives of Marxist philosophy in our country working in physics and neighboring disciplines simply do not understand the situation in modern science. At best, their knowledge corresponds to the level of science at the end of the last and the beginning of the present centuries."[37] Worse still, their appalling "scientific illiteracy" extended to the very cardinal principles of scientific knowledge. Tamm made it clear that the basic theoretical and experimental changes in the physical sciences required intensive concern with the epistemological, logical, and methodological questions of scientific knowledge, the area where competent scholars were in short supply.

Surprised and seriously perturbed by the unexpected turn of events, the philosophers sought to widen the scope of their operations. Particularly significant was the decision of the academy's Institute of Philosophy to invite noted scientists to their sessions for the purpose of forming a generally accepted response to the great ideas of modern physics. At one such session, held in June 1934 in commemoration of the twenty-fifth anniversary of the publication of Lenin's *Materialism and Empirio-Criticism*, two of the most prestigious physicists, Ioffe and Vavilov, delivered long papers on the philosophical foundations of modern physics, on the dialectical-materialistic interpretation of the revolutionary discoveries in science, and on the dynamics of the social functions of science.[38] Ioffe covered a wide range of problems related to the "development of atomistic views in the twentieth century"—the history of quantum mechanics—but he concentrated on Heisenberg's indeterminacy principle, which, in his opinion, offered the only realistic approach to the apparently "unrestricted freedom of electrons or atoms." He found it "strange"—and completely anti-Leninist—to place the label of idealism on such eminent scholars as Bohr, Heisenberg, and Frenkel, who had dedicated their lives to the search for a "full description of the properties of the atomic universe." Although many ideas of Western physics challenged both scientific tradition and common sense, they contributed, in Ioffe's opinion, to "a brilliant confirmation and enrichment of dialectical materialism."[39]

As an expert in luminescence, Vavilov chose the nature of light as the topic of his report and as the springboard for an analysis and full endorsement of Bohr's complementarity principle, which he presented as the most graphic illustration of the work of dialectics—the "unity of contradictions"—in nature.[40] For the elated Mitin, to have such leading physicists as Ioffe and Vavilov come out "openly and clearly" in favor of materialism and against "the waves of idealistic reaction" was an event of notable significance.[41]

To meet the mounting ideological pressure of the political authorities and their allies, some leading physicists resorted to the technique of making favorable but vague, rare, and unelaborated comments on key issues of philosophical import. Uninitiated in the intricacies of dialectical materialism, others found themselves in hot water either because of their careless use of philosophical metaphors or because of their insensitivity to philosophical guideposts marked by Stalinist selections from Lenin's *Materialism and Empirio-Criticism*. Despite rhetorical concessions, the leading physicists tended to agree with Leonid I. Mandelstam, an honored member of the Academy of Sciences, who, in his concluding lecture on the theory of relativity at Moscow University in 1933–34, stated that the philosophical meaning of Einstein's theoretical work was a special question that could be separated from physical questions.[42] At this time of reinvigorated Marxist criticism of Einstein's scientific contributions and epistemological views, Mandelstam told his students that all experiments showed that the theory of relativity was a "natural," not an "artificial," body of knowledge.[43] He made a point of informing his young listeners that, because of the extraordinary depth and breadth of its vision, the theory of relativity gave added strength to the full compass of modern physics.[44] Since he did not publish his lectures at the time, Marxist philosophers made no effort to make him a target of public attacks.

Physicists like Frenkel, Tamm, and Landau seem to have been inclined to a greater degree of discretion, generally relegating their support of the leading theoretical arguments of the pioneers of modern physics to parenthetical philosophical statements. But their sentiments were perfectly clear, for all that.

Frenkel, in particular, refused to acknowledge serious inconsistencies in Einstein's theoretical and philosophical views. He turned his arguments against those critics who claimed that Einstein tied his scientific ideas of relativity to a philosophical view of "relativism," a denial of the existence of absolute standards of objective truth. According to Frenkel, Einstein's theory of relativity could rightfully be called the theory of the "absolute" nature of scientific laws.[45] In an encyclopedia article on Einstein, he defended the general theory of relativity as "one of the greatest achievements of human genius."[46]

Vasilov was one of a small group of physicists who matched admiration for Einstein's scientific contributions with scathing attacks on physical idealism, particularly as built on Einstein's ideas. Never stingy in his praise of the theory of relativity as "the universally accepted basis of modern physics," he nonetheless showed little sympathy for Einstein's "idealism," a theme he tackled only on rare occasions. In an article published in *Priroda* (*Nature*) in 1934, he set the stage for an open disagreement with the "idealistic" subtexts of some of Einstein's ideas. Most of all, he disagreed with Einstein's claim, in "On the Method of Theoretical Physics," that nature was "the realization of the simplest conceivable mathematical ideas"—that pure "mathematical constructions" were the key to the understanding of natural phenomena.[47]

Einstein's statement, in Vavilov's view, offered an example of the "real danger" of the "idealistic path" chosen by a solid core of modern physicists. It provided an illustration of what Lenin had identified as "the forgetting of matter by mathematicians" and of the Kantian view of reason as governing the laws of nature. Vavilov did not challenge Einstein's recognition of the supreme role of mathematics in modern science; what he challenged was Einstein's emphasis on mathematical formulas as "pure constructions" of the human mind, unrelated in their origin to the accumulated wisdom of practical experience. The criticism of Einstein's mathematical idealism had not prevented Vavilov from noting just the year before (1933) that the theory of relativity and quantum mechanics had produced the most impressive examples of the power of mathematical extrapolation, and that mathematics had given the modern physicist an inordinately potent method for creating new theories.[48] In Einstein's mathematical idealism, not in Einstein's heavy reliance on mathematical procedures, Vavilov saw the main reason for "the practical sterility of the later phases in the development of the theory of relativity." In his view, physical idealism was a negative force on two counts: it gave support to capitalist ideologies dedicated to preserving the social status quo, and it helped create conditions that in the long run threatened to cause a general stagnation in science.[49]

Vavilov faced a dilemma of large magnitude. On the one hand, he did not conceal his admiration for Einstein's rich contributions to modern physics; on the other hand, he did not hide his concern with the comfort that Einstein's isolated philosophical utterances gave to the advocates of philosophical idealism. He "resolved" the problem by more than matching every criticism of Einstein's philosophical observations with frank statements on the unchallengeable value of Einstein's scientific achievements. By separating Einstein's philosophy from Einstein's science, Vavilov supplied subtle arguments in favor of protecting the scientific community from daily interference by Marxist philosophers.

Vavilov was a master in addressing himself both to physicists, most of whom placed the developments in their science ahead of the current concerns of dialectical materialism, and to Marxist philosophers, who favored Marxist orthodoxy over the achievements of science. In 1938, for example, he wrote that in Einstein's theory, space-time was an integral property of matter: "it depends on matter, it changes with matter, and it cannot exist without matter." He also noted that the theory had no room for "the idealistic notion of space-time" called for by Newtonian science.[50] But he carefully went on to observe that Einstein's theory of space-time needed additional work to become fully congruent with dialectical materialism. Asides like this, however, did not stop him from asserting that the atomic theory of the structure of matter, wave mechanics, and Einstein's theory of space-time were the three main paths taken by the ongoing revolution in physics. All three paths, he added, led to the triumph of dialectical materialism. Despite Vavilov's conciliatory rhetoric, the leading Marxist philosophers continued to make Einstein one of the main targets in their holy war on physical idealism, with the result that his standing in philosophical circles remained unclear and unsettling.

V. I. Vernadskii, a pioneer in the study of the biosphere, was one of the better-known scientists who extended Einstein's ideas beyond the realm of physics. Because he openly opposed dialectical materialism and criticized the Soviet curbs on the freedom of scientific inquiry and philosophical expression, a close-to-full text of his capital work, *The Philosophical Thoughts of a Naturalist*, would not be published until the age of perestroika.[51] In several papers that appeared during the 1930's, he made brief disapproving comments on both the Marxist philosophy of science and the Soviet political controls over the scientific community.[52]

Vernadskii was convinced that the space-time manifold constituted Einstein's main contribution to both the triumph of post-Newtonian physics and the enormous success of the twentieth-century revolution in science. Such basic ingredients of the general theory of relativity as the principles of general covariance and the equivalence of inertial and gravitational masses did not attract his attention. Vernadskii claimed that space-time as both a geometrical concept and a physical concept antedated Einstein's notion of space-time by several decades. In the advent of non-Euclidean geometry he saw the triumph of symmetry as an integrating principle of different configurations of cosmic space and a key for the understanding of Einsteinian cosmic harmony, to which he fully subscribed.[53] At one point, he claimed that the space-time of the living world was neither Newtonian nor Einsteinian, and that the space of both the living and physical worlds was "dissymmetrical," anisotropic, and nonhomogeneous.[54] Despite such negative statements, he is remembered as a leading scientist who recognized the enormous potential of Einstein's contribution to a revolutionary turn in the growth of scientific thought.

Soviet Reactions to the Einstein-Bohr Debate

Soviet scholars readily admitted that Einstein occupied two eminent positions in modern physics: he was the sole architect of the theory of relativity as an integrated body of theoretical principles, and he was one of the chief builders of the foundations of quantum mechanics. Peter L. Kapitsa, for example, specifically pointed to Einstein's famous equations expressing the quantum nature of certain types of radiation, which had been derived, according to him, from Einstein's studies of photoemission. He added that "the principles of quantum theory and quantum mechanics were predetermined by the discovery of photoemission."[55]

In respect to quantum mechanics, Einstein occupied a paradoxical position: he was not only a builder of that branch of revolutionary knowledge but also one of its severest critics. He attracted the most attention by his dissatisfaction with the methodological and epistemological foundations of quantum mechanics. His criticism found particularly strong expression in the much-heralded Einstein-Bohr debate that lasted for more than a decade and produced powerful reverberations in the community of physicists. The debate unfolded in two distinct phases. The first phase, which began in 1924, was dominated by Bohr's unwillingness to accept Einstein's quantum theory of light—and, in general, of the spatial structure of radiation.[56] During the second phase, which started in 1927, Einstein took over the position of attacker: he concentrated on the probabilistic character of quanta and on the role of measuring processes in defining micro-objects, which, he thought, limited the objective nature of physical reality. Guided by his allegiance to classical physics, Einstein wanted a quantum mechanics governed by the Laplacian law of causality and offering a more "complete" coverage of physical reality.

Soviet physicists and philosophers made no serious effort to take a position during the first phase, in large part because the international scientific community had little inkling of Bohr's views initially. For a time, Bohr showed his dissatisfaction with Einstein's theory of photons simply by ignoring its existence. During the second phase, the Einstein-Bohr dispute was carefully monitored in the Soviet Union and precipitated far-reaching scientific, philosophical, and ideological reverberations. Part of the continuous search for a Marxist version of quantum mechanics, Soviet reactions to the Einstein-Bohr dispute escalated into major conflagrations on several occasions. Predictably, the dispute refueled the campaign of the defenders of Stalinist orthodoxy to expunge bourgeois ideological residues from the revolutionary moves in modern physics.

In a provocative paper published in 1935, Einstein and his colleagues Boris Podolsky and Nathan Rosen claimed that the quantum-mechanical study of physical reality could not be considered complete. Quantum mechanics, they argued, did not satisfy the requirement that "every element of the physical reality must have a counterpart in the physical theory."[57] "Any serious consideration of a physical theory," they held, "must take into account the distinction between the objective reality which is independent of any theory, and the physical concepts with which the theory operates." They followed up with a clear and straightforward definition of physical reality: "If, without in any way disturbing a system, we can predict with certainty (i. e., with probability equal to unity) the value of a physical quantity, then there exists an element of physical reality corresponding to this physical quantity."[58]

More specifically, the three challenged Bohr's assumption that the wave function provided "a complete description of the physical reality" to which it referred. They claimed that it was possible to assign two different functions to the same reality and that, therefore, the quantum mechanical description of physical reality could not be complete. Their criticism was a delayed answer to Bohr's claim, made in 1932, that physical reality, as studied by atomic physicists, was fundamentally incomplete as long as its description was limited to "an objective existence of phenomena independent of the means of their observation."[59]

The Soviet scholarly community received this critique of quantum mechanics with mixed feelings. Marxist philosophers welcomed it as a relentless attack on the Copenhagen school for its flirting with neopositivist epistemology, which they identified as subjectivist idealism. The physicists were divided. Fock, who wrote the first Soviet textbook on quantum mechanics, stood firmly on Bohr's side: he thought that the authors' critique of the wave function was poorly designed, methodologically misdirected, and too deeply committed to the outdated principles of classical physics.[60] He noted that quantum mechanics represented a revolution in physics requiring a break with the deeply ingrained habits of thought about the Newtonian-Laplacian picture of the universe.

K. V. Nikol'skii was the most determined of the group of physicists who stood firmly against Bohr—so firmly that he attacked Fock's defense of him as a betrayal of materialism and of "Soviet physics" in its effort to rise above the idealistic aberrations embedded in Western science.[61] Directing his criticism at Pascual Jordan, whom he considered representative of the Copenhagen school, Nikol'skii argued that denying the objective existence of physical reality was tantamount to making physics part of psychology.[62] Nikol'skii introduced himself as a Soviet representative of Einstein's effort to separate quantum mechanics from Machian philosophy. Accepting the idea of fundamental differences between the macroscopic reality of classical physics and the microscopic reality of quantum mechanics, he commented favorably on the concerted search by the latter for new scientific perspectives.[63] On the philosophical level, however, he argued that quantum mechanics should deal exclusively with objects existing outside and independent of the human mind. He thought that only a reliance on statistical methods could create reliable foundations for an objective quantum mechanics, for those methods alone could treat microphysical reality as a concrete object, free of subjective admixtures.

Nikol'skii not only acknowledged the preeminent position that quantum mechanics held in modern physics, but also showed intense interest in clearing the way for its full philosophical and ideological reconciliation with dialectical materialism. In that effort, he found it advantageous to place Einstein on the side of materialistic philosophy and to accuse Fock and Tamm of claiming that the Copenhagen school, fastened to Machian philosophy, represented the only generally accepted orientation in quantum mechanics. He went so far as to identify a group of eminent Soviet physicists as "a branch of the Copenhagen school."[64] Fock responded by accusing him of trying to convert quantum mechanics into a branch of statistics.[65] Ironically, Nikol'skii could not muster much help for his arguments from Einstein's writings, which stood firmly on the position that the essentially statistical character of quantum mechanics owed to the theory's incomplete description of physical systems.

Nikol'skii's statistical orientation in quantum mechanics was clearly—and admittedly—an effort to accommodate a modern science to the philosophical principle of dialectical materialism. Fock's "Copenhagen approach," by contrast, was an effort to accommodate dialectical materialism to the needs and dictates of modern science. The mid-1930's witnessed the emergence of a third group as participants in the debate. Working quietly and carefully, these were the men like Landau, Ioffe, and Frenkel whose chief concern was to keep physics and philosophy fully independent of each other.[66] To that end, they were willing to go so far as to recognize two kinds of physical reality, one studied by physicists and the other by philosophers. Their basic intent was to restrain Marxist scholars from interfering with the professional work of the community of physicists.

Always ready to employ the full arsenal of abusive rhetoric, Maksimov was a front-runner in the Marxist philosophers' war against the Copenhagen school and its "Soviet outpost." He was especially annoyed at *Advances in the Physical Sciences*, which had invited Fock to write an introduction to the Russian translation of the Einstein-Poldosky-Rosen paper in the first place.[67] Fock, like Tamm and Frenkel, he said, manifested a marked tendency to echo idealistic pronouncements of Western physicists and to make comments unfavorable to materialism. He went so far as to warn that Fock and his group threatened to become the dominant factor in shaping public opinion in matters related to physics. Maksimov lumped them together with the "backward group" of Soviet intellectuals who had refused to become emancipated from the influence of "religious prejudices" and who behaved like the "exploiters and agents of capitalist culture."[68]

Soviet critics of the idealistic orientation of the Copenhagen school did not take full advantage of Einstein's criticism of Bohr's epistemological ideas. They knew that Einstein's occasional expressions of a "materialistic" view did not reflect his general philosophical orientation. In fact, Marxist philosophers concentrated more on attacking Einstein's "idealistic" inclinations than on defending his critique of the Copenhagen school. Einstein's subsequent publications, particularly "Physics and Reality" in 1936 and *The Evolution of Physics* (coauthored with Leopold Infeld) in 1938, attracted the attention of Marxist critics more as displays of Machian influence than as signals of materialistic leanings.[69] Marxist scholars were too involved in attacking the idealistic side of the Einsteinian equation to seek help from the materialistic side. All this did not deter Kol'man from siding with Einstein on the question of the objective nature of physical reality.[70]

Neither Einstein nor the Soviet Marxists cared for Bohr's complementarity principle. Einstein frankly admitted that he could not grasp its meaning.[71] "He did not like it," wrote Banesh Hoffmann. "It went counter to his instincts."[72] Soviet critics rejected it because it contradicted the law of dialectics, or in Marxist parlance, the synthesis of a thesis and an antithesis. As Bohr defined it, complementarity did not designate a synthesis of contradictory ideas. He was clear and forthright in insisting that, as in the case of other complementary but opposing modes of explanation, the particle and the wave were separate sides of the electron and could not be merged in the effort to describe it. In the words of Brian Pippard, Bohr held that "an electron (or proton, or any other small object) [was] neither wave nor particle, and must never be pictured as a hybrid of the two." Rather, they should be regarded as "alternative, incomplete, but complementary, modes of description, each to be used when appropriate, but never together."[73] The Soviet attack on complementarity carried two ideological messages: it protected the supreme reign of causality in nature, which was seriously challenged by the principle of complementarity, and it spearheaded the Stalinist search for the absolute uniformity of scientific explanations. In essence, the conflict between Soviet Marxists and Bohr was a conflict between monolithic and pluralistic philosophical commitments and habits.

The Physicists' Counteroffensive

In assessing the new developments from a broader Marxist position, Maksimov staged a savage attack on Tamm, Fock, Vavilov, and Frenkel, accusing them of idealistic distortions and dubious efforts to shape a worldview incompatible with Soviet ideology. He complained that the Soviet physics journals were "cluttered with idealistic rubbish, systematically imported from abroad."[74] Fock, he said, took an "idealistic position" when he wrote in opposition to Einstein's criticism of Bohr's epistemological views. In Maksimov's opinion, Tamm and his group manifested a servile attitude toward bourgeois ideology and should be removed from positions of leadership in Soviet scientific institutions. That step, he added, must be taken immediately because "the path chosen by physicists represented by Frenkel, Tamm, and Fock is not the path of the development of Soviet science." He urged "the masses of Soviet physicists, who fought for Soviet science and its practical application," to subject the views of the Tamm group to public condemnation.

Maksimov did not succeed in mounting a successful campaign against the alleged enemies of Marxist orthodoxy because the leading physicists were demonstratively unwilling to cooperate. In desperation, he allied himself with the physicist and academician V. F. Mitkevich. In that old-fashioned physicist who still believed in ether as a physical reality, who showed much distaste for modern efforts to expand the mathematical base of physics, and who, in general, was firmly convinced that physics had made no progress after Maxwell's mathematical codification of electromagnetism, he saw great promise for success.[75] Mitkevich did not disappoint him. In 1937, he asked N. P. Gorbunov, permanent secretary of the Academy of Sciences, for authority to organize a joint session of philosophers and physicists to discuss the ideological "evils" of "physical idealism."[76] Obviously counting on the Communist leaders for support, he intended the assembly to declare a holy war on the "idealists" Tamm,

Fock, Frenkel, Ioffe, and Vavilov. Mitkevich wanted to organize the session on the model of the anti-genetics session held a year earlier under the sponsorship of the All-Union Lenin Academy of the Agricultural Sciences, which had labeled Soviet geneticists "bourgeois scientists" and unpatriotic servants of capitalism and launched Lysenko on a meteoric rise in the academic hierarchy.

Gorbunov approved the project, putatively sponsored by the academy's Department of Physics and Mathematics, and appointed a special committee, headed by Mitin, to make the necessary preparations. It was at Mitin's suggestion that the committee selected the two leading critics of the theory of relativity to present formal papers: Mitkevich on the antimaterialistic features of physical idealism, and Maksimov on philosophical aspects of the principle of causality.

The session was never held; obviously, the political authorities were suspicious of the mechanist leanings that made Mitkevich too close an ally of Bukharin's philosophy. In inviting Mitkevich to speak at the proposed session, Mitin knew much about what he was against but little about what he was for. Mitkevich's project would probably have collapsed anyway because none of the leading physicists had even the slightest inclination to take part in an ideological debate. Unlike the community of biologists, the community of physicists was sufficiently united to prevent the rise of a "leader" solely dependent on authorities external to the world of scientific scholarship.

Backed by the authority of eminent representatives of Soviet physics, Ioffe decided to answer the leaders of the campaign against the alleged supporters of physical idealism in the Soviet Union. Directing his answer to the aggressively orthodox Maksimov and allying himself with Tamm, Frenkel, and Fock, who in his view were "materialists," not necessarily of a "dialectical" variety, Ioffe accused Maksimov of stretching the label physical idealism much too far and of trying to make nature subservient to his own laws of causality, which denied indeterminacy and complementarity, both confirmed by the experimental facts of modern science.[77] In his polemical exuberance, he charged, Maksimov worked on the monstrous task of allying a "reactionary" orientation in physics with the "progressive" commitment of the Soviet nation to socialist construction. As for Western physicists, Ioffe made a point of noting that Maksimov reserved his most virulent attacks for those well known for their liberal views and democratic instincts: Einstein, who was "a democrat and an anti-fascist"; Heisenberg, slandered by Hitler's propagandists; Bohr, attacked by the Danish press for his sympathetic attitude toward the Soviet Union; and Schrödinger, who emigrated from Germany as a resolute anti-fascist.[78] Maksimov, Ioffe pointed out with biting sarcasm, found himself in the same boat with Philipp

Lenard, whose criticism of Einstein's theories served as a prelude to his search for an "Aryan physics" suitable to the philosophy of National Socialism. Maksimov's writings helped discourage most physicists from tackling scientific questions of a purely theoretical nature and from entering the slippery path of epistemology. Most Soviet physicists now ignored Max Planck's dictum that modern physics was bound to discover not only "new natural phenomena" but also new insights into "the secrets of the theory of knowledge"—that modern physicists must combine work in the laboratories with a careful study of the wisdom of the great philosophers.[79]

Fock sought no help from the wisdom of great philosophers, but he did not hesitate to elaborate and systematize the main lines of Ioffe's counteroffensive. He combined a defense of modern physics with an attack on Maksimov's misunderstanding of dialectical materialism. Abstaining from logically involved and tedious arguments, he proposed that the philosophers give up trying to prove or disprove the scientific principles of the new physics. They must abandon their practice of ascribing idealistic impurities to various theories of quantum mechanics and the theory of relativity. Instead, they must concentrate on combating idealistic philosophies that abused the great discoveries of modern physics. They must protect physical theories from the corrosive influences of various idealistic schools in philosophy. In order to perform their duties efficiently, they must be thoroughly familiar with the intricacies and historical roots of the theories of modern physics. They must be able to differentiate between idealistic pronouncements made by individual physicists and the materialistic essence of their scientific theories. Above all else, they must not interfere with physicists engaged in scientific work.[80]

Philosophical critics of the Copenhagen school did not receive and could not have received much help from Einstein's thesis of the "incomplete" knowledge of physical reality offered by the theories and methodological tools of Bohr and Heisenberg. Einstein's criticism represented not a total and consistent denial of, but a minor departure from, the subjective and idealistic theory of knowledge. In "Physics and Reality," published in 1936, Einstein stated that "physics constitutes a logical system of thought which is in a state of evolution, and whose basis cannot be obtained through distillation by any inductive method the experiences lived through, but which can only be obtained by free intuition."[81] *Under the Banner of Marxism* wasted little time in attacking Einstein's "return" to the Machian theory of sensations and to the treatment of physical reality as a logical construction and "arbitrary creation" of the human mind.[82] But all was not lost. Einstein continued to show a skeptical attitude toward the principle of indeterminacy and toward the quantum-mechanical disregard of continuity in physical processes. Soviet philosophers also expressed satisfaction at the pronounced disharmony in Western interpretations of the relations of philosophical idealism to modern physics.

The End of the 1930's:
Old Dilemmas and New Conflicts

The tormenting complexity of the new physical theory, the rigidity of Lenin's views on the revolution in science, and the preponderant interest in laboratory research, all influenced Soviet scientists to stay as far away as possible from the theoretical and philosophical questions of modern physical knowledge. The cautious—and often deliberately oblique—writings of Ioffe and Fock indicated that the only feasible way out of the Leninist dilemma was to recognize that physical reality as treated by science and physical reality as treated by philosophy constituted distinct systems of theoretical thought. But during the 1930's, the situation in both physics and philosophy was too unsettled to encourage a serious effort in that direction. No scholar, whether physicist or philosopher, undertook to write a systematic and comprehensive treatise on the philosophy of physics. Interested in developing a Marxist philosophy of science, the Central Committee of the Communist Party appealed in 1938 to the scientists (primarily physicists and biologists) to concern themselves more extensively with the philosophical problems of basic theory.

As the 1930's came to a close, most scientists continued to avoid treating theoretical issues of a controversial nature. Philosophers, for the most part, now worked to consolidate their positions by joining ranks in the battle against physical idealism. They did nevertheless make some effort to temper their criticism by taking note of the great achievements of modern physics. Even Maksimov gave some attention to the triumphs of the theory of relativity and quantum mechanics and their potential affinity with the philosophical postulates of dialectical materialism. He considered various physical theories advanced in the West as specific components of larger dialectical unities. In modern physics, he saw a dialectical synthesis of such contradictory phenomena as the continuity and discontinuity of motion, the wave and corpuscular nature of light, and the causality and "indeterminacy" of physical processes.[83] He saw the future of physics in a general theory unifying the theory of relativity and quantum mechanics. At this time, Marxist philosophers gave wider compass to their recognition of the power of modern physics: undoubtedly, that impulse came in the wake of the news that the physicists had made key discoveries in nuclear fission.

Although the battle on the philosophical front had lost some of its intensity and emotional charge, the conflict between the physicists and philosophers remained unresolved. The physicists continued to make the philosophical norms of dialectical materialism more flexible and more adaptable to the needs of modern science. A handful followed the lead of Ioffe, Fock, and Vavilov, keeping up the pretense that they had contributed to the development of a philosophy of physics totally compatible with the spirit of Marxist theory. What they meant was that they had advanced the necessary philosophical arguments favoring a successful accommodation of dialectical materialism to the theoretical principles of modern physics. But most chose the path of Frenkel, Tamm, and Landau, retreating completely from the philosophical arena.

In 1937, the country marked the twentieth anniversary of the October Revolution. The journal *Priroda* contributed to the government-sponsored celebration by devoting an entire issue to reports on the twenty-year progress of Soviet science. Ioffe covered the rapidly growing and philosophically most volatile fields of physics. Obviously to avoid the wrath of Maksimov, Kol'man, and the increasing number of defenders of Marxist orthodoxy, he made no reference to the Soviet work on the theory of relativity or on any other facet of Einstein's science.[84]

The philosophers, on the other hand, worked on consolidating Leninist epistemological orthodoxy by preparing official rules for the interpretation of the multiple relations between physical theory, philosophy, and ideology. They continued to accuse various groups of physicists of having created unwelcome "outposts" of physical idealism in the Soviet Union. Mainly in defense of their raison d'être, they continued to claim that the full elaboration and consolidation of a dialectical-materialistic interpretation of the theoretical foundations of modern physics required much additional work. Maksimov, in an effort to widen his area of operations, undertook a close scrutiny of Louis de Broglie's philosophical ideas, only to conclude that the contributions of the French physicist were no less contrary to Marxist thought than those of Bohr and Heisenberg.[85]

In their search for a generally accepted mode of operation, the philosophers quickly agreed on one point of paramount strategic significance: they firmly rejected Fock's recommendation that they leave the scientific problems of modern physics to professional scientists. In 1939, Kol'man adopted a strategy of concentrating on the scientific—rather than on the philosophical—soundness of specific theoretical claims of Western physicists. He took pains to point out which quantum-mechanical ideas he considered incontrovertible contributions to science; but his main task was to attack "distorted" interpretations of those ideas by Bohr, Heisenberg, and other leaders of microphysics. For example, he charged the Copenhagen school with concerted efforts to make complementarity a universally applicable mode of scientific explanation, even though its usefulness did not extend beyond "a small circle" of experiments concerned with the interaction of instruments as macrosystems and atomic particles as microsystems. Bohr and Heisenberg, in their efforts to limit the role of causality in modern science, had translated a transitory weakness in methodology into a universal philosophy of agnosticism.[86]

Kol'man took much the same tack when he turned, in another article, to the theory of relativity. Here, too, he first briefly set out which of Einstein's theoretical principles he considered unquestionable contributions to science before launching into a diatribe on those he regarded as alien to dialectical materialism. He rejected Einstein's treatment of the speed of light as the maximum speed attainable by material objects and expressed categorical disapproval of the "Einsteinian" inference about the equivalence of the cosmologies of Ptolemy and Copernicus. In Einstein's denial of "general" simultaneity, he saw an outright negation of time as an objective phenomenon, a reality independent of the observer. He did not have much use for "Einsteinian" ideas of the "finite world" and of "the beginning and the end of the universe." He offered the following summary of his criticism of Einstein's scientific thought:

> The theory of relativity does not reach the depths of physical phenomena. It overlooks atomic processes and the role of individual particles of matter in explaining the causes of observable regularities. It studies physical phenomena by means of differential equations, which focus only on the continuity of matter. Concentrating on the processes that take place on the surface and disregarding the deeper causes of these processes, the theory of relativity is the product of a phenomenological approach to physical phenomena. The strength of the theory of relativity is in abstractions, the path to grand generalizations. But it also has a natural limitation that invites abuses. All efforts to build physics on the geometry of continuous space are destined to be failures, however remarkable the qualities of this geometry. Such efforts are groundless, metaphysical exaggerations, fed on the idealistic leanings of many theoretical physicists.[87]

Einstein's cosmic geometry, according to Kol'man, suffered from yet another flaw: it transformed the study of physics into a study of geometry. It exaggerated the positive aspects of the axiomatic method to the point of producing absurd results. Einstein erred in viewing geometry as an exclusive creation of the immanent forces of human reason and in relying on internal consistency as the only valid criterion of truth. He carried the separation of theory from practice to an extreme.

Kol'man's sweeping criticism of the idealism of Einstein's theory and its "erroneous" scientific assumptions did not go unanswered. In an article published in *Priroda* the same year, on the occasion of the sixtieth anniversary of

Einstein's birth, Fock, without mentioning Kol'man, reaffirmed the prevailing view of Soviet physicists when he stated that the idealistic leanings of Einstein's "philosophical rhetoric" did not in any way reduce the powerful contributions of the theory of relativity to the understanding of the objective laws of the physical universe. Contrary to the prevalent argument among Marxist theorists, philosophical idealism had no bearing on the nature and magnitude of the Einsteinian revolution in physics.[88]

Fock's was not the only voice challenging the rising chorus of Marxist critics of the theory of relativity. Lev Landau also observed the sixtieth anniversary of Einstein's birth by writing an article for the popular journal *Knowledge-Power* on the scientific soundness and vast scope of the general theory of relativity. The article appeared four months after Landau was imprisoned by the secret police on undisclosed political charges.[89] At the time of his arrest, he was preparing the first Soviet textbook on the special and general theories of relativity.*

Unabashed by the outcry from these eminent men, Kol'man expressed his further unhappiness with Einstein's recent assertion that accumulated practical experience did not form the cognitive basis for scientific abstractions, thereby contradicting one of the basic postulates of Leninist epistemology. In *The Evolution of Physics* (1938), it seemed to Kol'man, Einstein had exposed the idealistic cast of his mind when he asserted that "physical concepts are free creations of the human mind, and are not, however it may seem, uniquely determined by the external world."[90] What Kol'man did not say, however, was that in Einstein's view, external experience played a decisive role in determining the acceptability of new scientific knowledge on utilitarian, experimental, and observational grounds. External experience did not generate new scientific ideas but was the ultimate judge of their validity.

Kol'man's omission in fact says much about the Marxist philosophers' preoccupation with attacking modern physics on philosophical and ideological grounds. Marxist critics were so involved in tracing and criticizing the subjective side of physical phenomena as presented by the Copenhagen school in quantum mechanics and by Einstein in the theory of relativity that they dealt only sporadically with Einstein's recognition of the objective aspects of physi-

---

*Archival documents, made accessible five decades later, revealed that Landau was caught by the secret police in the act of helping to write an anti-Stalinist leaflet to be distributed at the 1938 May Day parade in Moscow. The leaflet called Stalinism a fascist distortion of the ideals of the October Revolution. In private conversations, Landau was known as a relentless critic of Stalin's increasing dependence on terror as a primary instrument of political control (Gorelik, "Fiziki i sotsializm," p. 49). He was released before the year was out on the intervention of Peter Kapitsa.

cal reality. They made little effort to explore the ideas that brought Einstein close to at least some of the principles of dialectical materialism. For example, they were much less interested in Einstein's firm adherence to strict determinism in the processes of nature, even though it supported a basic premise of dialectical materialism. In their struggle against the Copenhagen school's principles of indeterminacy and complementarity, they made little use of Einstein's defense of causality as the key to scientific explanation.

Their main argument against Einstein was ideological. Maksimov gave it a brief and most direct summary:

> No physical theory precipitated such an explosion of idealistic fantasies as the theory of relativity. Mystics, clerics, and idealists of all colors, including many serious scholars, became enamored with the philosophical conclusions of the theory of relativity. The more naive followers were especially attracted to the odd and miraculous consequences imagined to flow from the theory of relativity. . . . Idealists mustered all their forces for a struggle against materialism by tying space and time to philosophical relativism. Relying on the general theory of relativity, . . . they accepted the idea of the curvature and the finitude of cosmic space.[91]

Maksimov's statements carried transparent ideological messages: he identified physical idealism with the values of mystics and religious thinkers; he condemned relativism as a negation of universal and absolute laws of cultural and social evolution; he negated the idea of the finite universe presented in Einstein's cosmological paper in 1917 on the ground that it invited creationist challenges to the laws of nature; and he underscored philosophical materialism as the backbone of Marxist philosophy and official Soviet ideology.

In their attitude toward philosophical elaborations of the principles of modern physics, the leading physicists and philosophers were worlds apart. Most philosophers considered it their sacred duty to protect dialectical materialism from adverse ideological influences of modern scientific theories and philosophies of science. They argued that the theory of relativity could become an integral part of Marxist philosophy and Soviet science only after it had undergone major modification. Most leading physicists followed an opposite course: they were interested in protecting their science from the crippling influences of dialectical materialism. They thought that the conflict between philosophers and scientists could be reduced, if not fully eliminated, by narrowing the area of contact between the scientific and philosophical views of "physical reality." The fact that both sides in the bitter feud could receive public airing throughout the 1930's indicated that the political authorities were not quite ready to interfere decisively in this specific confrontation between science and ideology.

Despite the animosity of Marxist philosophers toward the perceived idealistic slant of Einstein's theories, interest in the world of ideas heralded by the relativistic approach to physical reality continued to grow. *The Principle of Relativity*, published in Russian translation in 1935, brought together basic comments on the theory of relativity by Lorentz, Poincaré, and Minkowski, and a series of Einstein's fundamental papers on the two theories of relativity. For unknown reasons, the German original, promptly translated into English, failed to include the submissions of Poincaré, whose ideas both antedated and postdated Einstein's formulation of the principle of relativity. The author of the review of the Russian translation in the journal *Socialist Reconstruction and Science* welcomed the inclusion of one of Poincaré's papers and noted that the combination of the genius of a mature Poincaré and the genius of a young Einstein was needed to make "invariance" both "natural and economical."[92] By "invariance," of course, he meant the very heart of the principle of relativity. Generally, however, Soviet physicists and philosophers placed limited emphasis on the role of Poincaré's theoretical ideas in the genesis of the theory of relativity. The main arguments of Einstein's "Physics and Reality" appeared in Russian translation soon after its publication in the *Journal of the Franklin Institute* (1936).

At the very end of the 1930's, Vavilov, speaking for his fellow physicists, gave an eloquent and blunt answer to Kol'man's and Maksimov's attacks on Einstein, without mentioning their names. In an unprecedented move, he suspended his practice of balancing positive and negative appraisals of Einstein's contributions. He even expressed a most favorable attitude toward Bohr's complementarity principle, considered by most Stalinist philosophers a flagrant attack on the universality of the law of causality. He could not have made himself plainer: Einstein's "materialistic" conception of space and time was not only a turning point in the history of modern physics but also a major victory for the Marxist philosophy of science. In Vavilov's words: "In Einstein's theory, space-time is an inseparable attribute of matter, and cannot exist without matter. We do not know of any space without material fields of force. Such is the fundamental thought of the general theory of relativity expressed in a concrete physical form."[93] (This article, we may note, was one of three in a collection commemorating the thirtieth anniversary of the publication of Lenin's *Materialism and Empirio-Criticism*. The other two essays were authored by Timiriazev and Mitkevich.)

The unsettled philosophical-ideological conditions in physics, particularly the aggressive campaign by Marxist philosophers against physical idealism, drove many Soviet physicists away from general problems of a theoretical nature. In 1937, the physicist D. I. Blokhintsev noted that, aside from G. A. Mandel's work on the "five-dimensional theory" and V. A. Fock's on Dirac's equations in the general theory of relativity, Soviet physicists had shown very little interest in exploring Einstein's scientific ideas.[94] Ten years later, things had not changed much. Vavilov could mention only two new original papers, Fock's "On the Motion of Finite Masses in the General Theory of Relativity" (1939) and "The Copernican System and the Ptolemaic System in the Light of the General Theory of Relativity" (1947), as solid theoretical contributions to Einsteinian studies.[95] The first paper, relying on approximations, offered a solution to Einstein's equations for finite spheric masses assumed to conform to the Euclidean notion of space infinity. It showed that Einstein's gravitational field equations embraced the equations of motion, a problem that Einstein and his associates had handled and solved independently at the same time.[96] That paper heralded Fock's "campaign" in the 1950's to reduce the general theory of relativity to a "general theory of space, time and gravitation."

The physicists' disinclination to explore Einstein's theory was due not just to the all-consuming concern of Marxist authorities with the ideological implications of its basic principles. They had to take account, to some extent at least, of the very limited possibilities of linking Einstein's scientific contributions to the practical problems of "socialist construction." During the 1930's, Soviet physicists like Bronshtein, Frederiks, Frenkel, Fock, Blokhintsev, and Landau limited their writing on the theory of relativity primarily to encyclopedia articles and popular surveys.

The Second World War: A Time of Conciliatory Strategies

The wartime alliance with the democratic countries of the West brought a tangible change in the philosophical treatment of the new physics. Attention now centered on the enormous extent to which its scientific contributions confirmed the basic propositions of dialectical materialism, rather than on the intellectual and ideological aberrations of physical idealism. Mitin expressed the new mood in a paper on the development of Soviet philosophy he wrote for the celebration of the twenty-fifth anniversary of the October Revolution in 1942. He portrayed modern physics as both the zenith of the new science and a magnificent confirmation of the correctness of the philosophical postulates of dialectical materialism. "The unveiling of the 'secrets' of matter," he said, "and the profound analysis of the new laws of nature have fully illustrated the cornerstone of the Leninist prediction of the inexhaustibility of the atom and the electron and of the transformability of all forms of matter." "Depending on the power of its inner impulse, new physics," he continued, "has overcome the enormous theoretical contradictions that emerged in the course of its vigorous growth. Here belong the contradictions between mass and energy, matter and motion, corpuscles and waves, and continuity and discontinuity."[97] The contradictions encountered by physical theory were, of course, graphic examples of the work of dialectics in nature.

In Mitin's analysis of the anatomy of the revolution in physics, the theory of relativity occupied a central position. In bestowing lavish praise on Einstein's contributions, Mitin was content to voice an optimistic note about the successes of Soviet scientists and philosophers in developing a uniform Marxist view of the theory of relativity, and in fashioning the only philosophically sound interpretation of Einstein's masterwork. "Soviet natural science and philosophy," Mitin asserted, "have been able to overcome and cope with the idealistic interpretation of the theory of relativity. Armed with the method of dialectical materialism, our science has been successful in giving a correct appraisal of the role and importance of that theory, in drawing consistent conclusions from it, and in formulating accurate interpretations of its substance bearing on both theory and the worldview."[98]

The theory of relativity, as Mitin saw it, did not deny "either the existence of the objective world or the objectivity of our knowledge of nature." It did not deny the absolute attributes of space and time and of the motion of bodies, for it recognized their objective existence and independence from the human mind. It left no room for arbitrary judgments in recognizing regularities in the processes of nature. The reliance on the views of observers and on frames of reference as points of departure in the theory of relativity did not deny the objective reality of natural processes. Space and time were relative only insofar as they were inseparable from the motion of physical bodies. In brief, despite all "idealistic abuses" and distortions, the theory of relativity represented "a step forward in the discovery of dialectical regularities in nature."[99]

Mitin's analysis was important only insofar as it described a specific phase in the Soviet philosophical interpretation of modern physics in general and Einstein's theory in particular. The basic characteristic of that phase was the obvious effort of Marxist philosophers to exalt the gigantic contributions wrought by modern science rather than to dwell on the upsurge of idealistic thought following in their wake. Behind Mitin's oratory was an admission that the best way to treat the conflict between the dominant Western philosophical interpretations of modern science and dialectical materialism was to pretend that it was inconsequential, or that it did not exist in the first place.

By 1944, Vavilov was able to speak glowingly of Einstein's success in making space and time indissoluble parts of the same reality and in identifying geometry as a physical science.[100] He did not neglect his ideological obligation to attack physical idealism as a symptom of the irreversible decadence of capitalist society, to be sure. But he did not select the same representatives of idealistic thought for attack that had attracted the attention of Stalinist philosophers during the 1930's. He, for example, refrained from involving Einstein in his war on idealism. In Einstein, he saw a towering figure in science, not an enemy on ideological grounds. Instead, he found in Arthur Eddington, the acerbic critic of materialism and determinism in science, an easy and particularly provocative target. Nor did he have kind words for Arthur Holly Compton's search for a meeting ground of science and religion in *The Freedom of Man* (1935).

In 1943, the Academy of Sciences marked the 400th anniversary of the Copernican revolution by sponsoring a conference on the far-reaching consequences of the new astronomy. Fock there submitted a paper directly challenging Einstein's denial of the utility of privileged coordinate systems in the general theory of relativity.[101] He was particularly unhappy with Einstein's idea of the equivalence of the Ptolemaic and Copernican systems in "relativistic cosmology," as set out in *The Evolution of Physics*.[102]

In a subsequent study, Fock clarified his position:

We have stressed repeatedly the fundamental significance of the existence of a preferred coordinate system. . . . Only if the existence of such a coordinate system is recognized as reflecting certain intrinsic properties of space-time can one speak of the correctness of the heliocentric Copernican system in the same sense as this is possible in Newtonian mechanics. If this is not recognized, or if the existence of preferred coordinates is denied, one is led to the inadmissible point of view that the heliocentric Copernican system and the geocentric Ptolemaic system are equivalent.[103]

On ideological grounds, Fock obviously thought that recognizing the two systems as coordinate would be a most unwelcome concession to the astronomical ideas of the medieval church. To avoid this situation, he introduced so-called harmonic coordinates, which recognized the unique features of individual coordinate systems by imposing supplementary conditions on their gravitational potentials beyond the reach of the Lorentz transformations.

In his anniversary paper on Copernicus, Fock reaffirmed and amplified his earlier claim that the real strength of the general theory of relativity was not in the mathematical elaboration of the principle of relativity, but in carrying Newton's theory of universal gravitation to its logical conclusion. The relativistic focus represented a weak component of Einstein's theory; the explanation of the riddle of gravitation represented its unsurpassable strength. Fock readily admitted that it was the theory of gravitation that placed Einstein among the greatest geniuses in the annals of science. Einstein was the first scientist to explain not only how gravitation worked but also what made it work.

Fock claimed that it was only as a theory of gravitation that Einstein's theory was in a position to affirm the superiority of Copernicus's cosmology over Ptolemy's. Infeld, in defense of the relativistic tier in the general theory of relativity, later claimed that Fock misread Einstein's (and his) reference to the two cosmologies as coordinate systems. If Einstein, according to Infeld, accepted their equivalence, he did so only in elaborating the mathematical structure of relativity theory without intending to attribute a "physical meaning" to it.[104] P. S. Kudriavtsev expressed a typical view of Soviet scientists when he wrote that the general theory's postulation of the equivalence of the two cosmic systems must be considered a "formal" claim, for "all historical, philosophical, and cosmological arguments were on the side of Copernicus's system." Obviously, he used "formal" in the sense of "mathematical."[105] No doubt, a sentence or two of explanation in *The Evolution of Physics* would have forestalled this particular criticism.

Fock's anniversary paper did more than pay homage to the father of the heliocentric system. It was also a clear and stern message to Stalinist philosophers that the time had come to replace the monotonous repetition of ideologically saturated philosophical exercises with a disciplined and technically competent inquiry into Einstein's ideas. Criticism should be welcome primarily as a way to elevate Einstein's thought to new heights, not as a device for trading slanderous philosophical clichés.

Stalinism After the War:
The Climax of Marxist Attacks

It was not until 1947, when the country began to gather momentum in massive efforts to rise from the ravages of the Second World War, that the kind of moderation Mitin had suggested came under comprehensive scrutiny and re-evaluation. The year began inauspiciously with the publication of M. E. Omel'ianovskii's slim volume *Lenin and Twentieth-Century Physics*, which gave a fleeting survey of the great discoveries of modern physicists from Roentgen to Dirac. Omel'ianovskii here reiterated the established Marxist views on the danger of the excessive mathematization of science and on the errors of the so-called energeticist orientation, and presented the theory of relativity as a confirmation of dialectical materialism. But he also made two observations that were absent from Mitin's report: he noted that in the West physical idealism was as strong as ever, and that Soviet physics had not made solid contributions to the advancement of dialectical materialism because it was torn by the chronic strife between idealistic and materialistic strains.[1]

The Theory of Relativity and Anti-Cosmopolitanism

On June 24, 1947, only a few months after Omel'ianovskii's book was approved for publication, Andrei Zhdanov, a member of the Politburo, gave the keynote address at a meeting of scholars, artists, literary figures, and ideologues gathered to "discuss" G. F. Aleksandrov's new book on the history of Western philosophy. In his bitter denunciation of ideological deviationism, Zhdanov devoted several passages to physics. While reaffirming Engels' dictum that materialism must accommodate itself to every discovery heralding a new epoch in natural science, Zhdanov demanded a relentless campaign against the representatives of modern "obscurantism" and "fideism," who advocated a return to the "Pythagorean mysticism of numbers," manipulated the discoveries of modern physics as the springboard for a revival of idealistic philosophy, and, disregarding Francis Bacon's warning, translated "the powerlessness of their science into a slander of nature." That group, as he saw it, included the followers of Einstein.[2]

Zhdanov's message was momentous and prophetic: the time had come to stop making concessions to the idealistic aberrations of leading Western scientists, to begin working on the creation of an ideologically pure and philosophically untarnished Marxist physics, and to pay special attention to the national roots of Russian science. The search for ideological purity and philosophical orthodoxy became closely associated with a campaign against cosmopolitanism, viewed as a servile attitude toward Western achievements in scientific thought. A stern warning issued by the journal *Bolshevik* identified cosmopolitanism as "the negation of patriotism, its contradiction": it preached "the complete indifference toward the fate of the motherland," and "it denied the existence of a civic and moral debt to one's own people and the fatherland."[3]

Anti-cosmopolitanism was a variation on the theme of Soviet patriotism, a rallying cry that had been pushed to the extreme throughout the war under the direct guidance of the Communist Party. A long succession of anniversary celebrations commemorated dramatic achievements in all fields of intellectual and technical endeavor. Early on in the war, the Academy of Sciences began seriously considering the establishment of a permanent center for the study of the history of science, with an emphasis on the development of scientific thought in Russia. Ambitious actions were undertaken to widen the base of historical studies of such national heroes in science as Lobachevskii, Mendeleev, and Pavlov, and, in general, to make the history of science in Russia appear richer, longer, and more impressive than heretofore.

A new look into the antecedents of Einstein's pivotal notion of the space-time continuum called for a recognition of Russia's primacy in setting the stage for the triumph of the general theory of relativity. A lengthy article published in 1952 exemplifies the new treatment. After a preliminary discussion of famous scientists and philosophers whose contributions marked distinct steps in the evolution of scientific thought, the author, V. I. Sviderskii, singled out Lobachevskii as the first scholar to introduce geometry as an empirical science and to trigger the vital series of developments that led to the triumph of the general theory of relativity.[4]

Anti-cosmopolitanism demanded a massive search not only for deeper and broader roots of Russian culture than previously conceded, but for "proofs" of the superiority of Russian achievement in individual fields of human endeavor as well.[5] Its main emphasis was on the originality and magnitude of Russian genius. Having become the leader of the newly formed "socialist camp," the Soviet Union, in the eyes of Zhdanovian philosophers, needed to bolster its national reputation as a country with a long cultural history and superior creativity. Another goal of the campaign was to document the deep and robust roots of materialism in the history of Russian thought. In all this, Russia's current and historical contributions to scientific thought received the most attention.

Einstein heard about the anti-cosmopolitan campaign soon after it became the main plank in the Zhdanovian war on ideological impurities in Soviet thought. In the fall of 1947, he published an open letter to the United Nations praising the nobility of its goals and lamenting the limited scale of its authority, which prevented it from taking a more active and effective part in securing international peace and eliminating the threat of nuclear warfare. He expressed the hope that the future would bring a "supranational" organization strong enough to maintain peace in the world.[6]

At the obvious prodding of the Communist authorities, four of the USSR's most eminent physicists promptly issued an open letter to Einstein objecting to the idea of elevating a supranational authority above the legal powers of individual states. Einstein's idea of a "world government" seemed to them a screen for imperialist designs to extend capitalist monopolies beyond the boundaries of individual states.[7] Zhdanovian anti-cosmopolitanism could not but stand opposed to Einstein's plan for a world government. In all this, the four scientists recognized the memorable heights of Einstein's achievement as a scientist and a humanist. In answering their letter, Einstein stayed close to his original argument that unlimited national sovereignty and the endless preparation for war were two sides of the same coin. With their only letter to Einstein, the four physicists satisfied two classes of peoples: the political authorities, by expressing an anti-cosmopolitan stand on the question of national sovereignty, and their fellow physicists, by making it clear that they held Einstein's scientific discoveries and humanitarian spirit in the highest esteem.*

In the war on physical idealism and cosmopolitanism, Marxist critics directed their heaviest armor at Einstein as the true progenitor of a sweeping wave of anti-materialist thought in modern physics. One way or another, they held him responsible for the "sins" of Milne, Eddington, Jeans, Jordan, and many other scientists accused of conducting a crusade against materialism in modern physics and in the philosophy of scientific thought. At the same time, they were unusually parsimonious in dealing with the substance of Einstein's

*The four Soviet physicists were S. I. Vavilov, A. N. Frumkin, A. F. Ioffe, and N. N. Semenov. All held key administrative positions in the Academy of Sciences.

scientific contributions. In 1947, for example, the *Great Soviet Encyclopedia*'s article on gravitation made only passing mention of Einstein. The author suggested two books for additional reading, both published in 1923. The reader was left with the unambiguous impression that after Newton very little true knowledge was added to the theory of universal gravitation. During Stalin's rule, Marxist philosophers did not write a single book presenting a comprehensive and systematic picture of the intricate web of Einstein's physical and epistemological arguments. They concentrated primarily on isolated ideas and principles inviting philosophical attacks on ideological grounds.

The immediate result of Zhdanov's declaration of war on cosmopolitanism and ideological deviations was the founding of the journal *Questions of Philosophy* (*Voprosy filosofii*), which like *Under the Banner of Marxism* before it (that journal had been discontinued in 1944) was meant to spearhead the effort to promulgate philosophical unity in the world of scholarship. The new publication became the leading philosophical mouthpiece of the Communist Party. The first issue featured Zhdanov's report at the June 1947 meeting and the comments of selected members of the audience. Carefully policed by the architects and guardians of the new ideology, the journal ran into all sorts of problems: heavy political pressures, administrative uncertainties, and difficulties finding trustworthy contributors from the academic community. B. M. Kedrov, one of the better philosophers of science, was accused of "cosmopolitan" indiscretions in his new book on Engels and modern natural science, and was quickly dismissed as the chief editor.[8]

At first, Soviet physicists paid little attention to the new ideological campaign. In a 1949 article, Vavilov, the new president of the Academy of Sciences, disregarded the Zhdanovian exhortations when he hinted that physicists should dedicate themselves to refining and amplifying, rather than to challenging and denigrating, Einstein's contributions to modern physics. Although he admitted that the general theory of relativity continued to be riddled with ambiguous ideas, he made no reference to Zhdanov's warning that it opened the gate to ideas embedded in mysticism and fideism.[9]

In 1948, L. D. Landau and E. M. Lifshitz published a second and extensively revised edition of their *Theory of Classical Fields*.[10] Quickly translated into English, the study was less than generous in personal references to Einstein but overflowed with ideas that were clearly drawn from his work. The authors went out of their way to observe that the general theory of relativity represented "probably one of the most beautiful of all existing physical theories" and to recognize that it was developed by Einstein in a purely deductive manner and only later was substantiated by astronomical observations.[11]

In 1950, Ioffe brought out *The Basic Ideas of Modern Physics*, the first effort in the Soviet Union to present a general historical survey of the evolution of modern physical theory. He was firm in his defense of the theory of relativity, which was "backed by incontrovertible proofs of its fundamental postulates, accumulated after 1905," and was "universally accepted by modern physicists."[12] Ioffe dwelt on the revolutionary significance of Einstein's ideas on the expansion of mass at velocities approaching the speed of light, on the equivalence of mass and energy, on photons, and on Brownian molecular motion.[13] At the same time, to live in peace with the Stalinists, he placed special emphasis on the contributions of Soviet scientists to modern physics and made a derogatory, but skimpy, reference to the efforts of "Heisenberg and Einstein, Eddington and Jeans," to transform science into a systematic study of experience—to make subjective data the main building blocks of physical reality.[14]

Ioffe was one of the most frequent targets of merciless attacks by Marxist ideologues. To protect physics from political authorities, he made concessions of an ideological nature, usually in formal announcements of philosophical texture, without doing tangible damage to the ongoing work in physical theory and experimental research. His perseverance in defending the guiding principles of the new physics contributed to the relative tranquillity and solid experimental work of Soviet physicists, particularly during the initial years of Stalinist rule. Before the war, says David Holloway, "thanks in large measure to Ioffe himself," physics "did not perish as biology had done; laboratories were not closed down in response to ideological criticism, and in spite of the emphasis on practical utility, research continued even if it did not promise immediate results."[15]

During the turbulent years of ideological purification and anti-cosmopolitanism, with attacks on him becoming more vicious and ominous, Ioffe could only keep quiet and wait for the storm to subside. In 1954, after Stalin's death, but with Stalinism still aflame, he published an answer to the critics of his book admitting to having made ideological errors of minor consequence. He had taken a "faulty" line, for example, in explaining the equivalence of mass and energy and in interpreting some of the unsettled principles of quantum mechanics. But he was categorical in rejecting the claim of one critic that the future of quantum mechanics could never be in a close alliance with the theory of relativity because of its doubtful scientific value.[16] Ioffe refused to retreat from his conviction that Einstein's theory represented a powerful and most promising turn in the forward march of scientific thought. He ended his comment with a double-edged conclusion: "My book does not offer a sharp criticism of Western physics. But I am convinced that it has conveyed a message to all careful readers that the development of this science has confirmed the philosophical principles of dialectical materialism and has not given any support to idealistic conceptions."[17]

Exposed to rapidly growing ideological attacks, which reached unprecedented intensity in 1949–52, the maligned physicists received notable support and sympathy from their professional community. In 1950, for example, at a time when Ioffe could do nothing to please Marxist philosophers, close to a hundred physicists contributed papers or editorial services to a festschrift in honor of his seventieth birthday. The volume, devoted strictly to scientific themes, was easily the most monumental expression of respect and gratitude a Soviet scholar had received from his peers.[18]

That same year, paying close attention to Marxist philosophical critics and their supporters among physicists, rather than to the general consensus of the scientific community, the powerful Lavrenti Beria, the head of the Soviet secret police and (since 1946) a member of the Politburo, ordered that Ioffe be relieved of his duties as director of the Leningrad Physical and Technical Institute, a model institutional component of the Academy of Sciences and a product of the eminent physicist's dedicated administrative work and profound understanding of major trends in the evolution of modern physics.[19] As president of the academy, Vavilov, who had the highest respect for Ioffe, had no choice but to carry out Beria's order. Two years later, however, the academy obtained permission from the government to appoint Ioffe director of the newly founded Laboratory of Semiconductors, in 1954 transformed into a full-fledged institute. It was only after Stalin's death in 1953 that Marxist attacks on Ioffe went out of style.

Einstein and Infeld's *Evolution of Physics* (1938) was another major work inviting ideological scrutiny. The Russian translation appeared in 1948, when the Stalinist campaign to purify Soviet ideology was in high gear. The book's explicit statement that the pivotal ideas in science were pure products of the human mind, unfettered by any kind of empirical substratum, was exactly the kind of pronouncement the current ideological crusade was directed against. As we saw, many found the authors' willingness to grant equal validity to Ptolemaic and Copernican astronomical views particularly objectionable. Perhaps by deliberate design, *The Evolution of Physics* reached a limited number of readers. A more accessible edition appeared in 1956.

Amid mounting criticism of Einstein's ideas, Fock repeated his elaborate plan to bring harmony to the relations between the theory of relativity and dialectical materialism. The Ioffe festschrift gave him an opportunity to clarify and confirm the view of the general theory of relativity he had originally advanced in 1939. That view differed tangibly from the ideas behind the current anti-Einsteinian campaign. Fock began his paper by expressing full agreement with Einstein's claim that the character of geometry was indissolubly linked with the distribution of masses and their motions. The link, Fock proposed, was mutual: "On the one hand, departures of geometry from Euclid's axioms are conditioned by the presence of gravitational masses, and, on the other hand, the motion of masses in the gravitational field is determined by the departure of geometry from Euclidean axioms. In brief, masses create geometry, and geometry determines their motion."[20]

Fock argued that though the general theory of relativity needed major restructuring, Einstein himself had provided the basic ingredients for that task. All that was needed was to shift the emphasis from the principles of the covariance of field equations and the equivalence of inertial and gravitational masses to "the hypothesis that physical space and time are described by Riemann's geometry presented in Einstein's gravitation equations."[21] That maneuver, in Fock's opinion, would give Einstein's equations more physical content.

At the time Fock wrote this paper, he was obviously more an admirer of the grandeur of Einstein's achievement than a critic of the general theory of relativity. Marxist philosophers did not know how to approach his revisionist views and preferred to leave him alone. They were partly satisfied that, in a quiet way, Fock agreed with Lenin that too much mathematics opened the gates to physical idealism. Fock, however, continued to be criticized for his favorable attitude toward the Copenhagen theoretical-epistemological orientation in quantum mechanics. No doubt, his restructuring of the general theory of relativity was an attempt to accommodate it to Stalinist interpretations of the Marxist philosophy of science.*

Most philosophers did not wait long to join the ideological crusade. After a short delay, they invaded the pages of *Questions of Philosophy*, the Party organ of the Academy of Sciences, and *Advances in the Physical Sciences*, a publication of the Institute of Physics. *Questions of Philosophy* found it important to reexamine Omel'ianovskii's book on Lenin in the light of the Politburo's ideological pronouncements. Expressing anger and dismay, the philosophers accused Omel'ianovskii of treating objective physical reality—for example, the relations between mass and energy—as nothing more than formal products of mathematical measurement. His failure to recognize privileged (or preferred) coordinate systems, they claimed, led him to the Machian conclusion that the conflict between Ptolemy and Copernicus was pointless. In Omel'ianovskii's

---

*After a brief stay in a Stalinist political prison in 1937, Fock tried to find a meeting ground between Marxist philosophy and the theory of relativity. He accepted dialectical materialism, but he never ceased to criticize the scientific ignorance of Marxist philosophers.

effort to formulate a philosophical interpretation of the theory of relativity, they detected a vague exercise in the Machian theory of knowledge. Especially in Omel'ianovskii's failure to acknowledge the independence of the development of Soviet physics, they saw a flaw of major proportions. Nor did they forgive him for not attacking Iakov Frenkel and other Soviet physicists whose thinking was "contaminated" by the idealistic thought of the bourgeois West.[22]

Maksimov, always ready to conform to the Politburo's signals, was eager to harness all his skills and militancy to eliminate the "unpatriotic" behavior of leading physicists. His ideological convictions and his temperament made him an ideal banner-carrier for the Zhdanovian forces. At this time, he was ready to side with the philosophers who opposed the idealistic leanings but not the scientific content of the theory of relativity. He was ready to concede that the mathematical apparatus of the theory of relativity was "irreproachable," and that its value for "the study of physical phenomena [was] vast."[23] Freed from the burden of idealistic distortions, the general theory of relativity, particularly its conceptualization of space-time, would confirm the basic principles of dialectical materialism.

Not interested in spelling out the strengths of the theory of relativity, Maksimov preferred to limit his work to pointing out the evils of physical idealism. This was a task called for by Zhdanov's patriotic crusade. Embedded in the theory of relativity, physical idealism, wrote Maksimov, was dangerous and injurious, not only because it impeded the development of science, but also because it was guided by the ideas of cosmopolitanism. "Physical idealism," as he assessed it, was "a movement politically opposed to communism, for it reflected, directly or indirectly, the ideology of the imperialist bourgeoisie." Maksimov went on to suggest that the physicists should be inspired by the efforts of the Lysenkoist forces to stamp out all traces of "servitude to bourgeois ideology." Like the Lysenkoists, Soviet physicists should publicly denounce colleagues supporting idealism and cosmopolitanism. He contributed to that effort by unleashing a savage attack on Frenkel and another leading physicist, M. A. Markov.[24]

In 1948, M. A. Leonov published the first textbook on dialectical materialism to incorporate the Zhdanovian assault on ideological deviations in science and cosmopolitanism. While acknowledging "kernels of truth" in the theory of relativity, he was especially eager to point out the germs of "fideism" and "mythological" delusions hidden in the fabric of Einstein's reasoning. In Einstein's formula of the equivalence of mass and energy, he detected an end-of-the-world scenario. Heavy reliance on Riemann's version of non-Euclidean geometry, which postulated a positive curvature of cosmic space-time, led Einstein to accept the idea of finite space and time and to invite the theological dogma of creation. As Leonov saw it, every challenge to the infinity of space and time flirted with the idea of creationism.[25]

Despite the ever-harsher criticism of the theory of relativity, a few recalcitrant philosophers continued to stand their ground as best they could. Most noted among the rare defenders of Einstein were I. V. Kuznetsov and G. I. Naan, both representing the new generation of Marxist philosophers of science in the Soviet Union, and much more familiar with the theoretical and methodological complexities of modern physics than their predecessors. Kuznetsov had a general grounding in theoretical physics, but his main interests were in the epistemological problems of quantum mechanics. Naan was deeply steeped in modern cosmology; in fact, he showed superb proficiency in both scientific matters and philosophical challenges related to cosmology.

When Kuznetsov published a monograph on Bohr's principle of correspondence in 1948, he obviously did not anticipate the extent to which it would infuriate the Stalinist ideologues. It mattered little that he opened by noting the fundamental agreement of the correspondence principle with the Marxist law of dialectics when what followed was his unrestrained praise of Einstein's scientific work. There was no doubt in Kuznetsov's mind that the notion of relativity elevated physics to new heights of achievement by broadening both its theoretical base and its range of application. The special theory brought together electrodynamics and Newtonian mechanics; the general theory united the special theory of relativity and the theory of gravitation. Einstein also laid the groundwork for relativistic quantum mechanics.[26]

Kuznetsov's monograph created a storm of hurricane proportions. The Zhdanovian philosophers and physicists, in a quickly called session sponsored by the Institute of Philosophy, furiously attacked him for his failure to concentrate on exposing the "evils" of physical idealism. Some critics used the occasion to accuse Kuznetsov of subservience to Western cosmopolitanism. Maksimov, ever ready to the attack, lamented Kuznetsov's failure to point out the superiority of physics affiliated with dialectical materialism over "Machian and other idealistic orientations."[27] One of the philosophers there, N. F. Ovchinnikov, mustered enough courage to claim that the Marxist theory of the interrelations of the absolute and the relative received solid confirmation from the principle of correspondence.[28] But he was one of the few in attendance to make any protest. Preoccupied with the idealism of the Copenhagen school, the critics made only indirect references to Einstein and the theory of relativity. There is no doubt, however, that the favorable references to Einstein were one of the major reasons for the attack on Kuznetsov's monograph. Whether because of these attacks or out of sincere conviction, Kuznetsov later became one of Einstein's severest critics.

In 1951, the Estonian scholar G. I. Naan responded to the Marxist attacks in kind. In an article in *Questions of Philosophy*, he castigated Maksimov and the growing number of opportunistic philosophers who wrote profusely and disparagingly about Einstein's alleged idealism. It was something of a miracle that the censors even allowed Naan's article to be published at all, let alone in the country's most prestigious philosophical journal. Naan made two claims: Einstein's principle of relativity was one of the great ideas of modern physics, and relativity as a physical notion should be kept separate from relativity as a philosophical notion. He challenged the widely circulated thesis that Marxist theory would do well without the principle of relativity as one of its cornerstones. He also argued that the future of Marxist philosophy rested in the inseparability of dialectical materialism from science.[29] As Loren Graham has pointed out, Naan argued that relativity was not a subjective phenomenon but was "inherent in the philosophical processes themselves."[30] Implicit in Naan's precise arguments was the thought that philosophers, however deft in crafting generalizations, should play no role in the validation of scientific knowledge. Marxist philosophers, he noted, should be advised to explore the multiple contributions of the theory of relativity to a confirmation of dialectical materialism. Unlike Maksimov and his group, who insisted that scientific theories should be modified to meet the directives of orthodox Marxism, or, if unmodifiable, should be rejected, Naan agreed with the consensus within the scientific community that dialectical materialism should be ready to absorb all advances in science automatically. He claimed that the writings of Maksimov and his ilk undermined dialectical materialism by showing that its most active defenders were unable to cope with the growing intricacies of physical theory.[31] Behind Naan's arguments was an unmistakable attack on the aspirations of Marxist philosophers to serve as a clearing-house for scientific ideas.

The lone voices of Kuznetsov and Naan were muffled by the fierce attacks on physical idealism and cosmopolitanism. From 1948 until Stalin's death in 1953, the journals concentrated almost exclusively on attacking the "anti-materialistic" ideology of the Copenhagen school and the theory of relativity.[32] They criticized physicists for their alleged efforts to transform mass into energy, to recognize motion without matter, to spread the "mystical" idea of the expanding universe, to sacrifice the law of causality to the principles of uncertainty and complementarity, and to make physics an ancillary extension of geometry. Omel'ianovskii lamented the failure of the Soviet community of physicists to produce a comprehensive interpretation of the theory of relativity "free from the influence of idealistic philosophy."[33] "The time has come," he wrote, "to make a systematic effort to fasten the theory of relativity to dialectical materialism."

The most significant outcome of the international symposium celebrating Einstein's seventieth birthday in 1949 was the monumental collection *Albert Einstein: Philosopher-Scientist*, edited by P. A. Schilpp. None of the Soviet physicists who were invited to contribute papers responded. Frenkel, the most frequent target of Stalinist attacks on physical idealism, wrote a paper for the occasion but was advised by unspecified Soviet authorities not to submit it. It happened that the celebration took place just when Communist attacks on the idealistic foundations of the theory of relativity and on "reactionary Einsteinianism" (*reaktsionnoe einshteinianstvo*) were accelerating. It was only after Stalin's death that Einstein's "Autobiographical Notes," the most valuable contribution to Schilpp's volume, appeared in Russian translation.

The Lysenkoization of Cosmology and Physics?

As the ideological war on physical idealism in general and on the theory of relativity in particular grew in intensity, an important event took place in August 1948. Supported by government and Party authorities, and inspired by Zhdanov's anti-cosmopolitan announcements, Trofim Lysenko contrived to get a carefully supervised congress of biologists in Moscow to outlaw genetics as a field of scientific inquiry and as part of the school curriculum. Lysenko informed the assembled scholars that his action had Stalin's full support. As the elaborate structure of that discipline was being dismantled, V. M. Molotov, representing the Politburo, and a group of philosophers gathered around the newly established *Questions of Philosophy* recommended that a similar ideological house-cleaning be undertaken across the full spectrum of science.[34]

Since Zhdanov had pounced on relativistic cosmology and theoretical physics as illustrations of the rising threat of idealism to Soviet science, the cosmologists and physicists were bound to be among the first groups of scientists to be officially directed to subject their disciplines to ideological cleansing. In Marxist circles, both physics and cosmology were recognized as occupying strategic positions in the pressing search for harmonious relations between the official ideology of the Soviet state and the philosophical foundations of ideas unlocked by the twentieth-century revolution in science.

Thanks to speedy preparations, the astronomers met in December 1948 under the sponsorship of the Leningrad section of the All-Union Astronomical-Geodetic Society.[35] Attended by five hundred persons, the conference concentrated on the ideological problems of cosmology. The three main papers focused on relativistic cosmology and its alleged incompatibility with Marxist theory. The idea of the expanding universe, pioneered by A. A. Friedmann, attracted critical attention for two reasons: by relying on the techniques of extrapolation, it interpreted the knowledge of our galaxy as the knowledge of the entire universe; and it suffered from "formalism"—from total dependence on mathematical operations not grounded in inductive data distilled from experience. Western relativistic cosmology was presented as a product of "speculation" elaborated by mathematics.

The ritual of the conference, as was customary in those days, would not have been complete without a sacrificial drama. The goats in this case were Ivanenko, Landau, and Lifshitz; the first was publicly chastised for considering the theory of the expanding universe a "triumph of materialism," the other two for presenting relativistic cosmology without critical remarks. In the three scholars' endorsement of homogeneous and isotropic space, the critics saw "a distortion of the objective and real world of astronomy."[36] The conference issued a resolution declaring that every favorable comment on the theory of "the finite expanding universe" was equal to expressing servility toward "the reactionary science of the Western bourgeoisie."[37] Cosmologists were ordered to create a new version of their science free of any implicit or explicit challenge to the idea of the infinity of cosmic space and time. In fact, they were specifically ordered to search for a materialistic interpretation of Hubble's red shift. In 1951, Omel'ianovskii, rapidly emerging as the most authoritative Marxist philosopher of science, had nothing but scorn for Friedmann's "pernicious" idea of the expanding universe. He commended Friedmann, however, for "refuting" Einstein's earlier idea of "a closed and stationary universe."[38] Small wonder that the physicists D. I. Blokhintsev and I. M. Frank asserted many years later that during the 1940's "our philosophers rejected categorically the astrophysical data supporting the notion of the expanding universe."[39]

Two weeks after the conference, in January 1949, the Communist Party ordered the Ministry of Higher Education and the Academy of Sciences to hold a national conference of physicists. A special committee was appointed to make preparations for the projected conference, which was expected to add more fuel to the battle against physical idealism and to reinvigorate the campaign against cosmopolitanism. Most of the members were affiliates of either the academy or Moscow University. The aim of the conference was to bring an end to the "ideological" deviationism of such leading Soviet physicists as Frenkel, Ioffe, Landau, and Tamm. A. V. Topchiev, the powerful general secretary of the academy and the chief representative of Communist authorities, served as the committee chairman. He made it clear at the outset that he expected the conference "to follow the model of the session of the All-Union Lenin Academy of the Agricultural Sciences" (that is, the Lysenko-sponsored congress held in Moscow in August 1948).[40]

The high academic status and extraordinary power granted to Topchiev represented one of the more significant indicators of the growing determination of the Communist authorities to expand and consolidate their direct ideological control over the activities of the Academy of Sciences. A minor expert in naphtha chemistry, but a very active, experienced, and dedicated Party worker, he was "elected" to full membership in the academy on direct orders from political authorities. As its "main scientific secretary," he was charged with seeing that the academic world fully followed the Communist Party's ideological lines. To legitimate his authority as a scientist, he added his name to selected papers produced by the members of the academy's Institute of Naphtha. A book he was credited with coauthoring, even though he had little to do with its writing, was given a Stalin Prize, a political move to bolster his scientific authority.[41]

The organizing committee and invited participants held forty-two meetings in search of "unity" in the interpretation of the basic issues of ideological purification and anti-cosmopolitanism as they affected physics. The more the members of the committee and their guests debated, the more obvious it became that the physicists were going to give much less than enthusiastic support to the Party's designs to engage them in the current ideological campaign. In the view of some of the university contingent, their colleague Maksimov thought the agenda ought to be shifted from a philosophical analysis of how the relativity and quantum theories related to dialectical materialism to direct "scientific" attacks on their specific principles.[42] His immediate goal was to give a distinct Soviet orientation to all the major branches of modern physics, a task to be undertaken by tested Marxist philosophers rather than by physicists, who had shown no enthusiasm for such an undertaking. There were open charges, emanating from philosophers and physicists alike, that Maksimov's attacks on the theory of relativity were attacks on figments of his own imagination. The meetings gradually degenerated into irreconcilable controversy, and the projected conference was canceled.

Some of the leading participants in the organizing committee's discussions adopted a strategy of making broad and vague concessions to Stalinist ideological demands, while at the same time acknowledging the vast debt of modern science to the contributions of Einstein and Bohr. When they criticized the proponents of physical idealism, they usually concentrated on the philosophical views not of Einstein and Bohr, but of Jordan, Jeans, Milne, and Eddington, who had written entire books against materialism, mustering arguments that in some respects were as outlandish as the statements Stalinist philosophers and their allies were making against idealism. One development left no doubt

in the minds of historians: the academic participants in the discussions "delivered a shattering blow" to the militancy of Maksimov and other defenders of
Stalinist ideology.[43]

"Academic Science" and "University Science"

The debates and maneuvers of the organizing committee produced sure signals of a conflict between the spokesmen for "academic science" and the
spokesmen for "university science." The latter group, consisting mainly of
Moscow University professors, exploited the anti-cosmopolitanism fervor by
making all sorts of references to the national purity of their own endeavors, as
opposed to those of the leading physicists in the academy, who espoused ideas
and behavior alien to Marxist standards.[44] According to G. E. Gorelik, "during
the 1940–1950s, the division of the Soviet community of physicists into two
parts was real. This split was most clearly expressed during the preparations for
an all-union conference scheduled for 1949."[45] S. T. Konobeevskii, dean of the
physics department of Moscow University, informed Stalin as early as October
1947 that the new "ideology" placed "university science in opposition to the
ailing academic science."[46]

The university members of the organizing committee argued that the first
duty of the planned conference should be to expose and criticize the manifestations of cosmopolitanism in the Soviet scientific community. Led by N. S.
Akulov, and relying on direct attack and innuendo, they chastised the distinguished members of the Academy of Sciences for not citing Russian scientists
in their papers and books. Their complaint had a personal message: they did
not find enough references to their names in the texts or footnotes of Ioffe,
Landau, and other leading representatives of "academic science."

As we have seen, with the exception of Mitkevich, who died in 1951, but was
inactive professionally during the 1940's, the leading physicists affiliated with
the academy were favorably disposed toward the theory of relativity and were
bitterly opposed to Maksimov's philosophical rhetoric. Some number of the
university physicists, by contrast, pressed long and hard for major surgical
measures to transform Einstein's ideas into a body of theory compatible with
dialectical materialism. It comes as no surprise that one of the most prominent
members of this "patriotic" and "anti-cosmopolitan" group was the extremist
Timiriazev. It was that relentless campaigner for a full rejection of the theory of
relativity who had helped the secret police prepare the dossier that had sent
Boris Hessen to a forced labor camp—and to his ultimate death at the end of
the 1930's. In Timiriazev's view, Hessen's defense and popularization of the

theory of relativity were crimes against the state. He publicly expressed his approval of Hessen's imprisonment.[47]

Nikolai Kasterin, who in 1919 wrote the first negative critique of the theory of relativity, figured prominently in the acrimonious conflict between the two institutions. Although he abstained from making direct attacks on Einstein in this period of anti-cosmopolitanism, he continued to work on the assumption that all physical problems could be resolved within the framework of the Faraday-Maxwell-Hertz tradition. All his research problems were cast strictly within the theoretical framework of classical physics. Many years before, in 1937, the academy had published a paper of Kasterin's—an effort, ambitious in intent but modest in substance, to effect a synthesis of "aerodynamics and electrodynamics" by relying exclusively on the mathematical and theoretical tools of classical physics.[48] Timiriazev and Mitkevich saw to it that the work was immediately translated into English and given wide international distribution. They hoped that Kasterin's new "synthesis" would gain recognition as a major victory for Newtonian physics and as a serious challenge to the theory of relativity. Not long after, in June 1938, with the blessing of the academy, a group of physicists and mathematicians held a meeting for the sole purpose of deciding whether Kasterin, on the basis of that submission, merited continued financial assistance by the academy. A large majority of participants expressed dissatisfaction with the anachronistic nature of the essay, which went into history as the last work published by the academy to support the idea of a return to classical physics. Selected criticisms by the leading discussants were published in the academy's bulletin, as was Timiriazev's defense of Kasterin's chief arguments.[49]

Even before that, Tamm had written a totally negative appraisal of Kasterin's scientific papers on aerodynamics, electrodynamics, and related fields. In his opinion, they were full of false assertions, absurd physical indiscretions, flaws in logic, and mathematical errors.[50] In the end, the academic council had recommended that the academy discontinue its financial support of Kasterin's research. It stated that Kasterin had not only ignored post-Newtonian science but violated a number of experimentally verified contributions of Newtonian science.[51] No doubt those rebuffs by the academicians were in the front of some of the participants' minds.

Kasterin managed to find new employment a few years later. In 1941, Moscow University appointed him professor of physics, completely disregarding the academy's negative appraisal of his scholarly work. The dean of the department admitted subsequently that Kasterin belonged to a group of physics professors who had ceased to be actively engaged in research, experimental or theoretical.[52] At that time, Moscow University introduced a policy, which lasted more than a decade, of not employing members and other affiliates of the academy to fill teaching positions. In 1944, Tamm and another eminent academy member, G. S. Landsberg, authors of well-received textbooks in the basic theory of electricity and optics, respectively, were among the candidates for teaching positions whose applications were turned down.[53] Immediately after the war, in the heyday of anti-cosmopolitanism, various Moscow University administrators and professors made a habit of attacking the academy as an "ailing" institution and a bastion of cosmopolitanism. Actually, Moscow University's physics faculty was in such disarray and decline that Peter Kapitsa himself made an appeal to the central government to do something about it.

No doubt, Timiriazev helped bring both Kasterin and Maksimov to Moscow University. With those transfers, Timiriazev hoped to make the university the national center of studies and ideological writings critical of Einstein's contributions to science and the modern worldview. He also hoped to make it the guiding force and organizer of opposition to the academy as a bastion of Einsteinian strength in the Soviet Union. However, his dream of creating an anti-Einstein center did not materialize. He and Kasterin had entered the nonproductive twilight of their professional careers, and Maksimov, who never had much support in the scientific community, became a controversial figure in Marxist philosophical circles. On at least one occasion, he was publicly criticized by the Communist press for cosmopolitan flaws in his interpretations of the history of science in Russia.[54] Timiriazev was almost bound to lose in any event because a solid contingent of Moscow University physicists favored Einstein's ideas.

Kasterin is generally remembered only as one of the very few physicists who refused to enter the challenging world of post-Newtonian science. Historians of physics in Russia usually give him credit only for his doctoral dissertation on the distribution of waves in a heterogeneous medium, completed in 1904 at Moscow University. Two articles published in the university's journal *History and Methodology of the Natural Sciences* showed a somewhat broader appreciation of Kasterin's work.[55]

Despite the inordinate intensity and immense scope of these postwar trends, despite the anti-cosmopolitan campaign, the attempts at ideological cleansing, and the heightened political oppression, the state did not succeed in Lysenkoizing physics. Quiet but consistent and persevering resistance by a majority of the leading physicists in the Academy of Sciences to Stalinist attacks on their discipline—and on science in general—played a major role in helping physics escape the fate of biology. Their job was made somewhat easier by the emergence of a new group of Marxist philosophers, typified by Kedrov and Naan, whose counterarguments helped frustrate Maksimov's attempts to bring his fellows into line at the organizing committee.

Einstein and the War on Energeticism

In the late 1940's, energeticism, a philosophical orientation inaugurated by Wilhelm Ostwald during the 1890's, became the target of calumnious attacks by the Stalinist defenders of ideological purity. Scoffed at by Lenin in *Material-ism and Empirio-Criticism*,[56] Ostwald's philosophy elevated energy to primary position among the building blocks of nature, relegating matter to a secondary and derivative position. A group of Marxist revisionists went even farther: they made the notion of energy the moving force of human society as well, thereby undermining the materialist foundations of Marxist sociology.

Marxist philosophers considered energeticism a specific expression of anti-materialism in the philosophy of modern physics. They argued that the revo-lution in physics brought about by quantum mechanics and Einstein's relativ-ity theories lured energeticism to many new positions and to many new forms of aggressive behavior in both science and philosophy. Indeed, the shifting of ontological emphasis from ponderable matter to energy, as an attribute of the field, was widely recognized as the dominant characteristic of modern phys-ics.[57] Einstein had explicitly stated that, according to the special theory of rela-tivity, matter and radiation were only particular forms of energy.[58] "Ponderable matter," he went on to say, had lost its reigning position in physics and was a particular form of energy. And Bertrand Russell had written in 1949, in *Human Knowledge: Its Scope and Limits*: "Mass is only a form of energy, and there is no reason why matter should not be dissolved into other forms of energy. It is en-ergy, not matter, that is fundamental in physics." [59] But for Soviet philosophers to recognize the primacy of energy would have meant violating the Leninist principle that there was no motion without matter, and, on the ideological level, rejecting universal causation as the sovereign explanation for the work-ings of nature and culture.

A. P. Aleksandrov, the head of the academy's Institute of Physical Problems and, subsequently, of the Institute of Atomic Energy in these years, later wrote that "soon after the end of World War II" he was asked to report to the Central Committee of the Communist Party, where several investigators questioned him on the high theory of modern physics. He found that the investigators held quantum mechanics and the theory of relativity in low esteem, obviously on ideological grounds.[60] At the time, he surmised, Marxist authorities were making preparations for a new wave of ideological pressures on the world of scholarship. They gave high priority to a war on the supporters of the idea of the "disappearance of matter," leaning on both the special theory of relativity and quantum mechanics. The time had come to attack the real and imagined supporters of energeticism and the alleged enemies of conservation laws. During the anti-cosmopolitan campaign, Marxist authorities considered energeticism and alleged attacks on conservation laws both pseudoscientific and unpatriotic.

Einstein's mathematical expression of the equivalence of mass and energy was interpreted as an outright endorsement of the ideas of the de-materialization of atoms and the annihilation of matter, a direct attack on conservation laws, and an invitation to creationist explanations in physics. Conservation laws attracted much attention not only as fundamental laws of nature, but also as the fundamental precepts of Soviet ideology: they cleared the ground for anti-creationism as a cornerstone of Soviet society. To be critical of the laws of the conservation of energy and mass was to be a disloyal member of Soviet society—keeping the door open for divine interference with the laws of nature, and, on the ideological level, recognizing religion as a system of socially beneficial values. Because of their prime ideological significance, conservation laws became a strategic topic of philosophical discussion.

Kuznetsov, who had suddenly abandoned his favorable view of post-Newtonian physics to become a leading Marxist critic of physical idealism, rejected every notion of a transformation of mass into energy and of energy into mass.[61] He rejected the first transformation on the basis that it violated the law of the conservation of mass, and the second on the basis that it violated the law of the conservation of energy. In claims on behalf of both transformations, he saw a surreptitious effort to negate the ontological primacy of matter, the cardinal principle of Soviet ideology. Ia. P. Terletskii was even more straightforward: he protested against Einstein's famous formula $e = mc^2$ as a violation of the law of the conservation of matter.

In 1952, *Advances in the Physical Sciences* subjected T. V. Kravets, a corresponding member of the academy, to severe criticism for his alleged advocacy of such "anti-Soviet ideas" as the transformation of matter into energy and of energy into matter.[62] Marxist critics found Einstein partially responsible for giving his mass-energy equation an "idealistic" interpretation. They recalled, in particular, two of his statements: his declaration that "mass and energy are ... alike; they are only different expressions for the same thing"; and his pronouncement that "both matter and radiation are but special forms of energy distributed in space."[63]

Marxist physicists showed no interest in dropping Einstein's mass-energy formula. They were interested merely in defining it in such a manner as to make it compatible with dialectical materialism. In that effort, their chief resource was Fock, who, in the face of the growing official pressure on the scientific community, took it on himself to come up with an ideologically acceptable solution to the problem. Fock explained that mass and energy presented different properties of matter that were not mutually transformable. He claimed that Einstein had established the mathematical proportionality, but not the physical identity, of mass and energy.[64] He did not accept Einstein's idea that "mass represents energy."[65] Nor did he accept the merging of the laws of the conservation of mass and energy into one law. Fock's main contribution was in dissuading the more zealous Marxist philosophers and physicists from starting a campaign to reject Einstein's principle of mass-energy relations *in toto*.

Since neither mass nor energy could be interpreted as a synonym of matter, the philosophers eventually realized that there was no need for a total war on Einstein's principle. Einstein made no reference to a possible equivalence of matter and energy, thereby eliminating any direct support for the idea of the disappearance of matter and creationism. The Soviet success in making the atomic bomb in 1949 provided convincing verification of the energy-mass equation and helped reduce the scope and intensity of Marxist attacks. Criticism of Einstein's principle was gradually reduced to semantic efforts to prepare dialectical materialism for compromises with selected ideas of the theory of relativity.

Ideologically conditioned to be on guard against the "decadent" influence of creationism, the archenemy of the Soviet philosophy of science, Marxist experts argued that matter was omnipresent and protected by conservation laws from dissipation and the threat of ultimate disappearance. A typical statement read: "The laws of conservation refute idealistic and religious 'theories' of 'the creation of the world,' 'the creation of matter from nothing,' and 'the initial thrust.'" The conservation of energy, mass, momentum, spin, and other natural phenomena provided different proofs for the permanence of matter in its multiple forms.[66] The law of the conservation of matter acquired a new meaning: it became a generic name for all conservation laws.[67] It became a philosophical, not a scientific idea. Gradually, it lost its philosophical meaning as well.[68]

The Climax of Anti-Einsteinian Sentiments

A combination of essays prepared for the canceled conference and newly solicited papers appeared in 1952 under the title *Philosophical Questions of Modern Physics*; it has come down in history as the infamous "Green Book." The volume marked a victory for the Stalinist defenders of Marxist philosophy and a major defeat for Soviet physicists, the Soviet scientific community, and the ethos of science. It signaled the iron determination of the defenders of Marxist orthodoxy to follow Lysenkoist models in transforming the entire network of Soviet science into an outpost of the Stalinist version of Marxism-Leninism.

Humbled and depressed, Sergei Vavilov, president of the academy, who did not live to see the publication of the aborted conference's papers, contributed the introductory chapter, which raised a number of burning questions on the relationship of physics to philosophy.[69] (The paper was extensively revised by unnamed persons, but for convenience, I have made him the author in all that follows.) Vavilov lamented the widely manifested inclination of Soviet physicists to avoid the philosophical problems of their discipline, in general, and the relationship of the basic theories of modern physics to dialectical materialism, in particular. He chastised physicists like Frenkel who responded to the criticism of their alleged idealistic indiscretions simply by avoiding philosophical issues altogether. To mount a successful campaign against philosophical idealism in physics, it was necessary to clarify such basic conceptions as "space-time, matter, mass, energy, charge, and spin." Physicists must rely on philosophy for help in their effort to explain continuous and discrete processes in the physical universe, the methodological role of mathematical models, and mathematical extrapolation as a tool of scientific explanation. Their concern with philosophy, moreover, should be two pronged: it should add new arguments to the current campaign against idealistic inroads into Soviet thought, and it should help Soviet physicists overcome their chronic hesitation to tackle theoretical problems of general significance. That hesitation, he observed, explained the low output of Soviet physicists in the domain of general theoretical ideas, the basis of "new and broader orientations." In "our excessive worship of foreign science," the main theme of anti-cosmopolitanism, Vavilov saw the key to his compatriots' reluctance to embark on original theoretical explorations.[70]

The contributors on the whole concentrated on the urgent need to put an end to idealistic violations in Soviet physics. Their attack on the idealism of Western physicists contained many old charges reinforced by new details and a heightened unwillingness to compromise. The gigantic figure of Einstein provided a particularly inviting target for a broad and critical examination, an assignment that Kuznetsov willingly took on. Kuzentsov, it will be recalled, was a militant member of the new generation of philosophers of physics who worked with exuberant determination to reverse the recent tendency—including his own—to underemphasize both the scope and the intensity of the antimaterialistic battle waged by the defenders of physical idealism. At this time, he chose to argue that Einstein's theory contained not only mistaken epistemological explanations but also erroneous scientific claims derived from faulty philosophical positions. He announced that "the unmasking of reactionary Einsteinianism in the realm of physical science [was] one of the most important duties of Soviet physicists and philosophers."[71] And he added:

> Einsteinianism . . . is one of the most active and most militant forms of "physical" idealism in the physical sciences of today. The Einsteinians are the inventors of the mystical "transformation" of matter into energy and of energy into matter. The Einsteinians have originated the wild fictions of the "finite expanse of the universe" and of "mathematical harmony" as testimony to "divine omniscience." The Einsteinians' attack on science has led to a revival of the worn-out myth of the miraculous "creation of the world" from a fanciful "proto-atom," set into motion by the divine finger.[72]

For those who might have thought it necessary to keep Einstein apart from the Einsteinians, Kuznetsov issued a stern warning:

> In all this, Einstein does not separate himself from the scandalous doings of the Einsteinians, but, on the contrary, sanctions and endorses their work by his authority. How low Einstein has fallen as a leader of the Einsteinians has been attested to by the following fact. At the end of 1948, a certain L. Barnett published a book in New York entitled *The Universe and Doctor Einstein*, an overt attempt to propagandize reactionary fideism under the guise of "a physical picture of the universe," which, he claimed, was based on up-to-date facts. This book would be indistinguishable from thousands of equally repulsive volumes published in the USA at the present time were it not for the fact that Einstein wrote a preface for it. In the preface, Einstein approved of the book by stating that it provided an apt characterization of "the present state of our knowledge in physics."[73]

To critics who thought that "the fallacious idealistic suppositions" of Einstein's philosophy were in no way related to "the sound materialistic pillars of Einstein's science," Kuznetsov answered that the "absurd" and "anti-scientific" conclusions drawn from a philosophical theory proved only that the theory was not sound—that it had nothing in common with science.[74] He took particular exception to Einstein's principle of general covariance—which states that "the general laws of nature are to be expressed in equations which hold good for all systems of coordinates, that is, are covariant with respect to any substitutions whatever (generally covariant)"—as a "universal principle" in the study of natural processes. That principle, he argued, originated not in experimental data but in the "fallacious idealistic assumption" that the laws of nature were "free creations" of the human mind—that the properties of physical phenomena were created by their relations to coordinate systems preselected by the observer.

While limiting his criticism of the special theory of relativity to individual propositions, Kuznetsov was ready to reject the general theory *in toto*. Influenced by Fock's earlier comments, he asserted that the general theory did not establish what it intended to establish in the first place: the equivalence of inertial and gravitational masses, as a basic postulate of the theory, was the result of erroneous reasoning, for it was based on the "subjectivization of space and time." Einstein's principle of the equivalence of inertial and gravitational masses, he said, applied only to the uniform and rectilinear motion of bodies in relation to each other. Moreover, such an equivalence, he noted (obviously under Fock's influence), did not exist outside relatively short intervals of time and very limited space. The equations of the general theory required an interpretation radically different from Einstein's, which suffered from the excessive "geometrization of the material gravitational field." Einstein's work on a unified field theory, he said, was doomed; after all, not much could be expected from efforts to unify electromagnetic and gravitational fields by formal constructions of "pure description," with no regard for concrete "physical ties." Einstein and his bourgeois allies were in debt to the "materialistic" development of modern natural science, rather than to "idealistic" aberrations of new philosophy, for their isolated positive contributions. But even when those scholars made positive contributions to science, they were "unable to advance fundamental theoretical generalizations based on the achievements of science and to draw all the consequences from the established scientific facts."

Kuznetsov crowned his vituperative and apocalyptic rhetoric with a statement that left no doubt about the clarity and resoluteness of his position:

> Critics of the idealistic orientation of Einstein's general philosophical statements have often argued in favor of preserving the theory of relativity as a "physical theory." Our analysis has shown that such an approach to Einstein's theory of relativity is incorrect. First, Einstein's general philosophical statements are not mere superficial hair-splitting, or, even worse, "unsupportable inferences." They are an integral part of the theory of relativity, and they determine much of its content. Equations displayed in this theory do not by themselves form a physical theory. . . . Second, every effort to "touch up" the Einsteinian theory of relativity, to "mend" it, or to "patch" it, misrepresents the real situation in that branch of physics. It disorients physicists. . . . The progress of science can be assured not by compromises and half-measures but by a total rejection of Einstein's conceptions.[75]

R. Ia. Shteinman, another contributor, charged that Einstein had not synchronized his theory of space and time with modern field theory, and that he encased his theory of relativity in mathematical formalism that was too rigid to allow for improvements. He also argued that Einstein oversubscribed to the philosophy of operationalism based on the principle that every physical theory was merely a statement on the interaction of physical phenomena established by measurement alone.[76] Shteinman charged that in his determination to construct a thoroughly mathematical picture of the universe, Einstein erred in not seeing space and time as "special forms of the existence of matter" and the field as "a variety of matter." For that reason, he failed to build a scientifically operative bridge between the theory of relativity and quantum mechanics—between space and time as *a posteriori* physical categories and the material structure and high-velocity motion of nuclear particles. Einstein, he said, could not demonstrate whether or not subatomic particles exceeded the speed of light. Shteinman repeated Vavilov's earlier statement that the links between Einstein's relativistic views of space-time and the quantum-mechanical interpretations of the microstructure of matter continued to be unexplored. He did not amplify his conclusion that only a consistent materialistic interpretation could return the theory of relativity to scientific sanity.

Kuznetsov and Shteinman went beyond the criticism of Einstein's theory of relativity: they introduced, but did not elaborate, a proposal for a substitute theory named "the theory of high-velocity motion." In fact, their proposal represented an attempt to produce a Marxist theory of relativity. They proposed to eliminate all references to relativity, to remove all mathematical operations considered to be void of physical content, and to take the role of observer out of the methodology of physics.[77] They behaved as if the dreams of Maksimov and Kol'man during the 1930's were well on the way to becoming a reality in the early 1950's.[78]

M. M. Karpov was a Marxist novice in command of only the most elementary knowledge of Einstein's work. That did not prevent him from offering a sweeping condemnation of Einstein's general intellectual orientation. Unlike Kuznetsov, he did not criticize the scientific parameters of the theory of relativity but limited his observations to matters of philosophy and ideology. Whereas Kuznetsov claimed that Einstein's theory was wrong on both scientific and philosophical grounds, Karpov argued that Einstein was a classic example of Western physicists whose scientific engagement contradicted their philosophical positions.[79] Einstein, according to Karpov, displayed four distinctive, and strongly objectionable, features of philosophical idealism. First, he saw the main source of scientific knowledge, not in the social and practical activity of man, but in a belief in "the inner harmony of our world," without which "there could be no science." Second, he accepted the idea of infinity on philosophical grounds but rejected it on scientific grounds. Third, he exaggerated the role of mathematics in physical theory. In his view, scientific concepts and laws, which furnished the key for the understanding of natural phenomena, were constructions of pure mathematics. The effort to formulate a unified field theory—to deduce a continuous field for all branches of physics from a single equation—could not but be a wasteful effort. Fourth, Einstein was a naive believer in "pure science," which prevented him from recognizing the role of social-class consciousness in shaping the course of scientific development. He did not believe in the *partiinost'* of science.[80]

After hearing about the denunciation of the theory of relativity by Kuznetsov and his colleagues, Einstein wondered, not without cynicism, why there was a need for scientists to work so hard to garner kernels of truth when the Soviet Communist authorities, guided by the principles of dialectical materialism, could deliver the whole truth by government decree. To give this argument an added sense of urgency, beauty, and sarcasm, he presented it as a short poem.[81] Behind his humorous response was a serious warning about the grave dangers to science of the interference of unknowledgeable outside authorities.

The Green Book marked the culmination of the Stalinist effort to achieve a unity of Marxist beliefs and scientific thought. The leaders of Soviet physics, accused of supporting idealism, were found guilty of unpardonable deviations from Stalinist philosophy. The critics reserved their harshest words for Iakov Frenkel. He was condemned for his alleged adherence to philosophical views contrary to Marxist thought, and more specifically, for failing to subject the principle of indeterminacy to strong criticism.[82] Paradoxically, he received one of the highest state prizes for his scientific contributions in 1947, the very year in which the attacks on him were at their most hostile.

Annoyed with the philosophical passivity of many leading scientists, the newspaper *Pravda* issued a stern warning on November 17, 1952: "still there are groups of professional physicists who avoid discussion and ignore all efforts to criticize idealistic currents in modern physics."[83] In the heyday of Stalinism, physicists who dodged philosophical issues were considered delinquent executors of their professional tasks. Philosophical neutrality had only one meaning: outright opposition to Marxist theory—and to the Soviet government.

Landau was among those who drew the wrath of Marxist critics for making a deliberate effort to avoid philosophical involvement. A contributor to the Green Book, after faulting Landau and Lifshitz for failing to provide an adequate philosophical analysis of their subject in their five-volume treatise on theoretical physics, saw in that ostensible effort to avoid philosophical discourse as a subtle expression of outright loyalty to physical idealism, particularly of the Copenhagen school variety.[84] Others found Landau guilty of favoring heterodox scientific ideas. He was said, for example, to view nuclear particles as "nonextended mathematical units." An article in *Questions of Philosophy* accused him of "poisoning" the younger generation of Soviet physicists by attributing too much importance to the principle of complementarity, by spreading the "false" notion of the expanding universe, and by giving primacy to the idealistic interpretation of the second law of thermodynamics. Such reasoning had led him to support the myth of "thermal death," even though Engels had dealt it a "mortal blow" many years ago.[85]

Abram Ioffe came under the same kind of attack, accused of ideological impurities as gross as those of Landau.[86] After Ioffe published his *Basic Ideas of Modern Physics* in 1950, the outcries against him became more frequent and more ominous. Critics especially resented his alleged inclination to view the evolution of modern physics as a self-propelled process, essentially independent of external influences, to gloss over the "negative" influence of idealistic thought on the scientific work of Einstein and the pioneers of quantum mechanics, and to ignore the efforts of Soviet physicists to harmonize the fundamental principles of their science with the guiding ideas of dialectical materialim.[87]

Ioffe's interpretation of Einstein's theories continued to elicit scathing responses from Marxist critics. He was found guilty of reducing the foundations of the theory of relativity to a purely logical analysis of procedures for the measurement of space-time, an approach that gave precedence to arbitrary choices of the "observer" over the objective "motion of bodies." Some critics noted that his argumentation favored the notion of mass as "a measure of matter." Omel'ianovskii claimed that Ioffe ignored the role of the "Machian bias" in Einstein's philosophical interpretation of the theory of relativity. Under Einstein's influence, Ioffe exaggerated the role of the observer in the theory of relativity and failed to recognize the primacy of relations between concrete objects. According to one critic, he did not draw precise philosophical conclusions from Fock's study of the general theory of relativity, which condemned physicists who made the equivalence of inertial and gravitational masses a universal law of physical nature and a basis for the "paradoxical" idea of the equivalence of the Ptolemaic and Copernican systems of the universe. Nor did Ioffe take note of the place of A. A. Friedmann's views on the expanding universe in the development of modern cosmology.[88]

Even the long-dead L. I. Mandelstam came in for a full share of vitriol. As noted earlier, Mandelstam was among the small group of scientists responsible for the quick incorporation of quantum and relativity theories into the mainstream of Soviet physics; and in his studies in molecular optics, spectroscopy, and nonlinear oscillations, he had helped create a strong Soviet interest in the theoretical base for modern radio techniques and automatic controls. In the course on the theory of relativity he taught at Moscow University in 1933-34, he had dealt exclusively with Einstein's scientific thinking, and so ought not to have been open to criticism for expressing ideas that ran counter to dialectical materialism. He explained his "philosophical" position in straightforward terms: "I have been particularly eager to emphasize . . . that all questions connected with the theory of relativity are physical questions above everything else. Its philosophical significance is a special question, but we are interested only in its physical significance."[89]

Soon after Mandelstam's death in 1944, the academy began publishing his selected works as an appropriate tribute to his professional dedication and scholarship.[90] Volume Five, which came out in 1950, contained his lectures on the theory of relativity. Unfortunately for his reputation, it included a remark implying that the philosophical meaning and the scientific meaning of the theory of relativity were unrelated questions. Although the statement was parenthetical and was printed from students' notes, it could not but arouse the ire of ideological purists. Since Mandelstam was dead and beyond personal recrimination, the authorities decided to rely on a more sedate mode of criticism: they asked the Institute of Physics of the Academy of Sciences to hold an intramural conference on Mandelstam's transgressions. At that conference, a procession of academic dignitaries, led by D. V. Skobel'tsyn, charged that Mandelstam's philosophical orientation had led him to erroneous thinking: he was a "spontaneous materialist" rather than a dialectical materialist. Even worse, his inclination to exaggerate the role of measurement and observation instruments in determining the nature of physical phenomena brought him closer to neopositivism than to any form of materialism. He was found guilty of making a deliberate attempt to skirt the philosophical problems of modern physics and of formulations sheltering idealistic aberrations.[91] His interpretations of simultaneity and space-time were considered alien to Marxist materialism.

The conference differed tangibly from typical ideological sessions both because it was conducted in an atmosphere of tranquillity and because several speakers paid homage to Mandelstam's solid contributions to science. The academician M. A. Leontovich went so far as to point out that Mandelstam's philosophical errors were negligible in comparison with the grand scope of his scientific work. If, according to Leontovich, Mandelstam showed inconsistencies in his interpretation of the basic principles of the theory of relativity and quantum mechanics, it was not because his thinking was influenced by idealistic thought but because it reflected the inner contradictions in the evolution of modern physics. "I think," he said, "that at the present time there are—and can be—no canonic explanations of the foundations of the theory of relativity and quantum mechanics."[92] Leontovich suggested that, instead of criticizing Mandelstam's scientific ideas, expressed some fifteen years earlier, they would do better to prepare a new study expressing the updated views on Einstein's physical principles. A similar conference was held under the aegis of Moscow University.[93]

Despite the relatively moderate tone of both conferences, it was clear that the days were over when theoretical physicists could take refuge in "spontaneous materialism" (as an easy and noncommittal substitute for dialectical materialism) and could avoid philosophical issues and their ideological implications. The temperate nature of criticism aired there did not satisfy Stalinist philosophers. The Kuznetsov group rejected moderation and lost no time in making Mandelstam a widely discussed and vehemently assailed foe of Leninist epistemology. The two conferences also demonstrated that, despite escalating ideological militancy and political threats, the scientific community continued to produce valiant individuals who put the interests of their science ahead of philosophical doctrines and ideological commands.

Marxist attacks on the special theory of relativity were not as sweeping and rancorous as those on the general theory of relativity. They were more scattered, produced less ill will, and operated within a much smaller sphere of theoretical thought. In one such attack, Ia. P. Terletskii, a physicist close to Stalin's anti-cosmopolitan campaign, wrote in *Questions of Philosophy* that Einstein, in his effort to improve techniques for measuring the simultaneity of separate events with the help of light signals, had failed to present an "objective" picture of the four-dimensional geometrical link of space and time and, in general, to make use of four-dimensional equations in mechanics and electrodynamics. According to Terletskii, Einstein did not relate general covariance to the laws of the physics of high speed. Instead, he presented them as conclusions based on a "subjective" choice of light signals.[94] It was somewhat paradoxical that Terletskii challenged Einstein's "operationalist" orientation, not the results of the special theory of relativity.

The Leading Interpretations of Einstein's Contributions

Determined efforts to create a Soviet theory of relativity fully subservient to Marxist-Leninist philosophy and official Soviet ideology were too ambitious and contradictory to produce the desired results. At the end of Stalin's reign, there were three Marxist (not necessarily philosophical) interpretations of Einstein's scientific contributions. Although not fully crystallized and symmetrical in the elaboration of theoretical subtleties, each orientation presented distinct attitudes and calls for action.

The first, articulated by Kuznetsov and Maksimov, saw Einstein's physical thought, particularly general relativity, as so deeply enmeshed in idealistic philosophy that it provided no basis for a sound and acceptable scientific theory. That interpretation was closely related to the strong anti-Einstein stand of the Timiriazev-Mitkevich-Kasterin wing of Marxist theorists in the 1920's and 1930's. It was bolstered by the anti-cosmopolitan and ethnocentric campaigns during the waning years of the Stalinist era.

By 1953, that orientation had run its course and had begun to show irreparable erosion. But some Stalinist crusaders were not ready to retreat from the battlefield or to take more conciliatory positions. In January 1953, Maksimov relied on the hospitality of *Questions of Philosophy* to trot out the same old arguments against Einstein's theory of relativity he had used for years. Einstein, he said, had enriched physics not with the theory of relativity, but with his mass-energy mathematical correlation, his study of Brownian motion, and his quantum theory of light. But in fact, he noted, the explanation of the equivalence of mass and energy was first formulated by the Russian physicist P. N. Lebedev in 1899. Einstein, according to Maksimov, made the Lorentz transformations part of his theory only after he gave them an idealistic orientation. The general conceptual framework of the theory of relativity must be rejected, he concluded, because it did not place physics on the path of scientific progress. It was the victim of a close alliance with Mach's subjective epistemology. "Einstein's views," he wrote, "take us back a step in both the theory of knowledge and methodology. Instead of materialism, Einstein and his allies preached and cultivated physical idealism, a metaphysical orientation favoring mechanistic views and mathematical formalism." As a general orientation in physics, he added, the theory of relativity was "false."[95]

The second interpretation came from the Marxist philosophers who looked optimistically at the possibility of separating Einstein's science from Einstein's philosophy and of creating a Marxist theory of relativity.[96] Those philosophers, represented most prominently by Omel'ianovskii, supported much of the ideological and philosophical campaign against Einstein spearheaded by Kuznetsov and Maksimov. "Relativity" as a general label for Einstein's theory, they agreed, should be discarded because, in their opinion, it was an umbrella for ideological and philosophical aberrations. They also echoed the continuing criticism of Ioffe's and Mandelstam's "idealistic" detours. But there were important differences between the two groups. Some members of the Omel'ianovskii circle were quick to point out that Kuznetsov and his followers found traces of idealism where none existed. And on one key question, they all contradicted Kuznetsov: the theory of relativity in their view was not only fundamentally sound, but also a dramatic step in the forward march of science and the infallible harbinger of a new epoch in the evolution of science. Although they recommended continued attacks on unwanted ideological-philosophical admixtures,[97] they entirely rejected Kuznetsov's suggestion of discarding the theory of relativity as an irreparable body of physical concatenations. Thanks to their many articles in *Questions of Philosophy*, the views of the Omel'ianovsii group reached a large segment of the interested public.

The third group of Marxist interpreters consisted of leading members of the scientific community and a few philosophers solidly grounded in science who viewed the ongoing preoccupation with physical idealism as an unwelcome distraction from the thorough exploration of the rich vistas of modern physics. They felt that in order to establish a symbiotic relationship with science, dialectical materialism must become an open philosophy, ready to adapt to the fullest degree to advances in science. The repressive atmosphere created by the Stalinist war on cosmopolitanism prevented them from forthrightly elaborating their views on the pernicious effects of ideological interference with the growth of science. When G. I. Naan suggested, in 1951, keeping separate accounts of the philosophical and physical substance of the theory of relativity to eliminate unproductive and debilitating ideological squabbles, he gave sustenance to the physicists' defense of Einstein's theory. At the time, however, few other philosophers echoed Naan's views.

Among the members of this group, no one better illustrates the pressure put on the leading physicists to conform to Stalinist Marxism, and the never-dying kernel of resistance to it, than Fock. In 1953, he resumed his criticism of the "inadequacies" of the general theory of relativity, which he had begun in 1939. Einstein, he wrote, made "extremely inconsistent" statements on the basic theory of modern physics. He repeated his thesis that the principle of the equivalence of inertial and gravitational masses, the pivotal premise of the general theory, had only local significance and was applicable only to weak and uniform fields and slow motion; therefore, it could not be accepted as a general principle. Without offering details, Fock claimed that Einstein's "misrepresentation" of the general theory of relativity was the result of a Machian bias built into his philosophy. Einstein's "misjudgment," in his opinion, was another example of how "in certain cases, faulty philosophical views" invited "errors in physics."[98]

These criticisms did not deter Fock from expressing the greatest admiration for the theory of relativity. Not allowing the "Machian bias," the main target of Marxist attacks on Einstein, to influence his thinking, he recognized the theory of relativity as one of the brightest pages in the annals of science. Since Mach's philosophy, in Fock's opinion, influenced only nonessential parts of Einstein's theory, it could be readily eliminated. In most of his contributions, Einstein allowed only "objective" facts to guide him and generally acted as a "materialist." Fock insisted that Soviet science would benefit more from efforts to refine the basic principles of modern physics than from attacking them on philosophical and ideological grounds. He argued deftly that all criticism of modern physics must be based on a thorough familiarity with scientific details, and he did not hesitate to attack Marxist critics—Maksimov and Kuznetsov, in particular—for their lamentable ignorance of the theoretical constitution of modern physics. He told the Zhdanovian crusaders that Einstein's theory of relativity was "the foundation not only for the full field of modern physics but for many technical advances as well." It had "fully met the criteria of practice—the basic measure of the correctness of a physical theory."*

Fock's chief supporter was A. D. Aleksandrov. A corresponding member of the Academy of Sciences and a visible member of the Communist Party, he earned solid academic credentials from his research on the transition from polyhedrons to convex surfaces. Like Fock, he did not hide his unreserved admiration for the theory of relativity, which he labeled "the greatest conquest of science" and "the titanic step on the path to understanding nature." [99] He too argued that despite its success in widening the horizons of scientific vision, the theory of relativity was not free from inner difficulties. But he was particularly eager to show that the writings of Soviet extremists like Kuznetsov and Maksimov were a declaration of war on physics no less than on alien ideologies. [100]

Aleksandrov agreed with Fock that dialectical materialism must grow with science rather than in opposition to it. He, too, interpreted Kuznetsov's and Maksimov's philosophical exhortations as efforts to pit dialectical materialism wholly against the most advanced developments in modern physics. Until the middle of the 1950's, few Marxist philosophers paid much attention to Fock's and Aleksandrov's warnings. While recognizing residues of materialism in the theory of relativity, most Marxist theorists were engaged in a war against the "agnosticism" and "solipsism" of Einstein's epistemological statements as direct challenges to dialectical materialism.

---

*Fock, "Protiv nevezhestvennoi kritiki," p. 25. What enabled Fock to make this public response was a letter that a group of prominent physicists, led by Landau, sent to Lavrenti Beria of the Politburo. Pointing out the potential dangers to the normal development of physics presented by the current attacks on Einstein, they pleaded for the authorities to allow *Questions of Philosophy* to publish Fock's rebuff of Maksimov's article. To everyone's surprise, the permission was granted (S. S. Ilizarov and L. I. Pushkareva, "Beriiai teoriia otnositel'nosti," *Istoricheskii Arkhiv*, 1994, no. 3, pp. 215–18).

Most remarkable of all, in the face of the Communist authorities' concerted effort to mobilize the scientific community in active support of their ideological war, were the few physicists, most notably Kapitsa, Tamm, and Landau, who remained conspicuously unwilling to participate. The silence of this small but powerful nucleus of resistance to Stalinist designs dramatizes the complete incompatibility of the critical and searching spirit of science with Marxist orthodoxy and the sacred beliefs built into dialectical materialism.

Turning Points

The death of Stalin on March 5, 1953, started a new era in the interweaving of Soviet ideology and science. The climate of moderation encouraged efforts to undo the damages of the Stalinist campaign to tie science to Marxist orthodoxy and Russian nationalism. Thanks to the Soviet physicists' dramatic achievements in nuclear science, they climbed to new heights of prestige and influence in shaping national science policy. They became a powerful force in a determined effort to protect science and the scientific community from the doctrinal excesses and rigidities of ideological controls.

Einstein's Detractors in Retreat

Stalinist philosophers were far from united in their reaction to the unambiguous signs of a post-Stalinist thaw. A strong majority acted promptly, but cautiously, to adopt more moderate ways of looking at Einstein's elaborate theoretical structures. Ernest Kol'man, one of the most dedicated defenders of Stalinist philosophy and ideology, for example, moved resolutely to accommodate his thinking to the new intellectual temper. In May 1954, a year after Stalin's death, he published an article on the controversial. issues of the theory of relativity. Many of Einstein's ideas still seemed to him to require a "materialistic reinterpretation" to become truly scientific, including the special theory of relativity, the mass-energy mathematical correlation, and the principle of inertial and gravitational masses. But at this point, Kol'man was less eager to criticize Einstein than to stress the gigantic proportions of his contributions to physics and to the modern world outlook. He even declared that the Machian slant in no way reduced the scientific grandeur of the theory of relativity. Among the many shining triumphs of the relativistic point of view in physics, Kol'man selected five for special emphasis. First, the theory of relativity confirmed Engels' statement, made in 1873, that the motion of all bodies was relative—the motion of a body was meaningful only when it was related to the motion of other bodies. Second, it disqualified such "metaphysical notions" as Newtonian space and time, ether, and action-at-a-distance. Third, it treated geometry as part of physics, insofar as it claimed that the attributes of space depended on the structure of matter and on mass. Fourth, in one important respect, it went far beyond Newton's world of science: whereas Newton "described" gravitation, Einstein "explained" it. Not divine omnipotence but the space-time curvature set the gravitational forces in operation. And finally, it established the unity of mass and energy and confirmed the law of the conservation of energy. Einstein helped give scientific expression to "the key claim of dialectical materialism that there is no matter without motion and no motion without matter."[1]

Kol'man repeated the old cliché that the physical content of the theory of relativity confirmed the basic principles of dialectical materialism. Semkovskii and Hessen had summoned up the same argument in the 1920's in the hope of mollifying Marxist zealots. Kol'man nevertheless went out of his way to note that, in so stating, he did not mean to imply that the theory of relativity should be considered a dialectical-materialistic view of the physical world—not even after all materialistic modifications had been rendered. "It is not and cannot be a dialectical-materialistic theory of the physical world because its vision is one-sided—it is interested only in the world ruled by the law of continuity."[2]

Some defenders of Stalinist orthodoxy, however, showed little inclination to tone down their criticism of Einsteinian idealism and its leading Soviet representatives. In a long paper delivered at the national conference of physicists held in Kiev in 1954, Kuznetsov fired a new salvo at Mandelstam, Fock, and Aleksandrov, accusing them of sabotaging all efforts to create a genuine Marxist theory of relativity. Mandelstam, he thought, erred in viewing space and time not as objective realities existing independently of human consciousness, but as subjective creations of the observer's choice of measuring techniques. Fock committed an even more grievous error: he viewed matter not as a basic and independent reality, but as a result of space-time relations. Fock did not know that space-time was a "form of matter." Aleksandrov erred in presenting the space-time continuum as an abstraction fully detached from the material world.[3]

All the same, even Kuznetsov was moved to change his position to the extent that, in this 1954 paper, he no longer advocated dismantling the theory of relativity altogether. Now he argued that it could be made stronger and more appealing by separating it from all Machian "metaphysical" and "idealistic" impurities—a Herculean task. Much work was needed to transform the theory of relativity into a system of scientific thought fully compatible with dialectical materialism. Kuznetsov clung tenaciously, though, to his fervent conviction that the main part of this assignment should be entrusted to philosophers, the guardians of Soviet ideology.

Kol'man begged to disagree. Sensing the growing popularity of the idea that the intellectual and social functions of science and philosophy should be clearly defined and rigidly separated, he asserted that the problems of modern physics, ranging from the structure of subnuclear objects to the macrophysical attributes of space and time, should be handled by physicists alone. Not surprisingly, Kuznetsov put forth a quick rebuttal. Dialectical materialism, he wrote, was much more than a passive recipient of scientific knowledge; it was an active and original contributor to science. What Kuznetsov failed to anticipate, however, was that the time was fast approaching when new policies would be designed to protect the scientific community from the degrading attacks of the custodians of dialectical materialism. Relying on the scurrilous rhetoric of the Stalinist stripe, Kuznetsov fought to preserve one of the most destructive weapons of bygone political authority.[4] The future was clearly on Kol'man's side.

Maksimov found a unique mode of adjustment to the new political situation: he went into retirement and withdrew from public debate. As a corresponding member of the Academy of Sciences, he continued to submit annual progress reports on his "activities" in scholarship. From his reports in the 1970's, it is clear that he never abandoned his war against the theory of relativity. Even at that late date, he was still calling for continued efforts to expose the "negative" influence of the "Einstein cult" on the development of modern science. Until his death in 1976, he fought the "philosophical relativism" built into Einstein's scientific legacy. None of his writings after 1954 appeared in print.[5]

In January 1955, *Questions of Philosophy* published a report on the current trend in Soviet thinking on Einstein's revolutionary ideas in physics.[6] The report gave a graphic picture of the gradual ebbing of orthodox Marxist criteria in evaluating the guiding principles of the theory of relativity. It made a passing attack on physical idealism for interpreting the theory of relativity in a "subjectivistic spirit" and negating "matter as objective reality," and for disregarding "the objectivity of space and time as the basic forms of matter in motion." It reaffirmed the currently popular notion that the theory of relativity was misnamed, a fact "previously acknowledged" by Minkowski and Einstein himself. But Minkowski had also stated that the question of renaming should be resolved by physicists, not by philosophers. In recording an unqualified opposition to writers of Maksimov's bent of mind, who would negate "all physical consequences of the theory of relativity," the report stated that such basic conclusions as the relative simultaneity of spatially isolated events were firmly embedded in modern physics. The report reaffirmed the great respect of the Soviet scientific community for Einstein and his theories. It noted that an over-whelming majority of the scholars who had been consulted viewed the theory of relativity as "one of the greatest achievements in physics" and as "the basis of the modern theory of elementary particles." The journal summed up the dominant attitude toward Einstein's scientific contributions this way: "De-velopments in science can lead to the establishment of new facts pressing for a reconsideration of the views based on the theory of relativity. The conclusions of this theory, however, have continued to be incontrovertible in the areas where they have been tested by the rich resources of physical experiment and technology."

Beginning with the mid-1950's, then, there was a perceptible change in atti-tude toward Einstein's theoretical thought. The general theory of relativity, much attacked in the age of Stalin, was now widely accepted not only by physi-cists but also by philosophers as one of the highest triumphs of science. Both groups tried to make up for the time lost under the pressure of successive Sta-linist attacks on even the name of the theory of relativity.

Einstein's Death

Einstein died on April 18, 1955. Soviet scholars joined the scientific communi-ties of all nations in mourning his death. The country's most eminent physi-cists did not hesitate to sing his praises, which also marked the end of attacks they had endured for their putative embrace of physical idealism and cosmo-politanism. One summed up the general sentiment by noting that it would be extremely difficult to find an idea in modern physics whose roots were not in Einstein's scientific legacy.[7] Ioffe, who saw Einstein as the brightest star in the skies of modern physics, made his appraisal both concrete and clear:

> For physicists in the whole world, and particularly for Soviet physicists, the question of the accuracy of the theory of relativity does not exist, for it was an-swered a long time ago: this theory is the most perfect generalization of the laws of motion formulated by modern physics. The task of physicists is to apply this theory to concrete problems, to enrich its substance, and to achieve a better understanding of its philosophical meaning.[8]

Tamm focused more on how the theory of relativity had pointed the way for other modern physicists. The Einsteinian conceptualization of space and time, he declared, had given meaning and a point of departure to one of the most creative stages in the evolution of physics. It could be argued, he observed, that the appearance of quantum theory at the beginning of the twentieth century owed more to Einstein than to its other founders; after all, the theory of relativity paved the way for a full transformation of physics. Einstein's theory of relativity eliminated the very roots of the classical orientation in physics and cleared the field for radically new theoretical and methodological perspectives.[9]

Fock, who was just then putting the finishing touches on his *Theory of Space, Time, and Gravitation*, a monumental revision of Einstein's conceptualization of the general theory of relativity, was particularly lavish in his praise of Einstein's genius. Einstein was the chief creator of the theoretical foundations of modern physics, he said, and "a figure of colossal proportions, standing among the greatest minds of all times."[10] In a paper published a few months earlier, he stated that Einstein's theory of gravitation was "undoubtedly one of the grandest achievements of human genius," which needed only to discard "elements and terms alien to it" to realize its full potential.[11] In his concerted effort to build a general theory of gravitation by shifting the emphasis from the general principle of relativity to the notion of the space-time curvature, Fock wanted to achieve two goals: to follow up a suggestion that Einstein himself had made in the 1920's and to elevate the discussion of the theory of relativity to safer scientific heights, free of ideological entanglement.

D. D. Ivanenko and Boris Kuznetsov added to the succession of commemorative articles with a long and systematic essay on Einstein's contributions to modern physics. Staying strictly within the framework of science, they concentrated on the two theories of relativity and their place in the mainstream of modern scientific ideas. The future of physics, they thought, lay in a joint study of "hypothetical gravitational fields," called for by the general theory of relativity, and of gravitons, offered by quantum theory.[12] The authors noted that they were disregarding Einstein's philosophy, for though they thought it was influenced by Hume's and Mach's ideas, it was unrelated to their immediate task of surveying the high points of Einstein's scientific thought. They clearly implied that an understanding of Einstein's philosophy was not germane to understanding the theory of relativity.

These eulogies heralded a new development in Soviet physics.[13] They marked an open rebellion of the country's leading physicists against men like A. A. Maksimov and I. V. Kuznetsov who had worked hard to discredit Einstein and to dismantle the complex body of his physical thought. They also gave powerful impetus to a burgeoning trend that favored a more conciliatory attitude toward the philosophical base of the theory of relativity and quantum mechanics as articulated by the dominant schools in the West. In refuting the philosophers of science who were critical of the main principles of Einstein's physical theory, Tamm and his colleagues did not intend to refute dialectical materialism; on the contrary, they based their arguments on the assumption of an essential agreement between the theory of modern physics and dialectical materialism. In all this, they had a single purpose: to end, or at least to soften, the interminable intellectually degrading and scientifically wasteful philosophical debates about the merits of Einstein's contributions and to clear the air for a freer, more balanced and comprehensive engagement in theoretical physics. They did not advocate separating physics from philosophy; their main interest was in entrusting the philosophical scrutiny of the theoretical principles of physics to those with proven competence in science and a realistic interest in philosophy. The time had come to heed the opinion held by many physicists that Marxist philosophy had failed to keep up with theoretical advances in modern physics.[14]

The Polish physicist Leopold Infeld, one of Einstein's close friends, made his contribution to the ongoing reassessment of Einstein's science, philosophy, and humanistic outlook. In his reminiscences, published in Russian translation in *Advances in the Physical Sciences* in 1956, he criticized the deeply rooted habit of Marxist philosophers to treat Einstein's thought as a classic example of neopositivist idealism. He asserted that to accuse Einstein of "pure idealism" was to display a mixture of ignorance and evil: Einstein, in his opinion, firmly believed in the objectivity of scientific truth, best expressed in his materialistic interpretation of the physical world.[15] Soviet philosophical commentators, on their side, did not entirely forget Einstein's ties with Mach's "idealism"; but most were at least now inclined to attach more weight to the "spontaneous materialism" of the theory of relativity—to the materialistic inclinations of Einstein's mind.

Attacks on Marxist Orthodoxy

At the end of 1955, the Academy of Sciences devoted a memorable session to the fiftieth anniversary of Einstein's special theory of relativity and its impact on modern physical thought. Landau opened the proceedings with a comprehensive survey of Einstein's work in both the theory of relativity and quantum mechanics. He made no reference to the relation of the theory of relativity to dialectical materialism. Quietly but firmly, Landau showed unmistakable signs of total alienation not only from the Marxist philosophy of science, but also from the Soviet political system. Whereas in the 1930's he labeled Stalinism a gross distortion of the humanistic ideals of Leninism, in the 1950's he was ready to admit, in private conversations of course, that he found Leninism no less objectionable than Stalinism.

Other speakers, covering specific problems related to all phases of Einstein's thought, agreed that every component of the theory of relativity had "confirmed the principles of dialectical materialism" and had "offered a large volume of concrete material for its future development." They also claimed that "every attempt to give an idealistic interpretation of the theory of relativity and its conclusions [was] the result either of a conscious distortion or of a misunderstanding of its meaning."[16]

These statements on the whole reflected the sentiment of physicists. Among philosophers, isolated defenders of Stalinist orthodoxy, not represented at the anniversary session, if they did not openly express their opposition to Einstein's "idealism," showed no signs of modifying the stand they had taken during the last years of Stalin's reign. But most shifted position to one extent or another. *Questions of Philosophy*, for example, in reporting on the academy's fiftieth anniversary session, was prepared to note that Einstein both completed Newton's scientific revolution and ushered in a new stage in the evolution of science. The journal reminded its readers that Lenin recognized the revolutionary proportions of Einstein's contributions to science.

In mid-1956, *Priroda* carried the de-Stalinized, scientific philosophical portrait of Einstein to the general reading public with an article that interlaced the high points of his personal life with his contributions to physics. The author, Ia. A. Smorodinskii, traced the history of physics from the end of the nineteenth century to the theory of relativity. Smorodinskii noted that the evolution of Einstein's thought was not a smooth succession of progressive steps. Einstein erred in bypassing quantum mechanics in his search for a unified field theory, which he limited to the search for a unity of electromagnetism and gravitation. He had also allowed "frequent carelessness and contradictions" to mar his judgments and to serve as a fountain of suggestive guides for "idealists in physics." But all this, Smorodinskii went on to say, did not diminish the brightness of light emitted by Einstein's fundamental achievement: the uprooting of the "metaphysical" idea of space and time. Nor did it prevent Einstein from playing a far-reaching "positive role" in the struggle of materialism against idealistic worldviews.[17] The article's main contribution was in presenting the general reader with a panoramic view of the vast range and diversity of Einstein's pioneering work in modern physics.

For several years after Stalin's death, the leading members of the scientific community made sustained attacks on the philosophical critics of the theory of relativity. The war on the self-styled defenders of Marxist orthodoxy reached a high point in 1957, when A. D. Aleksandrov publicly chastised I. V. Kuznetsov for his claim in 1952 that Einstein's theory of space-time was not an "objective fact" of science but the fabrication of a bourgeois mind. Thereafter, Kuznetsov, Maksimov, Shteinman, and their followers continued to draw scorching criticism for misinterpreting the basic principles of the theory of relativity.[18]

In October 1958, the All-Union Conference on the Philosophical Questions of Modern Natural Science convened in Moscow. In the keynote address, P. N. Fedoseev, a new breed of Marxist philosopher, stated that the time had come to eliminate forces that aimed to suppress scientific and philosophical controversies in the world of scholarship. To illustrate what he meant, he welcomed the dispute between the physicists Fock and Blokhintsev over the theoretical foundations of quantum mechanics, both boasting ardent and active supporters among their fellows. Whereas Fock saw quantum mechanics as a science of individual subatomic objects, Blokhintsev saw it as a science dealing with "ensembles" of subatomic objects. Both scientists claimed the compatibility of their specific interpretations with the guiding principles of "objective epistemology" and "dialectical logic" built into Marxist thought. What Fedoseev wanted to show was that dialectical materialism, as a Marxist philosophy of science, had become broad and flexible enough to provide room for contradictory scientific theories.[19]

The astronomer V. A. Ambartsumian, speaking on behalf of the scientific community, introduced his paper on cosmology with a plea for a concerted effort to prevent Marxist philosophers from placing barriers on the road to scientific progress. Until now, he said, they had hindered the Soviet scientists' search for new ideas; their excuse was that new ideas often challenged the philosophical premises of dialectical materialism. Philosophers took on the task of steering scientists away from the problems that they suspected invited "idealism."[20] A strong supporter of dialectical materialism, Ambartsumian was the logical person to point out the myopic policy that encouraged the rigid philosophical control of science and allowed Marxist philosophers to intimidate scholars favoring "dangerous" ideas.

Aleksandrov read a long paper on the general theory of relativity that generated a lively discussion and set the stage for the formation of a strong Soviet orientation in Einsteinian studies.[21] Accepting the theory of relativity as a general confirmation of the philosophical postulates of dialectical materialism, Aleksandrov asserted that the various schools of Western positivism could claim an affinity of principles with the theory of relativity only by offering a distorted interpretation of Einstein's ideas. He argued also that Soviet philosophers were habitually given to exaggerating the idealistic element in Einstein's philosophical thought and misrepresenting the objective principles of Einstein's scientific ideas. Philosophers, he wrote, exhibited a penchant for presenting Einstein's theory of relativity as part of a general philosophy of relativism. Obviously, Aleksandrov's criticism was addressed primarily to Stalinist philosophers, past and present. Philosophers, he contended, should acquire a deeper understanding of the theory of relativity, which he described as "a theory of space-time as an absolute form of the existence of matter."[22]

The 1958 conference prompted a certain accommodation between the physicists and the philosophers. It alerted the physicists to take a more careful look at Einstein's profusely exhibited epistemological ideas and to leave off their efforts to deny that the "idealistic" thought of Mach and Poincaré had become part of Einstein's thinking. Most philosophers now tried to show that Einstein's studies carried a solid component of "materialistic" ideas. Moreover, Einstein received plaudits for making a deliberate effort to give support to a general materialistic orientation.[23] Where, during the 1930's, the Stalinist defenders of Marxist orthodoxy saw Einstein's "Geometry and Experience" as a full-fledged personal manifesto of faith in idealism, zeroing in especially on his assertion that "as far as the propositions of mathematics refer to reality, they are not certain; and as far as they are certain, they do not refer to reality," the post-Stalinist philosophers now found grounds enough in that same essay to claim Einstein's fundamental affinity with materialism. Looked at in a new light, Einstein, recognized geometry as a branch of physics, based on logical constructions and "on induction from experience."[24]

The changed attitude toward the philosophical aspects of the theory of relativity led to a new division of labor between physicists and philosophers.[25] Both groups set themselves to establishing harmony between modern physics and dialectical materialism, but now they worked on different problems. The physicists who chose to engage in philosophy concentrated on the problems directly related to the epistemological, logical, and methodological problems of their science—to the construction of scientific theory, tools of scientific inquiry, and ladders of abstract thought. They dealt with philosophical explanations of the role of tensor analysis in advancing the general covariance principle, examined the philosophical parameters of the quest for a unified field theory, and searched for a new definition of the general theory of relativity that would give Einstein's contributions, both scientific and philosophical, clearer meaning and a sharper focus. In all these and many other problems of a similar nature, the physicists ceased to depend on the guidance and criticism of philosophers. They manifested a discernible tendency to avoid involvement in philosophical debates on ideological issues and to ignore the massive literature contemporary Soviet philosophers had produced. The more they wrote about the theory of relativity's affinity with dialectical materialism, the further they moved away from the guiding principles and ideological tenets of Marxist philosophy.

The philosophers, for their part, now concentrated on questions that had a direct bearing on the "scientific world picture."[26] The political authorities had given them fair warning not to carry the war against class enemies and alien ideologies to the domain of scientific debate. Their criticism of the philosophical thought of the leaders of Soviet physics and of the Western philosophers of science ceased to be arrogant and abusive. Noting the new trend, an old-fashioned philosopher publicly lamented that more and more books and articles on the philosophy of science were manifesting "a slackening of the struggle against idealistic distortions in modern natural science." P. L. Kapitsa recorded in 1957 that though "we in the editorial office of the *Journal of Experimental and Theoretical Physics* to this day are receiving articles which attempt to refute the validity of the theory of relativity, [such articles] are not even considered, clearly being anti-science."[27]

Despite extensive measures to reduce the intensity and scope of the ideological war, some philosophers continued to launch occasional verbal attacks on philosophical idealism, which they regarded as a powerful and ever-threatening vehicle of capitalist ideology. As late as 1959, for example, F. V. Konstantinov and his coeditors still held that Western idealists continued to preach agnosticism, which, "like religion," elevated the wisdom of divine authority by denigrating the power of human reason and its cognitive capacity. They singled out Poincaré's conventionalism as a pernicious orientation with a firm grip on "modern bourgois philosophy."[28]

Two events marked the clear decline—but not the full demise—of extremists of the Konstantinov stripe. The first was an Institute of Philosophy–sponsored conference held in December 1960 to discuss A. K. Maneev's new book, *A Critique of the Foundations of the Theory of Relativity*. Under the guise of a purely logical analysis of the basic scientific principles of the theory of relativity, Maneev concluded that Einstein's scientific conceptions stood on shaky ground and had little to offer to modern science.[29] Theoretical physics, as Maneev saw it, was very much in need of fundamental changes in its basic concepts, which should be fully divorced from Einstein's theory of relativity.[30] One leading discussant did not mince words, charging that Maneev was plainly oblivious to the fact that "during the five decades of its existence, the theory of relativity has received a mass of solid confirmations, has been verified by experiments, and for a long time has been the backbone of many branches of modern technical knowledge, including the entire theoretical and practical work in nuclear physics.[31] Several speakers defended Maneev—one of them lamenting the rapidly diminishing opportunity to discuss the question of whether Einstein was correct or not in the first place. But in the end, a decisive majority condemned the book as an ill-designed and misguided effort to dismantle a theory backed up by "basic confirmations" accumulated over a long period of time. The conference concluded that Maneev did not possess sufficient knowledge in modern physics to undertake a serious examination of the theory of relativity, and that he represented a lingering type of philosopher of science who worked against the best interests of both philosophy and science. That a summary of the proceedings of the conference was published in *Questions of Philosophy* served as a forthright and stern warning that works like Maneev's would face insurmountable difficulties in finding a publisher in the future. The 1960 conference signaled a new unity between physicists and philosophers, this time on terms set by the physicists.

The intent of the Institute of Philosophy was not to elevate the theory of relativity above all unfavorable criticism. Its sole aim was to prevent the philosophers from initiating scientific criticism and to separate philosophical discourse from scientific analysis. From now on, philosophers found it impossible to intimidate physicists, who demanded that separate tabs be kept on Einstein's science and Einstein's philosophy. The decision to criticize Maneev reaffirmed the inalienable right of the scientific community to serve as the sole authority in ratifying scientific knowledge. It also signified a major step toward unburdening the scientific community from ideological engagements interfering with the normal growth of positive knowledge.

Questions of Philosophy also figured in the second event. On March 12, 1962, in commemoration of the fortieth anniversary of the publication of Lenin's "The Importance of Militant Materialism" (an essay arguing that the future of Soviet society depended on the vigilant defense of Marxist theory and on the readiness to absorb the "progressive" achievements of Western culture), the journal published a lengthy editorial devoted to philosophy and natural science—and to dialectical materialism as the most reliable bridge connecting the two. If dialectical materialism was to "reestablish" its authority in modern thought, the writer stated, it must return to the Leninist tradition: it must shed the pathological residues of Stalin's cult of personality and terminate the sinister campaign against cosmopolitanism. The editorial's message was direct and authoritative:

> In their criticism of idealism in the creative work of prominent scientists in capitalist countries, many of our philosophers and natural scientists have neglected to note the progressive, materialistic moments in the worldviews of such eminent scholars as Einstein and many others. They have substituted superficial criticism, often consisting of simple name-calling, for a creative analysis of the real content of new scientific theories. It is imperative to reestablish a full measure of scientific objectivity and to consider the evolution of the views of scientific intellectuals in capitalist countries without making any concessions to idealism and mysticism.[32]

The time had come, the editorial noted, to take up Max Born's challenging invitation to study the avenues that united, rather than divided, Western and Soviet interpretations of the theoretical principles of modern physics.

Perhaps the clearest signal of all that the Stalinists' war on Einstein was over came from L. F. Il'ichev, a man whose expertise in Marxist theory had earned him full membership in the Academy of Sciences. He wrote in 1963: "The history of relativity has affirmed, rather than contradicted, dialectical materialism in its interpretation of space and time as forms of matter; it has given more depth to the idea of the indissolubility of matter and motion; and it has added precision to the notion of the interrelations and dialectical unity of the absolute and the relative. The objective scientific content of the theory of relativity has opposed relativism and all other forms of idealism."[33]

Moved by the tone of these philosophical debates, N. N. Semenov, a Nobel laureate, made a public statement on the evils of the Stalinist attacks on science. "Imagine," he told his colleagues at the academy's Institute of Chemical Physics, "what would have happened if we had rejected the principles of relativity and indeterminacy because our philosophers, typified by Maksimov, had told us that they were expressions of idealistic thought. This would have resulted in a disorientation of physics that, to give one example, would have erased every possibility of producing atomic energy and would have left us far behind the capitalist countries in answering this practical question."[34]

New Themes and Interests

The 1960's and 1970's witnessed a rapid growth and vast thematic expansion of Einsteinian studies. Particularly notable was the appearance and proliferation of the first Russian works on Einstein's professional evolution, family background, political activities, correspondence, and attitudes toward the nonscientific domains of culture. Pertinent general comments by such leaders in modern physics as Niels Bohr, Werner Heisenberg, Max Born, Wolfgang Pauli, and Arnold Sommerfeld on the broader meaning of the theory of relativity were now available in Russian translation, and elementary surveys of Einstein's science received a new lease on life. Philosophical concerns nevertheless continued to preoccupy the Marxist faithful, though many were at this point on a path away from Stalinism.

Take Vladimir L'vov, for example. In 1958, this former Stalinist who had made a virtual business of intimidating Soviet physicists and astronomers suspected of flirting with philosophical idealism,[35] published *The Life of Albert Einstein*. The first Russian biography offering a detailed description of the triumphs and dilemmas of Einstein's journey to immortality, it went into a second printing (of 75,000 copies) a year later. L'vov gave his readers fair coverage of the critical episodes in Einstein's long and eventful life, but only the most elementary glimpses of his contributions to science. He made a brave, and rather detailed, effort to disconnect Einstein's philosophical thought from the neopositivist bent of Machian epistemology. He did not forget to provide citations indicating the "predialectical"—but materialistic—frame of Einstein's mind.[36]

In 1959, Lev Landau and Iu. B. Rumer published a small volume on the essential facts of the theory of relativity aimed at satisfying a general public eager to become familiar with the Einsteinian revolution in physics. The authors made no effort to adhere to the Soviet tradition of casting a popular exposition of Einstein's physical ideas within a philosophical frame. In *Conversations on The Theory of Relativity*, published a year later, Boris Kuznetsov presented a rather more sophisticated general survey that was likewise aimed at the reading public. It, too, demonstrated that a serious discussion of Einstein's physical ideas required no philosophical involvement.

That work was followed by *Einstein*, a comprehensive biography, where Kuznetsov, in addition to scientific analysis, made generous use of humanistic analysis. Of particular note was his attempt to bring his subject closer to the pulse of Russian thought via an insightful, though somewhat strained, comparison of Einstein's and Dostoevsky's cosmic and moral philosophies.[37] The biography was hugely successful, with two further print runs after it came out in 1962 (in 1963 and 1967) as the sales climbed to 125,000 copies. Well-written and measured in philosophical judgment, the book covered all the key principles of the theory of relativity. In still another study, published somewhat earlier, Kuznetsov presented a general survey of the points of contact of the theory of relativity with quantum mechanics.[38] In that study, according to Siegfried Müller-Markus, Kuznetsov "took a remarkably unorthodox position" in treating the scientific texture of Einstein's arguments.[39]

The distinguished playwright Nikolai Pogodin used Einstein as the vehicle to probe one of the most critical dilemmas of the modern world: the ever-growing conflict between the use of the awesome discoveries of science for military purposes and the moral foundations of humanity.[40] His play *Albert Einstein* dwells on a fateful contradiction in Einstein's public life: his decision to join in an appeal to President Franklin Roosevelt for the go-ahead on the atom bomb and his dedicated work after Hiroshima in organizing a worldwide movement to outlaw nuclear weapons. Reminiscent of Bertolt Brecht's *Galileo*, Pogodin's drama depicted the rising discordance between the individual as the concrete carrier of moral responsibilities and society as the master and ultimate creator of science. In essence, Pogodin's drama is a celebration of the moral purity of Einstein's struggle for peace. In his reckoning, the uncompromising struggle against the threat of nuclear warfare, not the signature on the historic letter to President Roosevelt, revealed the true character and strength of Einstein's moral principles.

The year 1968 saw an important addition to the growing store of Einstein studies with the publication of U. I. Frankfurt's *Special and General Theory of Relativity*. The well-documented volume presented Einstein's theoretical principles within a scientific framework; it fully disregarded the relationship of the theory of relativity to the Marxist philosophy of science and, for that matter, to any philosophy. The study gave a careful account also of both the theory's historical antecedents and the major developments following on it. Frankfurt was successful in covering the work on the major ramifications of the theory of relativity. He was less successful in handling the arguments and modes of operation that gave the theory inner harmony and structural unity. The volume provided a welcome example of a historical approach to Einstein's scientific ideas. It is particularly noteworthy as the first Soviet effort to provide a comprehensive bibliography of studies on Einstein's scientific contributions.[41]

Numerous philosophical studies reflected the new rapprochement of the theory of relativity and dialectical materialism. They gave the reading public easily digestible commentaries on the theory's principal ideas and their role in shaping the modern scientific view of the physical universe. In showing the diversity of modern interpretations of Einstein's theoretical views, those studies exposed the futility of trying to place the theory of relativity into the rigid mold of dialectical materialism. They helped generate an awareness of the critical need to end philosophical interference with the professional work of the scientific community.

In 1962, for example, M. V. Mostepanenko published the first modern study of the multiple lines that seemed to him to establish a general unity of the theory of relativity and the fundamental principles of Marxist thought.[42] The essence of Einstein's theory, he wrote, was in "the materialistic depths of its picture of the world."[43] In his view, the Stalinist emphasis on the basic differences between the idealism of Einstein's theory and the materialism of Marxist thought must give way to a solid emphasis on their essential similarities. All the same, he contended that even when Western physicists leaned toward idealism, the internal logic of the development of their discipline compelled them to gravitate toward materialism.[44] Clearly, he meant to caution Soviet physicists to look for elements of materialism behind many idealistic statements.

Moved by the spirit of reconciliation, Omel'ianovskii went so far as to hail Einstein's propositions as the first development in science to show the key role of "dialectical contradictions" in the development of physics.[45] He added that the idea of dialectical contradictions marked the dividing line between modern physics and its classical ancestor. As had become fashionable, he made it clear that he spoke as a philosopher, not as a physicist. Omel'ianovskii brought the theory of relativity closer to dialectical materialism for the sole purpose of bringing dialectical materialism closer to the theory of relativity.

In 1965, the Academy of Sciences began publishing two collections of Einsteiniana: a four-volume set of Einstein's own writings and speeches, which was completed in 1967; and the fifteen-volume *Einsteinian Collection*, which was some twenty-five years in the making.[46] Those collections covered every phase of Einstein's life and work: opinions on his views in physics and philosophy, excerpts from his correspondence with close friends (such as Maurice Solovine, Max Born, and Louis de Broglie), valuable details from biographical studies, and insightful comments on the history of Soviet Einstein studies.

A succession of national and regional conferences on Einsteinian themes served a dual purpose: to highlight the expanding scope of scientific and philosophical studies of the theory of relativity and to inaugurate a new era in the relationship of dialectical materialism to relativistic physics. To a very large extent, those conferences chose to give special weight to cosmology, where the torrents of new information on exotic phenomena like quasars, pulsars, and black holes were taking physicists deeper into the mysterious world of the general theory of relativity and farther from the tested strengths of the scientific method. The philosophers of science could now boast of a solid corps of experts equally at home in physics and philosophy. Whether they studied the philosophical foundations of the Marxist Weltanschauung, the cold rules of the logic of scientific inquiry, or the techniques of modern historiography, the philosophers of science demonstrated a distinct interest in arguments and inquiries that helped them avoid direct ideological confrontation. They continued to criticize the idealistic bent of Western philosophy, but they were more interested in bringing dialectical materialism closer to Western ideas. By the end of the post-Stalinist thaw, they had brought dialectical materialism so close to Western expressions of philosophical idealism that it had lost much of its traditional identity and unity.

By the 1960's and 1970's, philosophers had come to rely on two strategies to demonstrate the essential affinity of Einstein's theories to dialectical materialism. One strategy was based on the general injunction that every verified scientific theory—including, of course, Einstein's—must be regarded as fully compatible with dialectical materialism. It required no philosophical tampering or outright modification. As the erudite philosopher P. V. Kopnin explained:

It is self-evident that any philosophy which tries to rebuff or to correct the facts of modern science is doomed to failure. [Speculative] orientation in science has outlived its usefulness, and there is no way we can resurrect it. Therefore, the strength of each philosophy is in its ability to accept science and its facts the way they are in their fullness and objective thoroughness. Any desire to retouch or correct the results of science in order to make them more appealing is the sign of a weak and vanishing epistemology. . . . Unfortunately, some Soviet philosophers favored such an approach to the theories of modern science, in general, and to Einstein's special and general theories of relativity, in particular: they accepted the fruits of science that agreed with specific interpretations of the positions of dialectics, and rejected the others. However, the experience of scientific thinking and the development of philosophical thought have clearly demonstrated the inadequacy of such an approach.[47]

In *The Theory of Relativity and Philosophy* (1974), E. M. Chudinov relied basically on the same strategy. Philosophy, he wrote, was entitled to suggest heuristic principles that gave scientific work a distinct style of organizing and presenting knowledge in tune with the intellectual climate of individual historical ages. The "style" included, among other things, the operating norms regulating the law of causality, the standards of theoretical simplicity and precision, and the imperative of "observability." Under no circumstances should philosophy be invited to supply science either with concrete facts or with scientific abstraction and theories. The relation of philosophy to science could be formal but not substantive: philosophy could aid science in formulating theories but not in selecting their content.[48]

Kopnin and Chudinov did not have in mind the general relations between science and philosophy, but rather the unique Soviet situation in which dialectical materialism had an exclusive monopoly on philosophical thought, and, particularly under Stalin, was granted unrestricted authority to interfere with both the substantive and the theoretical realm of science, while being firmly entrenched as a source of philosophical justifications of official ideology. Chudinov reminded his readers that in advancing the theory of relativity, Einstein did not depend on one philosophy but relied on "an entire series of epistemological hypotheses and principles presenting a philosophical subtext."[49]

The other strategy rested on the general observation, or assumption, that Einstein had in the course of time gradually retreated from Machian epistemological positions, and so had been brought closer to Marxist positions in the theory of knowledge. The proponents of that view, mainly philosophers, worked on tracing the principal steps in Einstein's alleged transformation with one goal in mind: to show that his changing position had helped narrow the differences between their respective philosophical views. Western historians and philosophers of science likewise recognized an evolution in Einstein's philosophical views in the direction of epistemological realism, but not necessarily of a Marxist variety.[50]

There is much substance in the repeated charge that the Brezhnev era, especially the 1970's, saw a marked retrogression in the post-Stalinist thaw. Indeed, some go so far as to identify the era as one of re-Stalinization. That is certainly fair enough in respect to economic affairs and internal political measures aimed at stemming the tide of dissidence. And neither philosophy nor science escaped entirely unscathed; there were some efforts to reconcile Stalinist orthodoxy with the moderating effects of the thaw and some official pressure to end the scientific community's attacks on Lysenkoism as a national disgrace. But there is no evidence of any official attempt to interfere with the ongoing de-Stalinization of Einstein studies, which continued to grow in depth and diversity. A basic philosophical and ideological affinity of Einstein's legacy with dialectical materialism continued to be taken for granted—needing, it was believed, only some elaboration of the finer points, and, perhaps, a rearrangement of theoretical principles, in the manner of Fock and his followers.

Einstein's Anniversaries

In 1975, the Soviet scholarly community took advantage of multiple landmarks in Einstein's life and career to add new perspectives to the study of his scientific legacy. That year marked the twentieth anniversary of Einstein's death, the seventieth anniversary of his special theory of relativity and photon theory, and the sixtieth anniversary of his general theory of relativity. The last, acclaimed as Einstein's greatest scientific achievement, attracted the most attention, particularly the tantalizing steps of trial and error that led to its final triumph and the subsequent efforts to resolve its inner difficulties and to give it more precision and added applicability.[51] An effort was also made to produce a basic bibliography of Einstein studies in the Russian language.[52]

A much more elaborate celebration took place in 1979, the centennial of Einstein's birth, which saw a wealth of studies by scientists all over the world in homage to the leading architect of the twentieth-century revolution in science. Not to be outdone, the Soviet Union staged a series of impressive events. Books, symposia, and scholarly papers came out in unprecedented numbers. A bibliography listed no fewer than 150 centennial publications. The participating journals ranged from *Communist*, always careful in placing a Marxist imprint on its published material, to *Questions in the History of Science and Technology*, devoted to the systematic study of the general impact of scientific thought on modern culture, to *Advances in the Physical Sciences*, alert to the ties between physics and philosophy. The Academy of Sciences and Moscow University held a joint commemorative session at which leading experts discussed Einstein's contributions to the modern physical picture of the world as refracted through the prism of dialectical materialism.

The pièce de résistance was *Einstein and the Philosophical Problems of 20th Century Physics*, a volume of essays drawn from *Questions of Philosophy*. The title notwithstanding, the collection concentrated on Einstein's scientific work in five areas: the probabilistic style of scientific explanation (the Brownian motion and quantum statistics); quantum theory (photoeffects and radiation); space and time (special and general theories of relativity); the structure and evolution of the universe; and the unified field theory.[53] The volume was notable primarily as an exhibit of the most dramatic effects of de-Stalinization on Einstein studies. It demonstrated the unity of Soviet scientists and philosophers in recognizing Einstein's ideas as the main catalyst of the twentieth-century revolution in physics—and in science generally. But it also demonstrated that there was no unity in the interpretation of the key components of Einstein's scientific legacy. The universal recognition of the vast magnitude of that legacy was matched by a refreshing and healthy diversity of ideas on its nature and implications.

In describing the heights of Einstein's achievement, the contributors to the anniversary volume—all scholars with established reputations—were concise and unceremonious. They were preoccupied with demonstrating their intimate familiarity with up-to-date Western interpretations of Einstein's physical principles and with submitting personal proposals for more advanced ideas in relativistic physics. The two most prominent in the latter respect were the mathematician-philosopher A. D. Aleksandrov and the physicist V. A. Fock. In what were straightforward reprints of earlier articles, they insisted that dialectical materialism could be blended with the theory of relativity by removing the principle of relativity from the central position of Einstein's physical ideas and by eliminating Einstein's mathematical formalism not directly connected with physical substance. The Fock-Aleksandrov correctives appealed more to philosophers than to physicists, a fair indication that the former did not fully abandon the long-nurtured hope of creating a Marxist theory of relatively, now declining steadily. The idea of a pressing need for a clearer and deeper understanding of the philosophical links between the theory of relativity and dialectical materialism attracted the attention of most contributors to the volume.

Philosophical links, predictably, were also the main concern for the contributors to the commemorative volume prepared by the Ukrainian Academy's Institute of Philosophy.[54] It consisted essentially of the papers presented at two institute-sponsored anniversary seminars held in Kiev in June and October 1979. The first seminar had concentrated on Einstein's views on methodological and epistemological questions of relativistic physics. The second had dealt exclusively with the influence of the theory of relativity on the guiding ideas, inner structure, and broader cultural ramifications of the scientific world. Published in 1983, the study went far afield in discussing the cosmic unity of science and the basic postulates of Einstein's humanistic philosophy. Where the Moscow symposium concentrated on past achievements of the theory of relativity, the Kiev symposium represented a bold effort to chart the new lines on the general cultural dynamics of Einsteinian physics. It also offered a cogent analysis of both the characteristic features and the role of creativity in science, of the relations of relativistic physics to nonphysical sciences, and of the humanistic underpinnings of a scientific world outlook.

As one might expect, Knowledge (Znanie), an organization devoted to the popularization of scientific knowledge and philosophy, joined in the centenary celebrations. In 1979, it published *Einstein and Modern Physics* in 46,000 copies.[55] This small collection of essays by different writers covered the early influence of Einstein's physical principles, epistemological views, and basic contributions to quantum mechanics. The authors concentrated on the depth and strategic position of Einstein's scientific achievements and philosophical involvement. Although earmarked for mass consumption, the book could only have appealed to readers who had no qualms about struggling through mathematical reasoning and philosophical subtleties.*

In the aftermath of the commemorative celebrations, many new topics were added to the already rich repertory of Einstein studies. A new book, for example, dealt exclusively with Einstein's practical inventions.[56] Einstein's views on the relation of science to the arts, ethical values, and philosophy received a closer, more modern, and more stimulating analysis. The epochal Einstein-Bohr debate of the 1920's and 1930's was subjected to a deeper, more flexible scrutiny. One author made a brave effort to apply Einstein's notion of space-time to the study of the "genetic psychology" of primordial human mentality. Nor did Soviet scholars overlook "the anthropic cosmological principle," particularly its relations to the evolutionary view of the universe, an idea tentatively built into relativistic cosmology.[57] Much discussion centered on the question of the general compatibility of that principle with materialistic philosophy.[58]

Omel'ianovskii, obviously moved by the magnitude of the centenary cele-

---

Priroda devoted its third issue of 1979 to the celebration ("Al'bert Einshtein, 1879–1979"). Ia. B. Zel'dovich offered an intimate portrayal of Einstein's achievements; I. Iu. Kobzarev described Einstein's relations with Max Planck; V. P. Vizgin analyzed the developments in physics that led to the theory of relativity; and A. D. Chernin presented the troublesome problems of relativistic cosmology.

brations, took it on himself to place the Einsteinian revolution in a Marxist context. His line of approach was to compare Einstein with Newton. On his analysis, Newton and the Newtonians aspired to create a closed and historical system of scientific axioms and principles explaining the structure and the dynamics of the universe—a system aimed at giving complete and permanently fixed answers to all relevant questions. Einstein, by contrast, created an open and historical system of physical explanations—a system that invited and encouraged continuous criticism and the search for new modes of inquiry and broader scientific horizons. The theory of relativity and quantum mechanics showed the preeminent role of revolutionary leaps in the historical flux of science. More than any other development in modern thought, the theory of relativity also demonstrated the decisive role of dialectics in the advancement of scientific learning. Omel'ianovskii hailed Einstein's theory as a dramatic display of "the law of dialectical contradictions," a synthesis of opposite theoretical programs and scientific traditions. He noted a striking difference between the subjective and objective sides of Einstein's creative impulse. Subjectively, Einstein favored a philosophical stance that found no use for dialectical materialism. Objectively, his heavy but unconscious dependence on the dialectical mode of reasoning made it possible for him both to create the theory of relativity and to anchor it in materialistic thought. In Omel'ianovskii's words: "The dialectical-materialistic theory of space and time as forms of the existence of matter has received significant confirmation from the theory of relativity."[59] But all this did not mean, he declared, that the general theory of relativity, in particular, was not in need of more depth and larger perspectives.

In *The Philosophical Foundations of the Theory of Relativity* (1982), D. P. Gribanov tried to pull together for the general public the central ideas of the vast literature produced in honor of Einstein's birth. Gribanov told his readers that Einstein could not be identified with any idealistic philosophical system or with the reigning schools of idealistic thought. Although not a dialectical materialist, Einstein did not participate in campaigns against Marxist thought; he could best be labeled "a spontaneous materialist and dialectician."[60] Einstein, according to Gribanov, accepted the idea of "the material unity of the world" and such of its basic correlates as the causal determination of natural and social phenomena, atomism, and the principle of the knowability of the world. His "cosmic religion" had no room either for a notion of God or for an elaboration of theological thought.[61] In Einstein's thinking, according to Gribanov, the concept of matter was closely linked with the notion of physical reality, which appeared in two all-inclusive forms: substance and field.

Gribanov believed that the growing effort in the Soviet Union to establish harmonious relations between dialectical materialism and the theory of relativity followed two distinct courses: a deliberate and thorough search for and emphasis on Einstein's ideas that appeared harmonious with materialistic positions; and a conscious effort to modify the guiding principles of Marxist thought in such a way as to bring them closer to the prevalent trend in Western—or "idealistic"—philosophy. The latter effort included a new move to define physical reality as a synthesis of objective nature and subjective perception. It also included the lavishly displayed willingness to cast the most favorable possible light on both the nonempirical mechanisms responsible for major advances in science and the imperative need to differentiate between the philosophical notion of matter as a reality existing independently of human consciousness and the scientific notion of matter concerned with the inner structure and dynamics of the building blocks of the universe.

The new Marxist definition of matter, as Gribanov presented it, stood for a dialectical synthesis of antithetical descriptions offered by the two masters of classical Marxism: Engels, for whom matter was a pure creation of thought and abstraction, and Lenin, for whom matter was an objective reality that existed independently of human perception.[62] Matter, as now seen, was both an objective (independent) reality and a creation of the human mind.

Gribanov did not identify Einstein as a dialectical materialist; what he claimed was that on several theoretical issues of cardinal significance, Einstein saw eye to eye with the Soviet interpreters of the Marxist philosophy of science. He reckoned that Einstein had little knowledge and less interest in the Marxist philosophy of science. He was ready to point out the areas in which Einsteinian and Marxist thought were far apart, but his primary interest was in the basic similarities between the two systems of theoretical principles.

How did Gribanov account for earlier attacks on Einstein by the guardians of Marxist orthodoxy? He offered a combination of random answers. The theory of relativity, he wrote, was presented in complex mathematical symbolism, which described objective reality only in an indirect fashion and thereby invited misinterpretations. Einstein gave much weight to the idea of the relativity of scientific truth. Particularly as presented in textbooks, Einstein's idea of space-time was not connected logically with the Marxist idea of the ontological primacy of matter. The situation was complicated by "terminological confusion," for which both the interpreters of the theory of relativity and Einstein himself were responsible. Einstein's ideological convictions displayed elements of confusion and inconsistency, a form of adaptation to the contradiction-ridden social conditions in which he lived. In all this, Gribanov made no effort to view the earlier attacks on Einstein as part of a Stalinist campaign to superimpose a rigid and uniform system of ideological controls over every phase of national life. By remaining silent, he absolved the Stalinist Politburo and Stalin himself of tampering with the most cherished values of science.

The expanding range of Einsteinian topics included an effort to examine the history of Soviet attitudes toward the theory of relativity and allied fields. The pioneers in this domain, however, were over-cautious and generally slow to take advantage of the opportunities opened by the thaw. They did not go beyond the developments on the philosophical front during the 1920's and early 1930's, which is to say, the scornful and sweeping attacks on Einstein and his views during the Stalin era escaped critical, detailed, and systematic historical study. Concentrating heavily on the bright side of past developments—on writing "progress reports"—the historians were inclined to attach little weight to the anti-Einstein tirades of the defenders of Marxist orthodoxy.[63] In "The Theory of Relativity and Soviet Science," K. Kh. Delokarov, a leading expert on the Soviet reception of Einstein's theories, mentioned I. V. Kuznetsov but ignored the fact that, in 1952, at the peak of Stalin's crusade for the ideological purity of Soviet science, he had wanted the theory of relativity thrown out *in toto*.[64] The first generation of post-Stalinist historians preferred to view the ideological critics of Einstein's theories as misguided and isolated individuals far removed from the pulse of Soviet science and philosophy. That those individuals were actually guided and protected by the highest echelons in the Communist hierarchy was generally overlooked. Reluctant to tread too deep into political waters, they failed to point out that the Maksimovs, Kol'mans, and Kuznetsovs, in their pursuit of ideological purity, had not only driven bright students away from the field of theoretical physics, but managed to alienate practically every scientist from the very doctrines they worked so hard to promote.

6 ∎

Accommodations to Einstein's Theory of Knowledge

In no domain of the Soviet confrontation with Einsteinism was the influence of the de-Stalinization process stronger and more penetrating than in epistemology, the branch of philosophy concentrating on the origin, nature, and limitations of scientific knowledge. The efforts to eliminate or drastically reduce the sharp and irreconcilable differences between the "subjective" bent of Einstein's theory of knowledge and the "objective" orientation of the official Soviet philosophy of science that characterized the Stalin age are the main focus of this chapter.

The Marxist Interpretation of Einstein's 'Pure Reason' and 'Intuition'

On several occasions, Einstein referred approvingly to the strategic place of pure reason and intuition in the advancement of scientific knowledge.[1] He was guided by the idea that scientific concepts were creations of the human mind, not of sensory links with the external world of nature and with the needs of society and its technology. He came close to Kant's notion of reason as a legislator of the laws governing nature.

The idea of pure reason did not appeal to orthodox Marxists, who preferred to look at science as a cumulative response of the scientific community to the technical and economic needs of society, and as a refinement and codification of empirical knowledge. In the 1930's, for example, Marxist doctrinaires subjected the Moscow school of mathematics, based on set theory, to merciless attacks for its alleged nonempirical and anti-technological orientation. Even Vavilov, who generally expressed liberal views in interpreting the relations of dialectical materialism to modern physics, argued in 1934 that Einstein's claim of a vital role for pure reason could not be accepted because it "isolated mathematics from matter," assumed that "reason makes the laws of nature," and, in general, endorsed philosophical idealism.[2]

Some thirty years later, Soviet philosophers were increasingly coming around to Einstein's thinking. Twentieth-century physics, according to M. V. Keldysh, president of the Academy of Sciences, showed that modern theory did not develop inductively, and that it could not be treated as the extension of an established theory or experiment: the physicist had learned that the revolutionary ideas in his science originated not in experimental data, but in mathematical deductions.[3] Implicitly, he endorsed the statement Einstein had made in his Herbert Spencer lecture at Oxford in 1933—that the general theory of relativity showed "in a convincing manner" the correctness of the view that the basic concepts of physics were, in a logical sense, "free inventions" of the human intellect, that is, inventions not derived from experiment or experience.[4] In the age of the post-Stalinist thaw, Marxist philosophers had a change of mind: they were now ready to admit that by relying mainly on mathematics, the most abstract of all sciences, and by depending solely on the mastery of its own logic, theoretical physicists had predicted the discovery of an "entire series" of particles.[5]

Einstein was convinced that intuition was the main source of the ideas from which pure reason made its deductions. He had no qualms about expressing his own indebtedness to that phenomenon; he, for example, credited intuition for his daring formulation of the relativity of simultaneity. Although flashes of intuitive creativity were products of nonlogical processes, they became parts of scientific thought only when they did not contradict the internal logic of the development of individual disciplines. Einstein's view of intuition was the same as Spinoza's: it had nothing in common with the metaphysical celebration of mysticism and irrationalism. Intuition, as Einstein envisioned it, not only did not stand in opposition to reason; it belonged to the more advanced spheres of human capacity for rational thought. In a way, it was an "extralogical" source of the rational components of larger orders of human thought. If there were other kinds of intuition, they were outside Einstein's interest. Like Kant, he believed that intuition not firmly linked with concepts was blind, just as concepts without intuition were empty.

Einstein summed up his arguments in two propositions: first, every theoretical system of physics was dependent on and controlled by sensory evidence; and, second, there was no logical way of proceeding from empirical data to general theoretical principles.[6] Where logic failed, intuition, a nonlogical process, moved into action. Intuitive insights, in order to become integral parts of the body of science, must not contradict sensory evidence. Einstein's emphasis on the creative role of intuition in scientific theory stimulated a lively interest in the affinity of scientific and artistic approaches to nature.

Traditionally, Marxist historians and philosophers of science were skeptical about the role of intuition as a research tool. For them, scholars who advocated an intuitive approach, like L. E. J. Brouwer in the study of mathematics, Henri Bergson in *Creative Evolution*, and Nikolai Lossky in the Russian school of ir-rationalism, "solved" things without ever making a comprehensive and sys-tematic study of the nature, sources, and strengths of intuition.

Intuition continued to be anathema to the Marxists traditionalists well past Stalin's death. For example, the *Great Soviet Encyclopedia* moved very little in its position between its first and second editions. The entry in the first edition (1935) stated explicitly that "dialectical materialism does not recognize intui-tion as a special form of knowledge."[7] Modern mysticism, it said, combined an affirmation of intuition with a negation of science. Some twenty years later, the encyclopedia recognized only the intuition linked with previously acquired "experience, habit and knowledge." In Western idealistic philosophy, the writer noted, intuition was the cornerstone of mysticism. He made no effort to define the role of intuition in the production of scientific knowledge.[8] As late as 1959, a team of philosophers steeped in conservative thought was still arguing the mysticism point: "Materialistic dialectics, acting in harmony with the facts of science, is guided by the idea that there is no mystical, supernatural, and in-tuitive knowledge. . . . We acquire all knowledge from the external world with the help of sensual experience and from thought built on sensation and per-ception."[9]

During the early 1960's and continuing into the early 1980's, the prodigious interest in Einstein's legacy contributed to a methodical exploration of the world of intuition as a legitimate mechanism of scientific creativity. Soviet writers, physicists no less than philosophers, now worked assiduously on iden-tifying, sorting out, and interpreting Einstein's scattered references to the phe-nomenon. They generally agreed with his firm belief in the paramount role of intuition in the advancement of scientific knowledge.

The article on intuition in the third edition of the *Great Soviet Encyclopedia* (1971) bore all the earmarks of the changing times. It acknowledged that in sci-ence, no less than in the arts, the search for truth might follow a path of intui-tion that did not depend either on the logical links between sense perceptions and conceptual thought or on language; rather, it occurred unconsciously and spontaneously.[10] Soviet writers disavowed any ties between intuition, as now defined, and metaphysical intuitionism; they went along with Einstein's argu-ment that one of the major functions of intuition was to link "complexes of sense experience" with "elementary concepts," a task beyond the competence of logical processes.[11] Whatever the mechanisms, perspectives, and potentials of linking processes, "the paths of science intersect the paths of art."[12]

P. V. Kopnin, one of the more discerning philosophers, observed that Einstein had no intention of denying the objective roots of scientific knowledge, but merely wanted to acknowledge the specific scientific role of intuition as a cognitive process unrestrained by the rigor and formal procedures of logical deductions.[13] Kopnin's statement and similar pronouncements were harbingers of a widespread and vigorous search for logical and epistemological arguments bringing dialectical materialism closer to the reigning orientations in the Western philosophy of science.

The evolution of Soviet Marxist views on the relation of intuition to scientific discovery thus passed through three phases: in the pre-Stalin and Stalin ages, intuition was not considered an integral part of the creative process in science; during the early phase of the post-Stalinist thaw, the emphasis was strictly on "the rational kernels" offered by intuitive creativity; and during the 1970's–80's, intuition was recognized as a nonlogical or translogical source of ideas integrated into logical frameworks of scientific knowledge.

"Creative intuition" was now treated as an arterial path to the acquisition of new scientific ideas of fundamental significance. But the products of intuition became parts of science only when they could be integrated into the logical patterns of established systems of scientific deductions.[14] Soviet epistemologists did not recognize intuitive insights that violated the dialectical principle of continuity in the growth of scientific knowledge. Moved by the excitement of the new epistemological discourse, the advocates of intuition simply strengthened the role of the subjective factor in the evolution of scientific wisdom.

Recording the rise of intuition to the heights of scientific methodology, M. I. Panov, a philosopher of mathematics, wrote in 1984: "Intuition has penetrated the modern methodology of science so deeply that without it, it has become impossible to describe the process of the acquisition of new knowledge and creativity."[15] The time had come, according to Panov, to give high priority to a systematic study of the "hidden laws" regulating the free creativity stimulated by intuition.

General Changes in Epistemological Views

Einstein wrote in his "Autobiographical Notes" that "out yonder there is this huge world which exists independently of us human beings and which stands before us like a great, eternal riddle, at least partially accessible to our inspection and thinking."[16] Lenin's *Materialism and Empirio-Criticism* teemed with statements on the world as an objective reality "independent of the minds of man and the experience of mankind."[17] By recognizing the objective sources of human knowledge both Einstein and Lenin stood on the side of epistemological realism; both agreed with Francis Bacon's statement that man could not hope to succeed "if [he] arrogantly searched for the sciences in the narrow cells of human understanding, and not submissively in the wider world."[18] But in answering the question of how the "eternal riddle" of reality existing "independently of us human beings" was captured by the human mind, the differences between the views of Einstein and Lenin were striking. For Lenin, human experience was a matter of adapting to objective reality, "the copying of which constitutes the only scientific methodology."[19] In the age of Stalin, Marxist philosophers were particularly eager to emphasize the direct and inevitable roads from accumulated human experience to the highest abstractions in science.

Einstein, by contrast, contended that accumulated experience was not the most important factor in the theoretical advancement of science. For him, mathematics was "a product of human thought" that was both independent of experience and responsible for major turning points in the evolution of scientific theory. Mathematical propositions referred to objects of human imagination, not to "objects of reality."[20] A scientific law, axiom, or theoretical orientation, he thought, owed not to experience as a reflection of objective reality, but to pure reason, imagination, and intuition. Einstein emphasized the subjective base of scientific progress; Lenin dedicated his book to a war against all subjective elements in scientific thought. Epistemological subjectivism in modern physics and the philosophy of science, as we have seen, was a casus belli for Maksimov, Kol'man, and other defenders of Stalinist Leninism.

The physicist M. A. Markov had displayed great courage in 1947 by challenging the Marxists' exaggerated emphasis on the objective and absolute nature of physical knowledge and by pleading the case for the vital role of the subject—or observer—in the cognitive construction of physical reality.[21] In taking such a line, he was obviously under the spell of the epistemological strategy of the Copenhagen school in quantum mechanics. That strategy, contrary to the tack taken by dialectical materialism, was clearly expressed in Heisenberg's dictum that "natural science does not simply describe and explain nature; it is part of the interplay between nature and ourselves; it describes nature as exposed to our method of questioning."[22] Markov's statement, subjected to harsh criticism in Marxist journals, came at the time when subjectivism in the theory of knowledge, particularly in the philosophical discussions centered on the theory of relativity and quantum mechanics, was held to be a cardinal sin by the crusaders against cosmopolitanism. To accuse a physicist of favoring a subjective view of physical reality was to accuse him of desecrating Soviet patriotism, dialectical materialism, and the politics of the Stalinist government.

Soviet ideology, as presented by the Stalinist forces, contradicted Bertrand Russell's firm conviction that no particular politics followed from epistemology. Especially in the Stalin era, Marxist epistemology served as a link between ideology and Soviet politics. This was the time when agnosticism, solipsism, and subjectivism were the chief targets in the attacks on creationism. This was also the time when Marxist ideologues claimed that the chaotic diversity of Western epistemological schools was a telling sign of the functional disharmony and irreversible decay of the world of capitalism.

Markov was the only Soviet scientist daring enough to write positively of the subjective as a component of physical reality during the Stalin years. But once the thaw set in, the topic became the central concern of a long line of philosophers of science and philosophically oriented physicists. Epistemological views changed so rapidly, in fact, that in the 1960's only a minuscule group of die-hard Stalinists questioned the important role of the subject in shaping the world of scientific knowledge.[23] Both philosophers and physicists went so far as to credit Einstein with rediscovering "Marx's materialistic formulation of the interaction of the object and the subject in cognition."[24] The subjective component of physical reality usually referred to the reliance on laboratory equipment, mathematical apparatus, and "intuitive leaps." It also referred to coordinate systems, frames of reference, and observers, the main engines in building a great and complex physical theory on the principle of relativity. The objective component of physical reality referred to natural phenomena existing outside and independently of human consciousness. Now it was readily acknowledged that there were physical objects—such as quarks, partons, and black holes—whose existence was confirmed by theory alone, not by experimental proofs.[25] The knowledge of the very existence of these and many other objects was built on subjective rather than on objective grounds.

Although Soviet scholars and ideologues continued to criticize the excesses of the Western "idealistic" philosophy of science, that did not interfere with their efforts to reconcile dialectical materialism with philosophical orientations deeply engaged in the subjective aspect of physical reality. They agreed in principle with Imre Lakatos's "activist" theory of knowledge, according to which, "we cannot read the book of Nature without interpreting it in the light of our expectations or theories."[26] Omel'ianovskii resented only Lakatos's alleged unawareness of the "activist" orientation of dialectical materialism.[27] In the post-Stalin era, Western philosophies of physics and dialectical materialism differed more in the formal aspects of their presentations than in the substance of their ideas. The following statement, made in 1977, clearly demonstrates that even committed Marxists were ready to accept much of the "subjectivism" of Western physics—particularly of the Copenhagen school:

> The role of the subject in the development of physics is . . . in the selection of the objects of inquiry and of basic theoretical assumptions, in continuous improvements in measuring instruments, in the preparation and execution of experiments, in the formulation of new physical ideas, . . . and in the construction of new theoretical pictures of the world. The modern trend in the growth of physical knowledge seems to indicate that in the future, subject . . . will become more active in responding to the growing role of theoretical designs for new knowledge, particularly in the advancement of conceptual and cognitive tools of physical science.[28]

Soviet philosophers continued to take note of the influence of Mach on Einstein. But where they had once never missed an opportunity to attack Einstein for his Machian leanings, most were now inclined to explore and emphasize the differences in the epistemological thinking of the two giants. The philosopher K. Kh. Delokarov found it important to note that Einstein was attracted to Mach's generally negative view of the mechanistic orientation in science rather than to the specific aspects of his epistemological thought. Einstein shared with Mach a general target of criticism, not a specific elaboration of epistemological arguments.[29] At this point, Soviet philosophers showed much more concern for distinguishing Einstein's thought from Machian and other neopositivist epistemologies than for linking it to dialectical materialism. They operated on the assumption that by abandoning neopositivist positions, Einstein had come closer to the Marxist theory of knowledge.

Einstein had announced his changed thinking as early as 1936 in a letter to his friend Maurice Solovine:

> In Mach's time a dogmatic materialistic point of view exerted a harmful influence over everything; in the same way today, the subjective and positivistic point of view exerts too strong an influence. The necessity of conceiving of nature as an objective reality is said to be a superannuated prejudice while the quanta theoreticians are vaunted. Men are even more susceptible to suggestion than horses, and each period is dominated by a mood, with the result that most men fail to see the tyrant who rules over them.[30]

By 1949, his thinking had evolved to the point where he found "Mach's epistemological position . . . essentially untenable."[31]

These and similar statements led V. A. Khramova to conclude that Einstein's philosophy "had nothing in common with Machism."[32] She warned that Einstein's habit of using Machian terminology provided no real grounds for assuming a basic similarity in the philosophical orientations of the two physicists. And she argued that the same applied to Henri Poincaré. Even when Einstein appeared to be making concessions to Poincaré's conventionalism, as S. G. Suvorov had suggested, he stayed at the level of formal rhetoric that gave no signs of a retreat from his firm belief in the objective essence of physical knowledge.

The debate about Mach's place in Einstein's thought led P. S. Dyshlevyi and F. M. Kanak in 1977 to draw the following conclusions:

> There are good reasons . . . to believe in the sincerity of Einstein's announcement that Hume and Mach (the skepticism of the former and the anti-Newtonian attitude of the latter) spurred him to reexamine the principles of classical physics. This announcement, however, should not be viewed as an indication that the philosophy of Hume and Mach influenced Einstein in his theoretical inquiries or that Einstein was a subjective idealist. Einstein's world outlook, seen as a totality of his views on the questions of methodology, has nothing in common with idealism and positivism.[33]

Compare this statement to the standard assessment of Einstein's subjective philosophy in the Stalin era, as articulated by I. V. Kuznetsov in 1952:

> Scientific concepts and theories are valuable only insofar as they have an objective content, which depends neither on man nor on mankind. A discovery of objective truth is the goal and the justification of every physical theory; such theory has no other justification.
>
> . . .
>
> [Einstein] understands the laws of nature, not as expressions of the real ties between objects and between various characteristics of objects, which exist outside and independently of [human] consciousness—of observers and measurement procedures—but as the method for a comparative review of measurement results. . . . This is a typical subjective-idealistic approach that has nothing in common with and is irreconcilably opposed to science.[34]

In the post-Stalin era, Marxist philosophers continued to see, but not to emphasize, inconsistencies in Einstein's philosophical views. However, they limited those so-called inconsistencies to idealistic lapses in complex arguments of a basically materialistic nature. They were now encouraged to stress "the general materialistic tendency" of Einstein's philosophical discourses, expressed primarily in the emphasis on the dialectical aspect of natural processes.[35] Einstein, Bohr, Born, Schrödinger, and de Broglie, once condemned as the leading representatives of physical idealism, were now held to be among the leading—though not always consistent—critics of physical idealism.[36] Marxist philosophers now argued that the extensive reliance on research tools, the subjective apparatus of physics, made it possible for modern physicists to attain objective truths fully reflecting absolute reality—the world existing outside and independently of human consciousness.[37] They conceded that scientific theories, in contrast to unprocessed collections of empirical data, represented subjective reflections of objective reality. The politics and the philosophy of the post-Stalin thaw encouraged them to be more tolerant of the epistemological views of the leading physicists in the West.

The new interpretation, as one might expect, did not win universal acceptance. The Stalinist mentality lingered on in some circles. Mitin, for example, flatly refused to acknowledge that a change of position in Soviet philosophical circles had brought Marxist thought closer to the reigning epistemological ideas in the West. He argued that the real change took place, not in the Soviet Union, but in the West, which had come closer to materialistic positions in the philosophy of modern physics.[38] He interpreted the alleged withdrawal of Einstein, Bohr, and Heisenberg from the positions of subjective idealism as new proof of the philosophical superiority of dialectical materialism.

Many Soviet critics now agreed with Heisenberg's remark that from the very beginning, Einstein had showed a strong inclination to advance an epistemological view of physics that would open the gate for "an objective description of nature, independent of man." They did not go so far as to agree with Heisenberg's statement that Einstein, fearful of "discrediting 'ideological superstructures,'" stuck too closely to "the reality of matter" and to the objective existence of space and time.[39] Nor did they stop referring to idealistic lapses in Einstein's epistemology.

Fock and Aleksandrov also departed from the new mainstream. They made an elaborate effort to show that the theory of relativity could be constructed without reference to the subject (observer) in the analysis of empirical data and in the elaboration of theoretical structures. Equally firm was their opposite conviction that there could be no quantum mechanics without a close look into the interaction of the object and the subject.[40] They claimed that Hans Reichenbach, Adolf Grünbaum, and Mario Bunge, among others, held similar views in the West. Referring to the interpretation of quantum mechanics as a science of the interaction of micro-objects and observation instruments, judged by most physicists as a recognition of a subjective orientation, Einstein noted: "What does not satisfy me in that theory, from the standpoint of principle, is its attitude toward that which appears to me to be the programmatic aim of all physics: the complete description of any (individual) real situation (as it supposedly exists irrespective of any act of observation or substantiation)."[41]

Some Soviet philosophers of science, following Einstein's lead, thought it advisable to treat objective reality and physical reality as different ideas. As Dyshlevyii put it, objective reality did not depend "either on man or on mankind," whereas physical reality, consisting as it did of "the totality of theoretical constructs," depended on the choice of physical theory.[42] Although Einstein, in his famous debate with Bohr, had not drawn a precise line between the two categories, he deserved credit, in Dyshlevyi's view, for formulating "a fundamentally new position in the methodology of physical knowledge, according to which the empirical base of physical theory is made up, not of nature as such, but of nature as seen through the prism of the practical (experimental and theoretical) activity of the researcher."[43] In that scheme of things, objective reality became a philosophical category, and physical reality a scientific category. As far as some Soviet scientists were concerned, the notion of objective reality was useless.

Most Marxist theorists recognized all along that Einstein's philosophical discourses represented a mixture of idealistic and materialistic principles. During the Stalin era, they dealt almost exclusively with the sins of Einstein's idealism, rooted in Mach's epistemology. In the post-Stalin age, the situation was the exact opposite: Marxist philosophers continued to mention Einstein's idealistic transgressions, but they were interested primarily in the materialistic base of the theory of relativity. Most important of all, they stopped viewing idealism as a philosophy without a "kernel of truth."

If Einstein did move away from Mach's suggestive philosophical thought, where did he end up? E. D. Chudinov, one of the more astute philosophical commentators, argued that Einstein came close to Kant's epistemological positions, an interpretation shared by several other critics. In Chudinov's view, Einstein's epistemological orientation was deeply rooted in the Kantian idea that man perceived phenomena through the prism of logical categories.[44] Under Kant's influence, Einstein's theory of knowledge gravitated strongly toward rationalism. It depended on empirical evidence, not to draw inductive conclusions, but to verify logical hypotheses.

From 1970 to the beginning of perestroika in 1985, Einstein's gradual but decisive rejection of Machian neopositivist ideas became a topic of intensive investigation. But though Soviet critics were now ready to admit that Stalinist philosophers had wrongly rebuked Einstein for his alleged dependence on Mach's theory of knowledge and were ready to reinterpret Einstein's philosophical ideas, they showed little inclination to do the same for Mach.

New Attitudes Toward Einstein's
Mathematical Idealism

In the post-Stalin era, the Marxist theorists' views on the origins and orientations of mathematical principles—and of physical idealism and mathematical
idealism—underwent a radical turnabout. Although they did not abandon
Engels' idea of the empirical roots of mathematics, which linked that science
with the material substratum of the world, they broadened their general outlook to allow pure intuition a paramount role in the advancement of mathematics and the full detachment of mathematical abstractions from empirical
moorings. In brief, they worked diligently on locating their epistemology
squarely in the middle between the extreme positions of "metaphysical materialism," which fully ignored the place of the subject in the construction of scientific knowledge, and various schools of idealism, which found little use for
the object.[45] They continued to criticize mathematical idealism and philosophical relativism, but their swords were duller, their arguments less un-
bending, and their general offensive more subdued than in Stalin's day.

Lenin's warning that too heavy a reliance on mathematics was a sneaky way
of inviting idealistic aberrations into the work of science lost most of its credibility and ideological urgency. Its place was gradually taken by Marx's calmer
and more realistic statement that the higher the level a science occupied on the
scale of methodological and substantive mathematization, the higher its place
on the ladder of professional perfection. The sweeping acceptance of Marx's
judgment, however, did not mark a complete erasure of Lenin's injunction.

Lobachevskii was now praised not only as the founder of the first non-
Euclidean geometry, but also as an early pioneer in axiomatics—in setting up a
system of logically interlaced geometrical propositions that reached far beyond
the limits of experience. He also received credit for having created a system of
mathematical principles in the pure interest of science rather than in response
to some immediate and practical need of society. The general theory of relativity, as now viewed, marked the beginning of a new era in physics because it presented the first body of physical thought to be constructed on a purely axiomatic basis, without appeal to experience and experiment. It legitimated Poincaré's idea of the possibility of a pure geometrical description of nature.[46]

Elie Zahar, a British philosopher of modern physics, has noted judiciously
that "Einstein wanted both to geometrise physics and to turn geometry into an
empirical science."[47] Gaston Bachelard, the eminent French interpreter of "the
spirit of modern science," puts it more bluntly: in post-Newtonian science,
"physics has become a geometrical science and geometry a physical science."[48]
Where in the Stalin era, Marxist scholars almost completely overlooked Ein-

stein's contribution to making geometry an empirical science, they now made a point of stressing and praising that facet of his work.

Some philosophers continued to adhere to the traditional injunction that certain major mathematical operations of the theory of relativity were unrelated to physical content—that Einstein carried the idea of the mathematization of physics to an extreme. So did a few physicists, represented most forcefully by A. A. Logunov.[49] But there was general agreement on both sides that mathematics gave physical research much more than technical assistance in processing empirical data as raw material for the formulation of the laws of nature. As the general theory of relativity showed conclusively, it also gave a point of departure and research designs to the study of physical processes. The general theory of relativity was a product of involved mathematical operations that created the principles of the inertial and gravitational masses and of the space-time continuum. The structure of the latter, according to Chudinov, was determined by a symmetrical metric tensor and by equations that were invariant in relation to a group of continuous transformations.[50] The function of empirical facts in physics, as Einstein saw it, was not only to provide raw material for mathematical formulations of the physical laws of nature, but also to test mathematically deduced hypotheses presenting regularities in the work of nature. A growing tolerance of nonempirical ideas (the creations of "our mere imagination") as the primary motor of scientific progress showed that the custodians of official ideology were ready to abandon a stronghold of Stalinist controls over the scientific community.

Einstein recognized the triumph of "the axiomatic method" in physics. He believed that "nature is the realization of the simplest conceivable mathematical ideas," and that by relying on purely mathematical constructions "we can discover [the concepts and laws] which furnish the key to the understanding of natural phenomena."[51] Soviet critics acknowledged Einstein's major contributions to the growing role of mathematics in modern scientific methodology. But they were equally adamant in claiming that, despite the revolutionary scope of the axiomatic method, it could not alone order scientific knowledge and build the background for new advances. In suggesting less than exclusive authority for the axiomatic method, they relied on Kurt Gödel's rule that axiomatic systems could not produce proof for their completeness and incontrovertibility.[52] Despite these reservations, however, they recognized the enormous fruitfulness of the axiomatic method as a potent tool of modern science.

Scientific reality has three dimensions: empirical data, mathematical formalism, and theoretical constructions. Stalinist philosophers had followed the Baconian line of reasoning: they considered the empirical base the principal component of scientific thought, assigning auxiliary or derivative positions to mathematical formalism and theoretical constructions. Einstein rejected that configuration, assigning primacy to theoretical formulations. In his opinion, theory determined how scientists saw the empirical substratum and what kind of mathematical apparatus they chose. Post-Stalinist philosophers did not quickly accept Einstein's ordering. But with the ideologues in retreat, they gradually found themselves in a position to undertake a sweeping reevaluation of Einstein's philosophy of mathematics. By the end of the post-Stalinist thaw, most philosophical challenges to Einstein's mathematical idealism had become a relic of the past. By then, too, all the scathing public criticism to which many eminent Soviet mathematicians had been subjected for their alleged leanings toward mathematical idealism—issued not just by Marxist philosophers but also by some Marxist mathematicians—had come to a full stop.

The Einstein-Bohr Debate Reconsidered

In the age of the post-Stalinist thaw, the Copenhagen version of quantum mechanics, no less than the theory of relativity, acquired full respectability in Soviet scientific and philosophical literature. By the end of the 1960's, there was nothing of major consequence in the Copenhagen school's theoretical system that was not considered entirely compatible with dialectical materialism. The crowning event came in 1976, when the Institute of Philosophy published *The Principle of Complementarity and Materialistic Dialectics*, a collection of essays that fully integrated complementarity into the Marxist philosophy of science.[53] Now viewed as a *new* philosophical category of dialectical materialism, complementarity was championed not only as a special tool the Copenhagen school used to resolve a critical problem of microphysics, but as a conception capable of describing the work of nature in general. The volume, for example, devoted separate chapters to the role of complementarity in biology, biophysics, space exploration, and geography.[54]

In I. S. Alekseev's view, Marxist theorists were now in a position to claim that Bohr's philosophy of science was in certain respects more "progressive" than Einstein's. He claimed that the philosophical background of the methodological differences between the two philosophies corresponded to the differences between "dialectical and pre-Marxist materialism."[55] In support of his assessment, he recalled Marx's inference that prior to his recasting of materialism, its chief defect was to regard reality exclusively as an object shorn of all subjective elements.[56] Einstein "fell behind" because he did not agree that experimental activity was part of physical reality.[57]

The search for a dialectical synthesis of opposing philosophies in the East and the West was strengthened in the Soviet Union by ambitious efforts to recognize the signs of unity among the clashing theoretical orientations in science. The time had come, for example, for in-depth inquiries into the threads unifying Einstein's and Bohr's philosophical and methodological thought. No longer was there a need for deciding who was correct, Einstein or Bohr; an increasing number of physicists and philosophers thought that both were correct. As a commentator noted, the debate did not produce a loser. "It was not a dispute between two enemies," he wrote, "but a duet of two great masters."[58] The debate was now perceived as a gigantic step forward in clarifying and strengthening the theoretical foundations of quantum mechanics.

All commentators agreed with Einstein that his debate with Bohr was limited to a number of key philosophical questions and did not affect his high respect for quantum mechanics as a body of scientific knowledge. In an article published in 1948, Einstein had added new fuel to his critique of the epistemological slant of the Copenhagen school, but he had also made it clear that "under no circumstances" was he prepared to deny that the theory of the Copenhagen school represented "a significant and in certain regards even a decisive step in the progress of physical knowledge."[59]

On his last visit to the USSR, in 1961, Bohr returned the compliment. Speaking to an overflow audience of Moscow students and professors, he made the following assessment of Einstein's contributions to a long series of philosophical disputes on the theoretical and methodological foundations of quantum mechanics as articulated by the Copenhagen school:

> Today, when Einstein is no longer with us, I should like to note how much he had done for quantum mechanics with his persistent and indomitable striving for perfection, architectural harmony, classical completeness of theories, and unified thought, on which the picture of physics could be built. By detecting contradictions at every forward step in physics, apparent outcomes of previous steps, he stimulated the progress of quantum mechanics.

In Alekseev's view, Einstein's ideas represented both the zenith of the Newtonian scientific orientation and "the first step" in the rise of quantum mechanics. On methodological grounds, he wrote, modern physics owed a huge debt to Einstein for his introduction of frames of reference as the most essential tools of the scientific analysis of physical reality. In his opinion, there was an easily discernible analogy between Einstein's "frames of reference" and Bohr's "instruments."[60]

Bohr's formal addresses and informal talks with Moscow physicists stimulated a lively interest in comparative historical studies of quantum mechanics and the theory of relativity. The result was an array of impressive and detailed studies on such vital topics as Einstein's contributions to quantum mechanics, the points of contact between the theory of relativity and quantum mechanics, and the Bohr-Einstein debate.[61] In Fock's telling, frank talks with Bohr had strengthened the ongoing effort of Soviet scholars to reduce the distance separating their philosophy of science from Western theories: in the course of them, Bohr had modified some of his more extreme views on causality and physical reality, bringing them closer to dialectical materialism and to Einstein's views.

P. A. Aronov and B. I. Pakhomov noted that, in a way, the Bohr-Einstein debate produced a grand scientific and philosophical synthesis of the pivotal components of the new physics—and of science in general. Referring to the outcome of the debate, they could not resist the temptation to interject Engels' statement that every epoch-making discovery in science produced a new and more perfect form of materialism. They made no effort, however, to explain their dramatic claim that the Bohr-Einstein dispute raised dialectical materialism to a higher—dialectical—level of advancement.[62]

By the beginning of the 1980's, there was no major epistemological orientation in the Western philosophy of science that did not attract conciliatory attention in the Soviet Union. The steady growth of epistemological diversity contributed to a perceptible disintegration of dialectical materialism as an ordered, original, and consistent philosophy of science. The crushing of the structure of dialectical materialism went hand in hand with clear signs of a developing philosophical pluralism, at first limited within the loose framework of the Marxist philosophy of science. Those signs were among the most fateful indicators of a rapid disintegration of official Soviet ideology as an active social and political force.

Approaches to the General Theory of Relativity and Cosmology

In the turbulent years after Stalin's death, the mainstream of physicists continued to consider Einstein's general theory of relativity sound in all its fundamental principles. But two major groups departed from that view. One, led by Fock and Aleksandrov, devoted Marxists both, found it unnecessary to look for help outside the framework of Einsteinian thought, but wanted to restructure the theory into a unique Soviet version fully compatible with dialectical materialism. The other, led by Ivanenko, considered the theory in need of major repair before it could be brought together with quantum mechanics into a unified field theory. This chapter examines those two schools of thought in some detail. It also explores the major controversies in the field of cosmology arising in the general theory.

Toward a Soviet General Theory of Relativity

No component of Einstein's work provoked more scientific, philosophical, and ideological debate in the Soviet Union than the general theory of relativity. After all, this intricate body of principles represented not only the highest and most comprehensive level of Einstein's scientific achievement but also a quintessential challenge to Marxist epistemology. It became the fountainhead of scientific principles opening the gates to the study of such previously unattended cosmic phenomena as black holes and pulsars, mysterious objects called for by Einstein's theory of gravitation but not readily susceptible to empirical testing.

In the proliferation of studies dealing with the general theory, Fock's monumental study of 1955, *The Theory of Space, Time, and Gravitation*, was most conspicuous for the depth of its analysis and the unorthodox nature of its conclusions.[1] Quickly translated into English and widely noted in the West, Fock's study offered an elaborate mathematical effort to make space-time the central physical notion of the theory of universal gravitation anchored in the general theory of relativity. The book was an elaboration of the thesis that Einstein's theory of gravitation was not a mathematical derivation from the general principle of relativity, but a physical derivation from the elaborate four-dimensional structure of space-time. In facing the riddle of space, Fock, very much unlike Einstein, found use not only for Riemann's non-Euclidean geometry but also for Lobachevskii's. Fock argued that only Lobachevskii's geometry, reinforced with Friedmann's idea of isotropic space with a uniform mass density, could be effectively applied to "the regions of space that include many galaxies."[2]

Fock was one of the most important Soviet contributors to both quantum mechanics and the theory of relativity.[3] In quantum mechanics, he made substantial contributions to the expansion and refinement of the new mathematical apparatus.[4] Paul Dirac, in his authoritative work on quantum theory, gave Fock credit for "the most convenient representation for describing states of the harmonic oscillator, a cornerstone of the theory of radiation."[5]

In 1938, Einstein, Leopold Infeld, and Banesh Hoffmann had reported a new discovery related to the general theory of relativity that, in Hoffmann's words, "enhanced its already extraordinary beauty and revealed an aspect of it unmatched by other theories."[6] Based on elaborate and intricate calculations, they had found that a Newtonian approximation of the equations of motion could be derived from the gravitational field equations.[7] The effect was to prove that, in the general theory of relativity, the principle of inertia did not have an independent meaning but was reducible in its entirety to the laws that identified the gravitational field by its dependence on the position and motion of material bodies.[8] The final report was so extensive and involved that Einstein decided to publish it in skeletal form. In the same year, Fock, working independently and relying on different mathematical operations, made an identical discovery and was able to present his procedures and findings in a compact paper.* After Friedmann's epochal suggestions on the expanding universe, Fock's discovery was the most important contribution to the general theory of rela-

---

*Fock, "O dvizhenii konechnykh mass." Fock saw the Einstein-Infeld-Hoffmann paper for the first time when he read the page proofs of his contribution, to which he added a postscript explaining the distinctive features of the two approaches. Emphasizing the relative simplicity of his approach, he concluded that each approach had a special strength. He noted that, after having read the published paper of Einstein and his associates, he had made no changes in his study.

tivity to come from the Soviet scientific community. In his paper, Fock hinted at what would become a sustained effort, in the 1950's, to show that Einstein's general theory of relativity was much in need of transformation from "a formal mathematical theory to a physical theory."[9]

Einstein claimed that his general theory of relativity was built on two principles: the principle of general relativity, identified as the general covariance of differential field equations, and the principle of the equivalence of inertial and gravitational masses. In Fock's view, developed in *The Theory of Space, Time, and Gravitation*, those principles did not form the basis of Einstein's theory of gravitation: the principle of general relativity did not have physical content, and the principle of the equivalence of inertia and gravitation was approximative and purely local. Both were heuristic devices in search of substantive truth. In Fock's words:

> "General relativity" does not exist in nature. . . . As a name for the Einsteinian theory of gravitation, the term "general relativity" can be viewed not in a physical but in a formal-mathematical sense—in the sense of "general covariance." The physical essence of Einstein's theory is not in its covariant form but in its presentation and exact mathematical formulation of the close ties between the gravitational field and the geometry of space and time. The idea that geometry may depend on physical processes occurring in space and time was known to B. Riemann, but it was only Einstein who established that the phenomenon of gravitation plays a decisive role here and presented exactly formulated mathematical ties between gravitation and geometry. This formula is the essence of Einstein's theory, the work of a genius.[10]

Since the general theory of relativity, as Fock viewed it, did not rest on the principle of covariance, it could not be considered an extension of the special theory of relativity. Only if interpreted as a concrete body of knowledge held together by physical principles, rather than by mathematical symbolism of the most abstract kind, could the general theory of relativity be identified as a true scientific interpretation of the gravitational field.[11] The persuasive power and logic of Fock's argumentation and the skeptical attitude of Marxist circles toward the idea of relativity prompted many supporters of Einstein to look for a more appropriate name for his leading creation.

As Fock envisaged it, Einstein's theory of gravitation embraced two ideas: the idea of the unity of space and time in a single four-dimensional manifold of infinite metric (as postulated by the special theory of relativity in 1905) and the idea of the unity of space-time and gravitation (as postulated in the general theory of relativity in 1916). He did not try to downgrade Einstein's general theory of relativity; all he wanted to do was to shift the pivotal importance from one set of Einstein's principles to another set. By placing the primary emphasis on the formal-mathematical aspects of his theory, Einstein, according to Fock, did not recognize the true greatness of his achievement. Like Nietzsche's Wagner, he was a genius who misunderstood himself.

Fock's proposed transfer of emphasis influenced his followers to argue against the need for assigning the observer a role in physical and philosophical inquiry. That suggestion was clearly intended to keep the subjective element out of physical analysis and to avoid any challenge to the objective moorings of Leninist epistemology. Iu. B. Molchanov, a noted Marxist philosopher of science, found no difficulty in eliminating the subject (observer) from relativistic physics, but he thought that in quantum mechanics the problem was much more complicated.[12]

Referring specifically to the principle of general relativity, Ia. P. Terletskii supported Fock's views. He thought that the principle of relativity was not the basic substance of the theory of relativity, and that the principle of the independence of the laws of physics from the choice of inertial systems of coordinates was only a precondition for establishing geometric links between space and time, the principal concern of the theory of relativity.[13] The philosopher Kh. M. Fataliev was quick to note that the main content of the general theory of relativity consisted, not of the covariance principle, but of the propositions that the attributes of space-time depended on the concentration and motion of masses, and that the metric depended on the gravitational field.[14] Fock received welcome support from A. Z. Petrov, a leading expert in post-Einsteinian advances in the mathematical apparatus of the theory of relativity. The general theory of relativity, according to Petrov, was "the theory of the dependence of space-time on the motion and distribution of matter in the broadest meaning of the term."[15] Its name should be "the general theory of gravitation" or "the theory of the gravitational field."[16] In his view, the true essence of the general theory of relativity as a general theory of gravitation was best described in terms of a space-time continuum, along the lines of the Riemann manifold—and not, as in the Einsteinian formulation, as a combination of the principle of the equivalence of gravitation and inertia and the principle of the general covariance of field equations, according to which "the general laws of nature are expressed in equations that hold true for all systems of coordinates."[17]

To give his reshuffling of the two sets of principles dramatic effect, Fock sought help from a paradoxical comparison:

The name of "the general theory of relativity" is similar to the name of "the West Indies" in the sense that both are mistakes made by their discoverers. The label "the general theory of relativity," however, is more dangerous than the label "the West Indies": at the present time, no one is likely to place the West Indies in India, but many still believe that the concept of relativity is the essence of the general theory of relativity. To avoid this kind of misunderstanding, it is better to use a label that does not invite arbitrary (or erroneous) interpretations. For this reason, we shall call Einstein's theory of 1915 the "chrono-geometric theory of gravitation," or simply, the "theory of gravitation."[18]

Fock professed that his effort to recast the general theory of relativity did not diminish his "greatest respect" for Einstein's genius and preeminent place in the annals of science.[19] He reminded his readers that "the entire theory of atomic energy [was] based exclusively on the correlation of mass and energy as postulated by the theory of relativity."[20] It was obvious, however, that his major ambition was to free the theory of relativity of all ideas opposed to the materialistic footing of Marxist philosophy and ideology. Soon after the publication of Fock's book, the leading physicists at the Academy of Sciences called an informal meeting to discuss its theoretical orientation and general merits. Leopold Infeld, who had coauthored *The Evolution of Physics* with Einstein, happened to be in Moscow at the time and was invited to take part in the discussion. The main speakers—Infeld, Landau, Tamm, and Ginzburg—thought that Fock's revisions of the general theory of relativity were both misdirected and unnecessary. But Infeld's general impression, that "all the Soviet physicists were opposed to Fock's understanding of relativity theory," was clearly off the mark.[21]

It must have come as something of a shock to him, in the circumstances, to learn that Fock's ideas had found some support in the West. G. J. Whitrow, for example, agreed with Fock that the general theory of relativity was basically a theory of gravitation and proposed that it should accordingly be called "Einstein's gravitational theory." He thought that since Einstein's persistent effort to build a unified field theory on the principle of the general theory of relativity had ended "in failure," there was no reason to assume that the general theory—unlike the special theory—was true.*

But in the West, as in the Soviet Union, there were also many detractors. John C. Graves, an American student of gravitation, acknowledged the correctness of Fock's emphasis on "the fundamental difference" between the special theory of relativity as a theory of flat space-time and the general theory of

---

*Whitrow identified Fock as one of the group of eminent relativists who had some doubts about the universality of the principle of the equivalence of inertial and gravitational masses: "Academician Vladimir Fock of Leningrad accepts the validity of the principle for weak uniform fields and slow motion . . . but has severely criticized Einstein for giving the principle 'a widened interpretation by taking it to imply the indistinguishability of fields of gravitation and acceleration and asserting that from the point of view of the principle it is as impermissible to speak of absolute acceleration as it is to speak of absolute velocity'" (*Time, Gravitation and the Universe*, pp. 119–20; Whitrow is here quoting from Fock's *Theory of Space*).

relativity as a theory of curved space-time. But he saw no justification for Fock's rejection of the label "relativity" as a quintessential description of Einstein's theory. He argued that the general theory of relativity was much more universal than Fock was ready to admit.[22] A similar criticism of Fock's position came from the Soviet philosopher M. V. Mostepanenko, who wrote that the theory of relativity was much more than a theory of space, time, and gravitation—that it was the foundation for the full range of problems treated by modern physics. Despite these and similar criticisms, Fock was widely acclaimed for his attempt to give firmer footing to the general theory of relativity.

V. I. Sviderskii, in one of the first reviews of *The Theory of Space, Time, and Gravitation*, praised Fock's effort to separate Einstein's theory of gravitation from the "subjective" leanings of the principles closely affiliated with the notion of general relativity. He recognized the study as the first basic work on the general theory of relativity to appear in decades, and as a welcome synthesis of Einstein's scientific ideas and Marxist dialectics.[23]

Fock's studies stimulated a vigorous involvement of Soviet scholars in a reexamination of the foundations of relativistic physics, especially the theory of the gravitational field. In 1958, Sviderskii published *Space and Time*, a widely read book representing the first effort by a Soviet scholar to offer a historical survey and a synthesis of Marxist views on space-time. The book offered a tightly woven discussion of the absolute and relative, continuous and discrete, and finite and infinite attributes of space and time. In one notable respect, it constituted a major departure from the work of Stalinist philosophers of science: it dwelt more extensively on the strengths of the Soviet studies of space and time than on the weaknesses of Western studies. The book reflected the fluidity and vacillation of the views of Soviet philosophers on relativistic cosmology at the end of the 1950's. In general, however, it claimed a complete unity of the theory of relativity and dialectical materialism.[24]

Following Fock, A. D. Aleksandrov, in a paper presented at the Moscow Conference on the Philosophical Problems of Modern Natural Science (October 8, 1958), contended that Einstein's principle of relativity was an epistemological rather than a physical notion, and that the principle of general covariance was a "mathematical requirement of heuristic nature, rather than a law of nature." Space-time, he argued, was an absolute reality, not a relative phenomenon.[25] In his words: "As the only absolute form of the existence of matter, space-time is the main and primary object of the theory of relativity; it is not a relative manifestation of phenomena viewed through distinct frames of reference."[26] In a later rendition of the same paper, Aleksandrov sought help from John Synge, who stated in his *Relativity: The General Theory* (1960): "The geometrical way of looking at space-time comes directly from Minkowski. He protested against the use of the word 'relativity' to describe a theory based on an 'absolute' space-time."[27] Aleksandrov agreed with Fock that "the general theory of gravitation" was the most appropriate name for the general theory of relativity, and that Einstein deserved credit for discovering a new world of cosmic relations.

The essence of the general theory of relativity, as Aleksandrov saw it, was neither in the general principle of relativity nor in the arbitrary choice of coordinate systems, but in the inner structure of the space-time continuum.[28] The general theory differed from the special theory, not in elevating the application of coordinate systems to higher levels of geometry, but in providing a more elaborate analysis of the structure of space-time.[29]

Aleksandrov, no less than Fock, wanted to take relativism out of Einstein's physics. By relativism he meant generally a purely subjective and agnostic view of both the physical and the cultural world. He admitted, however, that the relative, as Einstein used the word, presented only a specific manifestation of the "absolute," or the "invariant."[30] As the eminent physicist Iakov Frenkel had stated in 1930, the "theory of absolute laws " summed up Einstein's physical ideas more adequately than "the theory of relativity." Aleksandrov did not forget to remind his readers that an identical suggestion came from Hermann Minkowski, the chief mathematical sculptor of the special theory of relativity. Another critic observed that only a notion of space-time as absolute reality could meet the scientific standards set forth by dialectical materialism.[31] Aleksandrov, like Fock, was guided by the idea that eliminating the idea of relativity would make the role of observer in Einstein's theory obsolete and would help reinforce the objective side of the methodological equation.

Still, Aleksandrov, like Fock, never failed to emphasize the gigantic proportions of Einstein's scientific legacy and the fundamental congruence of the theory of relativity with dialectical materialism.[32] In general, he was interested more in exploring the points of contact between the two than in conducting a war on real or imagined manifestations of idealistic philosophy and capitalist ideology. One cannot help feeling that he was interested above all in removing what he considered artificial and distracting barriers from the passageways linking Marxist philosophy with Einstein's physical theory.

Aleksandrov expressed a strong interest in bringing an end to the repeated philosophical claims that Einstein's interpretation of his own physical theory was marred by a deep-seated positivist bias. He noted in one of his better-known papers:

We do not have the slightest inclination to consider Einstein's theory flawed, positivist, and idealistic. We are categorically opposed to the assertions made by a number of authors that Einstein's theory is "a Machian theory of relativity."... The theory of relativity is, first, a theory of space-time, and, second, like every thoroughly scientific theory, a body of knowledge that is neither reactionary nor idealistic. This does not mean that we deny that there is a need for giving this theory more depth and for correcting positivist and idealistic interpretations encountered in the literature.[33]

Aleksandrov's argumentation displayed a deep-seated ideological motive. An active and dedicated member of the Communist Party from the time he joined in 1951, he was widely known as an energetic crusader against ideological deviations in mathematics and physics. In painting a picture of the theory of relativity as a "theory of the structure of absolute space-time," he depended heavily on scientific metaphors and logical arguments operating under the shadow of a philosophical umbrella held together by Lenin's theory of knowledge, the main source of ammunition for the relentless attacks on the strategic premises of physical idealism.

Aleksandrov and Fock thought that it was the duty of both the scientific community and the philosophical community to establish harmonious relations between the theory of relativity and dialectical materialism. To that end, it was necessary, in their opinion, to expand the capacity of dialectical materialism to absorb new scientific ideas and to modify the theory of relativity by reducing its excessive dependence on mathematical formalism devoid of physical content. To many alert contemporaries, that undertaking appeared as a new effort to create a distinct Soviet theory of relativity. The time, however, was decidedly turning against the efforts to create unique varieties of Soviet science.

In their reinterpretation of the primary principles of the general theory of relativity, Fock and Aleksandrov wanted to achieve two goals: to reduce some of the key differences between Einstein's general outlook and dialectical materialism and to shift the emphasis from philosophical to scientific grounds. When they expressed strong distaste for relativity as a scientific concept and for mathematical operations detached from physical content, they acted as loyal supporters of dialectical materialism and official Soviet ideology. In *The Theory of Space, Time, and Gravitation*, Fock made a point of acknowledging his philosophical indebtedness to Lenin, from whom he borrowed selected ideas and philosophical metaphors but not the fervent manner of argumentation. Despite that declaration, the massive study showed no direct philosophical involvement.

In a way, Fock and Aleksandrov continued the Maksimov-Kol'man-Kuznetsov tradition in choosing Einsteinian themes for critical reassessment. Both groups criticized Einstein's reliance on extensive mathematical formalism, his introduction of the observer as a mandatory methodological device, his emphasis on relativity as a sovereign approach to the scientific study of physical reality, and his assumption of the equivalence of the Ptolemaic and Copernican systems. There was, however, a major difference between the two orientations. The Maksimov group expressed its critical arguments in philosophical idiom and harsh innuendoes. Fock and Aleksandrov, by contrast, relied heavily on carefully constructed and tactfully handled scientific arguments.

Volume 31 of the second edition of the *Great Soviet Encyclopedia* (1955), still tightly controlled by political authorities and ideological supervisors, produced unmistakable proof that the Communist Party favored the Fock-Aleksandrov orientation. Fock was asked to write a general physical survey of the theory of relativity, and Aleksandrov to provide a philosophical commentary.[34] Fock asserted explicitly that Einstein's extension of the principle of relativity to accelerated motion was incorrect, and that the Soviet interpretation of space-time (which is to say, his and Aleksandrov's interpretation) as the foundation of the modern theory of gravitation represented a higher stage in the evolution of Einstein's scientific legacy.[35] He also stated that the new theory of space, time, and gravitation was far more complicated than Einstein had anticipated. While recognizing the scientific preeminence of Einstein's non-Euclidean view of cosmic space, he did not fail to assign Lobachevskii a prominent position in the pantheon of illustrious pioneers in the long evolution of ideas crowned by the union of the theory of space-time and the theory of gravitation.[36] There was no doubt in Fock's mind that the greatest and decisive contribution to that union came from Einstein.

It is clear why the Communist authorities would have seen the Fock-Aleksandrov orientation as coming closer to the Marxist dream of an original Soviet theory of relativity than any other orientation then on the horizon. It subordinated the relative method of modern physics to the absolute laws of official Soviet ideology. It heeded the Leninist warning that under no circumstances should mathematical operations be allowed to "dissolve" the real content of physical nature and become a gateway to idealism. It satisfied Russian national pride by assigning Lobachevskii's geometry a much more active role in the general theory of relativity than was customary in the West. It emphasized the close ties between the physical notion of space-time and the law of causality as the chief mechanism of the scientific study of nature. And it was presented by a man of high standing in the scientific community. Omel'ianovskii, the doyen of the Marxist philosophers of science, worked carefully on supplying the new orientation with philosophical metaphors adorned with citations from scientific literature.[37] Serious Western attention to Fock's monumental study could not but strengthen his position at home. Unaware of Fock's work, Synge had published an equally monumental study expressing views on the general covariance and equivalence principles very similar to Fock's.

As viewed by Marxist philosophers, space and time were both absolute and relative. They were absolute insofar as they were "universal conditions" of all forms of existence and were independent of changes in concrete forms of matter in motion. They were relative insofar as, in their concrete nature, they were conditioned by the physical properties of matter in motion.[38] Whether viewed as absolute realities or relative phenomena, they were easily related to the materialistic slant of Marxist thought. The dialectics of absolute and relative attributes of space-time, the essence of the Fock-Aleksandrov orientation, was offered as a compromise between Marxist ideological structures and Einstein's physical principles. Particularly as Aleksandrov saw it, the basic emphasis was on space-time as an absolute physical notion; he recognized the relative aspect only to assign it a minor position in his theoretical scheme.

In 1961, A. Z. Petrov published *Space-Time and Matter: An Elementary Survey of Present-Day Relativity,* a short work whose clear intent was to endorse and popularize the Fock-Aleksandrov line of reasoning.[39] Both Fock and Aleksandrov published (or republished) a series of articles on the same theme in *Questions of Philosophy,* and in various widely publicized collections of essays on the relations of philosophy to the natural sciences.[40] The fourth volume of the *Philosophical Encyclopedia,* released in 1967, featured Aleksandrov's article on the theory of relativity.[41] The largest and most widely noted national conferences on the theory of relativity, held in Kiev in 1964 and 1966, dealt almost exclusively with the space-time and gravitation complex and its place in modern cosmology.[42] Although other orientations were represented at the conferences, there could have been no mistake in anyone's mind, from the wide airing given to the Fock-Aleksandrov views, that they were the preemptive favorite. Nor would there have been any misunderstanding about the far-reaching support for that orientation among the leading articulators of Marxist philosophy, such as Omel'ianovskii, Mostepanenko, and Fedoseev. Those philosophers felt that at long last a Marxist theory of relativity had reached full flower. The supporters of the Fock-Aleksandrov orientation as a new Marxist theory could now claim that by dethroning the general principle of relativity, Soviet scientists were in a position to avoid Einstein's "error" of accepting the Ptolemaic and Copernican astronomical views as equivalent coordinate systems.

Despite the wide initial publicity given to the Fock-Aleksandrov views, particularly in Marxist circles, no official or other effort was made to suppress the supporters of competing orientations. Indeed, one has the distinct impression that the views close to Einstein's configuration of relativistic principles contin-ued to be prevalent. None of the numerous proposals for major or minor revi-sions of the scientific body of the theory of relativity detracted from the univer-sal recognition of the brilliance of Einstein's genius and the extraordinary power of his scientific arguments. In 1967, Aleksandrov admitted that Einstein built the general theory of relativity, as a theory of space-time continuum and gravitation, on such a solid foundation that no new theory was likely to chal-lenge its scientific preeminence in the foreseeable future. Fock and Aleksan-drov stopped referring to their orientation as a distinct Soviet or Marxist the-ory of gravitation; they considered their work a Soviet contribution to the re-finement of the granite structure of scientific principles formulated by the genius of Einstein.

In 1967, the country celebrated the fiftieth anniversary of the October Revolution. The Academy of Sciences contributed to the festivities by prepar-ing detailed progress reports on the national effort in the advancement of sci-entific knowledge. For full coverage, physics alone required two bulky vol-umes. The chapter on the general theory of relativity, carefully crafted by A. Z. Petrov, created the clear impression that the Fock-Aleksandrov recasting of Einstein's theoretical principles represented a high point in the evolution of the general theory of relativity. Petrov was careful to note, however, that systematic work on the verification, elaboration, and refinement of the general theory of relativity had just begun.[43]

From 1959 to the beginning of perestroika in the mid-1980's, textbooks and general surveys of Marxist philosophy showed a tendency to pay more atten-tion to the notion of space-time than to the cluster of principles directly related to the idea of general relativity. S. Sh. Avaliani, a Georgian philosopher of sci-ence, justified the practice by claiming that, after all, "the study of the philo-sophical aspects of space and time has always been the fundamental problem of the philosophy of natural science."[44] There was also a pronounced tendency to consider the Einsteinian notion of space-time an integral part of the Marxist philosophy of science. So much was this conception taken for granted that no need was felt to make extensive personal references to Einstein. The section on space-time in *The Foundations of Marxist Philosophy* (1959), edited by F. V. Konstantinov and others, made only one personal reference to Einstein. The corresponding section in *Dialectical and Historical Materialism* (1985), edited by A. P. Sheputin, was equally parsimonious.[45]

In adhering to the official version of dialectical materialism, these and similar textbooks dispensed with Einstein's "relativity" as an idea incompatible with the Marxist philosophy of science. In specialized scientific and philo-sophical literature, however, many scholars took the opposite road: they tried to find a safe, though somewhat restricted, place for relativity in both physics and philosophy. L. B. Bazhenov went so far as to assert that "most physicists" did not accept Fock's critique of the general principle of relativity.[46] From all appearances, that was a fair assessment of the situation.

For all the Fock-Aleksandrov orientation's evident success and impressive arguments, it encountered critical difficulties. It appeared at a time when the scientific community, with the government's tacit but cautious agreement, favored abandoning the search for distinct Soviet sciences, a relic of the Stalin era. The new trend, at first not clearly expressed, favored a conscious search for the unity of Western and Soviet science, showed increasing tolerance toward the diversity of theoretical interpretations, and allowed for a much broader flexibility of ideological imperatives.

There were unambiguous signs from the start, for example, that few leading physicists accepted the Fock-Aleksandrov orientation. Landau and Lifshitz, in their article on gravitation for the second edition of the *Great Soviet Encyclopedia* (vol. 43, 1956), chose to make no mention of Fock's views on the general theory of relativity.[47] Fock, undoubtedly the most gifted Soviet mathematical physicist, waited until 1960 to receive the Lenin Prize for scientific achievement. There was an element of irony in the prize: he did not earn it for *The Theory of Space, Time, and Gravitation*, but for a 1957 collection of articles on quantum mechanics originally published in 1928–37. Obviously, the physicists critical of Fock's ideas on the general theory of relativity had the last word in the committee allocating the most prestigious national prize.

For a decade or more after Fock published his masterwork, his star rose in the scientific community. But beginning in 1967, the Fock-Aleksandrov orientation experienced a loss of momentum. In the mid-1960's, Fock made the unexpected decision to stop writing about the theory of relativity. Perhaps he took to heart two developments in the country: the growing opposition to his views not only among physicists but also among philosophers; and the unpublicized decision of the Communist Party hierarchy to discourage the search for a Marxist theory of relativity. The starkest measure of his fading influence is the *Great Soviet Encyclopedia*. In the second edition, published in the mid-1950's, the article on the theory of relativity had been written by Fock and was generally limited to an exposition of his ideas. In the entry for the third edition, published in 1974, all traces of Fock's interpretation had disappeared save for a strong emphasis on space-time and the inclusion of his name among the authors cited at the end.[48]

The Search for a Unified Field Theory

The physicists who followed Ivanenko in trying to combine relativistic and nonrelativistic theories into a unified field theory did not constitute a distinct school as such, nor did they show much interest in epistemological questions. They depended on imagination no less than on an intimate familiarity with the newest developments in physical theory. They operated on the assumption that both quantum mechanics and the theory of relativity had much work to do in straightening out their inner structures and epistemological underpinnings before they could take significant steps toward unity. As a rule, they were well grounded in and contributed to both grand branches of modern physics.

As a first major step toward a unified system of theoretical thought and methodological principles, Ivanenko recommended close and careful scrutiny of the growing body of specific arguments in favor of making Einstein's explanation of gravitation the central factor in the study of the microphysical universe.[49] Work on a unified field theory, he wrote in 1959, must concentrate on the principles integrating the "form" of space-time shared by the quantum explanation of nucleons and the general theory of cosmology.[50] He also thought that a nonlinear interpretation of quantum-mechanical equations—an effort that had engaged the attention of the French physicists Louis Victor de Broglie and J.-P. Vigier—presented the most logical first step in the search for the theoretical unity of quantum mechanics and the general theory of relativity, the equations of the latter having been "essentially nonlinear."[51]

Ivanenko suggested that physicists in quest of a unified field theory would find it advantageous to search for precise answers to two major questions: are different parts of the universe dominated by different particles and antiparticles, and was there a need for a new cosmology that would take into account the possibility of some parts of the universe expanding and others contracting?[52] A positive answer to the first question, he noted, would require a major recasting of quantum-mechanical methodology, a positive answer to the second would require a major recasting of relativistic cosmology. In their search for a unified field theory, physicists must consider the possibility that the general theory of relativity did not necessarily apply to the universe as a whole but only to parts of it. In Ivanenko's opinion, the space curvature of individual parts of the universe might be either positive or negative; he even ventured a guess that our part of the universe, because of its low density, was characterized by "negative curvature of the Lobachevskii type."[53] Einstein's theory of general relativity had no room whatever for Lobachevskii's non-Euclidean geometry.

Ivanenko's unorthodox ideas did not stop him from concluding that "the great Einsteinian conception of geometricized gravitation [would] continue to serve as a reliable basis for, and a stimulating source of, inquiries into the total picture of physical reality."[54] To increase its effectiveness, however, the general theory of relativity was in much need of improvements in its mathematical apparatus and a widened scope of experimental verification.[55]

Ivanenko did not do much active work in physics after the 1950's. In the years of the thaw, he was more effective as a competent and discerning commentator on current Western and Soviet efforts to enrich relativistic cosmology than as a master of new theoretical bridges spanning the theory of relativity and quantum mechanics. What he lacked in deeper concentration on the burning questions of physics was compensated for by his unusual ability to keep a watchful eye on up-to-date developments in the area where quantum mechanics and the theory of relativity came into close contact.

Ivanenko never swerved from his position. As late as 1979, he still insisted that the gravitation energy problem could not be regarded as definitively solved; it continued to be a "difficulty" requiring further clarification, "maybe even by means of some extension of the principles or formalism of general relativity."[56] He noted that relations between cosmological and atomic phenomena continued to be veiled in mystery. In mathematical theories of gauge fields and torsion, he saw sound efforts to open new paths to a unified field theory not issuing from the general theory of relativity. He also showed that Soviet physicists were not only aware of these developments, but made contributions to them as well.

Ivanenko did not share the view of most experts that the immediate future would be dominated by fine tuning and enriching the general theory. Ready to acknowledge the gigantic proportions of Einstein's scientific achievement, he felt constrained to point out that he found "some serious difficulties" in it. To relieve his uneasiness, he recalled that Newton's "ideas on absolute space and time, the axiomatic laws of motion, [and] the phenomenological law of gravitation" had drawn their share of criticism without bringing their creator any loss of honor within the scientific community.[57] Ivanenko was aware of the golden rule in the history of science that without criticism there was no science in the first place.

The Divided Views in Cosmology

Cosmology, a branch of astronomy in which the hard facts of science flowed together with philosophical imagery and poetic creations, invited more controversies than any other science. A. L. Zel'manov described it as a discipline made of three kinds of ingredients: empirical data (mainly of an observational nature), theory and laws of physics, and philosophical principles.[58] The growth and consolidation of the Einsteinian legacy was easily the mainstream of Soviet cosmology.

Einstein's original equations called for a static universe whose curvature was related to the mean density of matter in space and to an improvised cosmological principle. The new, and generally accepted, relativistic cosmology, as formulated by Georges Lemaître, combined Friedmann's idea of an expanding universe with Hubble's red-shift observations. The universe, according to the Lemaître model, is nonstatic, homogeneous, and isotropic. The emergence of relativistic cosmology, according to Akbar Tursunov, marked "the formation of cosmology as an independent discipline."[59] No modern discovery, in Ginzburg's view, contradicted the laws of physics supporting relativistic cosmology.[60]

We need not rehearse here in full the fate of astronomy at the hands of the Stalinists. As noted earlier, the entire field underwent a series of ideological expurgations that completely dislodged several national traditions in the science. In the end, a large proportion of the country's astronomers, perhaps as many as 20 percent, were imprisoned.[61] In the circumstances, most astronomers who wrote on the subject opted to ignore Friedmann's cosmological ideas. In 1940, M. S. Eigenson, in an article on the current state of cosmological studies, referred to the red-shift and to relativistic cosmology but did not mention Friedmann.[62] V. N. Petrov, in an article published the same year, risked a confrontation with the authorities by giving an optimistic and generally favorable account of the influence of the general theory of relativity on modern astronomy, but he, too, chose to ignore Friedmann's contribution.[63]

Stalin's death did not bring immediate recognition to Friedmann's ideas on either the scientific or the philosophical-ideological level. It is not all that surprising that the second edition of the *Great Soviet Encyclopedia*'s entry for cosmology (1953) should say nothing about his contribution to the triumph of the idea of the expanding universe, but spoke only of his confirmation of Lobachevskii's notion of the infinity of cosmic space and his "idealistic responses" to various questions of philosophical relevance.[64] But to find that he was still effectively a nonperson as late as 1960 is truly stunning. He was not mentioned in the monumental *Astronomy in the USSR During Forty Years, 1917–1957*, published that year for the specific purpose of demonstrating the great Soviet achievements in the astronomical sciences. In fact, that work did not cover cosmology at all. The editor apologized for the omission, placing the blame on "a series of causes."[65] The astronomer B. A. Vorontsov-Vel'iaminov was strangely silent too. Although he devoted several chapters to cosmology in his *Essays in the History of Astronomy in the USSR* (also published in 1960), he did not make a single reference to Friedmann.[66]

But Friedmann did eventually get the acclaim he deserved. As one of the post-Stalin thaw's many consequences, it ultimately opened the gates for a reconsideration of his cosmological contributions. Presumably, his widespread recognition in the West helped to move the Soviet scientists to take another—and more favorable—look at their pioneering compatriot, the more so because the name Friedmann was prominently displayed in nearly all recent Western biographies of Einstein and in general surveys of the rise of modern cosmology. In the West, Friedmann was firmly established as one of the most original and resourceful founders of relativistic cosmology.[67]

The full integration of the theory of the expanding universe into Soviet cosmological thought came in 1963, a year in which *Advances in the Physical Sciences* devoted an entire issue to Friedmann and the USSR Academy of Sciences held a jubilee session in his honor.[68] In 1966, the "Classics of Science," a series of reprints of experimental and theoretical studies responsible for the progress of modern science, published a volume of his most noted studies.[69] One of his key cosmological papers, originally published in the German journal *Zeitschrift für Physik*, was translated into Russian for the first time. The volume also included glowing reviews of Friedmann's contributions by Kapitsa, Fock, and Zel'dovich. In a way, these comments were pleas for a broader and firmer integration of Friedmann's ideas into Soviet astronomical thought. The tome on astronomy in the Academy of Sciences' ten-volume history of science in the USSR offered a special chapter on cosmology that went a long way toward recognizing Friedmann's contribution to astronomical studies, the general theory of relativity, and the prestige of Soviet achievements. In 1967, *Relativistic Astrophysics*, a grand and sophisticated survey of recent developments in physical approaches to the structure and the evolution of the universe, described his hypothesis as one of the greatest triumphs of twentieth-century science.[70]

Even A. Z. Petrov, that passionate partisan of Fock and Aleksandrov, spared no words in exalting Friedmann's achievement. In *The Development of Soviet Physics: 1917–1967*, a collective work observing the fiftieth anniversary of the October Revolution, he wrote:

> In his widely known study published in 1922, the Soviet scholar A. A. Friedmann presented a beautiful theory of general-relativistic cosmology. This study gave, in essence, a sketch of the basic ideas of cosmology: a hypothesis of the homogeneous distribution of matter in space and, as a consequence, of the isotropic nature of space-time. . . . This theory is particularly important because it offers a generally accurate explanation of the fundamental significance of the red-shift effect. The solution of field equations based on these hypothe-

ses—the Friedmann solution—has become a model for all cosmological theories.[71]

Petrov claimed that Friedmann's contribution was more in suggesting a sound path to the continued growth of cosmology than in making that science a complete and integrated body of knowledge. He went on to imply that the future of Einstein's scientific legacy in the Soviet Union was in the diversity of scientific views and in unburdening cosmology of rigid ideological assignments. Petrov's own major scientific contribution was in preparing an invariant classification of all possible gravitational fields and in participating in the development of an invariant-group method for the solution of certain categories of Einstein's field equations.

The new enthusiasm for Friedmann's ideas was part of a larger development: the rapidly growing popularity of cosmological views based on the general theory of relativity. In Volume 13 of its third edition, the *Great Soviet Encyclopedia* recognized that among the proliferating new cosmological theories, preeminence still belonged to the orientation built on Einstein's general theory of relativity and on the model of a homogeneous, isotropic, and nonstatic universe.[72] Although a majority of Soviet scientists continued to support the idea of cosmic isotropy, even at that late date (1973), there were clear rumblings about the research potentials of the notion of cosmic anisotropy. Marxist philosophers, in particular, tried to draw a line between "absolute" and "relative" attributes of the space-time continuum.[73]

Despite the favorable reception of relativistic cosmology, most experts thought that many questions—for example, the origin of quasars and pulsars—required more precise explanation and a new look at its Einsteinian basis. As early as 1955, A. L. Zel'manov had claimed that the widespread tendency to extend the physical attributes of the Metagalaxy to the entire universe and to all phases of its evolution was the main reason for the conflicting ideas on the expanding universe—a qualification that made it possible to accept the idea of a universe that was infinite and yet expanding.[74] By then, some number of cosmologists had come to recognize that the idea of an expanding universe did not necessarily imply that all the galaxies were involved.[75] But even with an increasing consensus in favor of that notion, most cosmologists still saw the future of their discipline in the elaboration and refinement of Einstein's ideas.

In its gradual rise to preeminence in Soviet astronomical thought, relativistic cosmology participated in a long but relatively quiet war with the Biurakan orientation, named after a leading astrophysical observatory in Armenia. Founded and dominated by V. A. Ambartsumian, a member of the USSR

Academy of Sciences, the Biurakan school represented a bold and comprehensive effort to blend the historical orientation of dialectical materialism with the phenomenal discoveries of modern astronomy. Ambartsumian, for example, allowed for the possibility of black holes, the grand riddle of relativistic cosmology, but he was quick to add that their existence did not explain the activities of the so-called nuclei of galaxies. He did not go beyond the cursory statement that "eruptions and fragmentations of superheavy bodies of extremely high density" were the natural sources of galactic activities, supported by "all known facts."

Where the big-bang theory, firmly built into relativistic cosmology, had room for only one explosion in the evolution of the galaxies, and the steady-state theory left room for none at all, the Biurakan theory viewed the evolution of the galaxies as an unceasing succession of explosions. It favored the "unsteady" state of galaxies. Repeated explosions in the superdense nuclei of galaxies produced entire "associations" of cosmic bodies. In contrast to the prevalent view among modern cosmologists, Ambartsumian saw the evolution of the universe as a transition from superdense and massive bodies to objects of lesser density and massiveness. He opposed the condensation theory of cosmic evolution. In *The Structure and Evolution of the Universe*, Zel'dovich and Novikov gave Ambartsumian credit for anticipating analogous activities in galactic nuclei and quasars, but thought that most of his challenges to the dominant view in relativistic cosmology lacked sufficient factual support.[76]

Friedmann's theory, in Ambartsumian's view, was correct but limited: it explained only a fraction of the key cosmological problems.[77] The models of the universe constructed on the basis of Friedmann's theory, he wrote, represented only the first steps of a mathematical description of the expanding Metagalaxy. In his opinion, those models were too narrow and too simple to satisfy the momentum of modern cosmological advances.[78] He argued vehemently against making cosmology a branch of "applied physics."[79] The forthcoming scientific revolution would be sparked by a new observational astronomy and by a cosmology based on the anisotropic view of the Metagalaxy, which required a new and more complex mathematical apparatus. There was a good chance, he thought, that the new cosmology would be guided by the idea that all parts of the universe were not governed by the same laws of nature. Ambartsumian boasted that his "evolutionary" cosmology anticipated such recent discoveries of cosmic phenomena as quasars, pulsars, and black holes. Its current state of development, he thought, required that general astronomy be devoted primarily to a search for "empirical regularities," a stage preceding the concern with high theory.[80]

Ambartsumian displayed a marked predisposition to accept all popular ideas in cosmology, but not until he modified them to fit his grand theory. He went so far as to claim that modern cosmology was in its entirety a relativistic discipline, even though his general theory contradicted Einstein's basic cosmological ideas and Friedmann's notion of the expanding universe. Despite his professed great respect for both Einstein and Friedmann, he made only a limited effort to explore their ideas.

Ambartsumian made the most of a belief in the full compatibility of his cosmic views with dialectical materialism. He went by the rule that "the evolution of the universe is cast within the framework of a struggle of dialectically contradictory tendencies."[81] On many occasions, he asserted that his strict adherence to the methodological and theoretical principles of dialectical materialism gave his studies sharper vision and more depth. Marxist ideologues, not surprisingly, appreciated the historical bent of his ideas and were lavish in their praise of them. Omel'ianovskii did not hesitate to present Friedmann's homogeneous and isotropic model of the universe as a mere prelude to the triumph of Ambartsumian's "extremely nonhomogeneous" model.[82] *Communist*, the theoretical organ of the Central Committee of the Soviet Communist Party, hailed Ambartsumian's cosmological theorizing as the triumph of a comprehensive effort to employ the dialectical method in the study of the historical dynamics of celestial bodies; in fact, it was thanks to his achievements that cosmology had become a dialectical science.[83] During the 1960's and 1970's, Ambartsumian (usually helped by his associate V. V. Kaziutinskii) contributed articles on the growth of Soviet astronomy to symposiums observing important dates in the history of the Soviet Union. In all his writings and addresses, he rarely failed to refer to the Biurakan conceptions as a triumph of Soviet cosmology.[84] He had the temerity to claim that his theory of the transition from the more dense to the less dense states of matter was as important as the development of quantum mechanics in the history of modern physics.*

Unlike the Marxist philosophers, the scientific community on the whole found Ambartsumian's cosmological notions laughable. Ginzburg, for example, cited the astronomer Iu. N. Efremov's biting comment on Ambartsumian's cosmological enthusiasm: "During the quarter of a century of its existence, the Biurakan conception of the formation of stars has not made a theo-

---

*Ambartsumian's research associate Kaziutinskki was asked to write the article on astronomy for the academy's grand enterprise, *The History of Philosophy in the USSR*. His selection was clearly calculated to create the impression that Ambartsumian's ideas represented the most promising current in Soviet astronomy. Kaziutinskii, however, did not try to show that the "orthodox" or "classical" (that is, pre-Ambartsumian) orientation was contrary to the principles of dialectical materialism (Evgrafov et al., eds., *Istoriia filosofii v SSSR*, vol. 5, part 1, pp. 625–33).

retical step forward, nor has it received observational confirmations or advanced new arguments against the classical conceptions. There is no need for it, no scientific arguments support it, and it is a topic of debate only on popular scientific and philosophical levels."[85] Ginzburg wanted a "materialistic philosophy" that imposed no limitations on the choice of models of the universe. The laws of cosmic evolution, he said, must be established on the basis of "astronomic observation and modern physics," not on philosophical—that is, ideological—grounds.[86] Pleading for "healthy conservatism," Ginzburg suggested that it would be most advantageous to stay with the current physical and astrophysical theory at least until its potential was fully realized.[87] Most specialists agreed that its peak of development was still to come.[88] Ginzburg admitted that the general relativistic orientation in cosmology should allow for a diversity of models of galactic evolution.

According to G. M. Idlis, the general theory of relativity did just that. It allowed for the construction of multiple cosmological models of the homogeneous and isotropic world, which might be presented as either finite or infinite and as either static or nonstatic.[89] In their total effect, cosmological orientations served as attractive vehicles for popularizing the most sublime achievements and challenges of modern science and for bringing new victories to the scientific worldview. Einstein's space-time manifold, gravitational fields, and description of matter, the stock in trade of modern cosmology, made the universe more complicated but also more harmonious and more intelligible. They combined to create the architectonic principles of unity overlaying the oceans of diverse thought.

Zel'dovich and Novikov spoke for a majority of the scientific community when they stated that the foundations of modern astronomy were laid by the general theory of relativity, which needed only further refinement and additional modes of application. "Until now," they observed, "no data has presented a serious challenge to the adequacy of the application of the theory of relativity to the entire universe."[90] By giving science a new view of the gravitational field, based on a non-Euclidean geometry and a relativistic interpretation of the space-time continuum, Einstein had given cosmology a lodestar and a path to inner unity.[91] His insights promised to bring together the powerful intellectual resources of the two reigning currents in modern physics, the theory of relativity and quantum mechanics. It made it possible for cosmology to become more than a study of local models of the universe.

M. V. Mostepanenko echoed the most popular opinion in both his own community of philosophers and the community of scientists when he asserted that no question of the structure of the universe could be asked and answered without the help of the theory of relativity.[92] The post-Stalinist unity of the the-

ory of relativity and Marxist philosophy had been achieved mainly by the retreat from dialectical materialism, the underpinning for the Stalinists' assertion of ultimate authority in the domain of science. Defenders of philosophical
orthodoxy like Maksimov and Kuznetsov, who had made the attack on the scientific side of the Einsteinian equation a sacred ideological duty, were silenced
and rapidly forgotten. Under pressure from the scientific community, they
ceased to search for a Marxist physics, their most cherished activity in the Stalin
age. Similarly, Ambartsumian's theory of continuous cosmic eruptions or
revolutions lost ground not only because of its lack of documentation but also
because of the swiftly declining interest within Marxist circles in creating a
Marxist cosmology or a Soviet cosmology, an unwelcome residue of Stalinist
ideals.

Freed of the straitjacket of philosophical orthodoxy, Soviet physicists threw
themselves into the study of the general theory of relativity as actively as any in
the world. Einsteinian scientific research proceeded in two major directions:
toward a reconsideration and realignment of primary concepts and principles
and toward the creation of a more modern and comprehensive mathematical
apparatus—some of it of topological vintage—transcending the limitations of
Einstein's and Hilbert's equations and the classical tools of tensor calculus. On
the substantive level, the most important work was done in the areas of relativistic cosmology and astrophysics. In a *Brief History of Time*, Stephen W. Hawking, Lucasian professor of mathematics at Cambridge University (a position
once held by Newton), mentioned Ia. Zel'dovich, A. Starobinskii, E. Lifshitz,
and I. Khalatnikov among the major contributors to the explanation of cosmological problems "from the big bang to black holes." He was particularly
impressed with their original ideas on the quantum-mechanical approaches to
black holes and on the relations of Friedmann-like models to the big-bang
theory.[93] Although each of the many attempts to modernize and enrich Einstein's ideas had its partisans, most experts thought that the Einsteinian paradigm was unscarred by the ravages of time and erosion and saw no threat to its
supremacy anywhere on the horizon. Work on the problems presented by the
inner logic of Einstein's legacy was the order of the day.

8 ▪

Einstein's Humanistic
Influence

The broadening of the scope of Einsteinian studies was one of the more important and interesting developments of the post-Stalinist thaw. The period saw a lively discussion and rich literature on the relations of science to aesthetic creativity, moral norms, and religious thought, much of it stimulated by a new exposure to Einstein's views on society and related matters. In the Stalin era, almost all writing about Einstein had been limited to scientific and epistemological messages. Very little was written on Einstein's character, sentiment, and general outlook as a unique mirror of the cultural values of the age. In fact, it was not until the late 1950's that a full-scale biography of Einstein was written by a Soviet writer.[1] The exploring of that side of Einstein is the subject of this chapter.

Broader Cultural Dynamics of Science

It is not too much to say that Einstein's writings and influence proved crucial in the post-Stalinist quest for fresh clarifications of a broader cultural base of science—of the relations of science to such grand systems of cultural expression and values as ethics, belles-lettres, and philosophical thought. In all these mammoth enterprises, Einstein supplied cogent suggestions for a humanistic scrutiny of the place of science in modern culture. One of the basic functions of the modern concern with the "science of science" was to undertake a systemic and comprehensive study of the full cultural matrix of science. In an obvious effort to strengthen the humanistic attributes of dialectical materialism, post-Stalinist philosophers of science were determined to go far beyond epistemological boundaries. They agreed with Nietzsche's dictum that "philosophy reduced to epistemology was a philosophy which cannot pass its own threshold."[2]

C. P. Snow's essay on "two cultures"—a plea for a more thorough and intimate interaction of scientific and nonscientific (humanistic) "cultures"—stimulated much favorable comment in the Soviet Union.[3] It reinforced the burgeoning effort to study the broader cultural matrix of science, especially the links between science and various forms of aesthetic expression. There was a sudden increase in both humanistic approaches to science and scientific approaches to humanistic philosophy. "Science and culture" became a popular theme of scholarly discussion. For the first time, the Soviet public could read in Russian translation Werner Heisenberg's musings on science and art and on their common striving for universality in the worlds of values and ideas.[4] It could also read about Kepler's excitement in detecting an aesthetic quality in "the harmony of spheres," Niels Bohr's discourses on science and art, Erwin Schrödinger's comments on science, art and play, and Wolfgang Pauli's meditations on emotionally charged symbolic images as preludes to scientific ideas. All these writings had a common refrain: the close cultural bonds of science and art. All provided ample evidence of a critical need for an extensive examination of the humanistic base of science.

That examination had been complicated by what was considered a brewing crisis in science, the reigning component of Soviet society and culture. According to an authoritative—and somewhat exaggerated—comment, Soviet science and technology "often cease to be integral parts of culture; very often they lock themselves in their own problems, methods and results, creating their own methodology, philosophy, and culture, which inevitably lead them to alienation and isolation from the general values of humanity and from culture as such."[5]

In the post-Stalin period, Soviet ideologues and scholars produced many comparative studies on differences and similarities between scientific and artistic cognition, in which the Stalinist idea of the cognitive superiority of science over art soon gave way to the notion of full parity. At the same time, interest in pure aesthetic values increased rapidly and became a major academic concern in the humanistic approach to art.

As might have been expected, Einstein's humanistic observations and aphorisms played a major role in the growing concern with the unity or complementarity of science and art. Einstein himself provided ample material for that discussion. He readily acknowledged that all his artistic leanings developed hand in hand with his "pursuit of science," and asserted that all the prominent scientists he knew viewed science and art as manifestations of the same creative process.[6]

It was especially toward the end of the post-Stalinist thaw that all this became a subject of study. Thanks in large part to changes in Marxist thought that recognized the active role of the subjective factor in scientific knowledge and ended the traditional overemphasis on ideologically guided epistemological approaches to artistic creativity, scientists, philosophers, and literary critics began busily exploring the secrets of intuition and imagination as creative mechanisms in both art and science.[7] They explored the scientific uses of such aesthetic qualities as symmetry and harmony and, in general, the common cultural and psychological attributes of science and art.

Soviet scholars were encouraged by Galileo's statement that "truth and beauty are the same thing" and by Paul Dirac's remark, made in a Moscow lecture, that "the beauty of a scientific theory is the only reliable indicator of its truthfulness."[8] No longer did they hesitate to comment favorably on the Pythagorean school in ancient Greece, whose most distinguishing general features were the internal harmony of scientific knowledge and the close interdependence of mathematical symbolism and the aesthetic unity of the cosmos.[9] In the elegance of mathematical expression, the simplicity of the laws of nature, and the projection of cosmic harmony they saw, as Einstein did, an impressive blend of science and art. The universality of Kepler's creative work attracted them for the first time as a majestic picture of an aesthetically perfect cosmos expressing "a musical-numerical harmony." The role of science as an inquiry into the aesthetic attributes of cosmic harmony became a particularly attractive topic of historical studies.

In the relation of science to art, Einstein saw more than a specific expression of values of a high humanistic order. As Soviet commentators noted, he also saw a unique heuristic tool for consolidating existing knowledge and opening new vistas. In Einstein's opinion, they wrote, an effective theory of physical reality must rest on "external confirmation" and "inner perfection," the former referring to the verification of theoretical claims by empirical facts, and the latter to the "naturalness" and "logical simplicity" of the basic concepts.[10] The aesthetic aspect of inner perfection not only added a humanistic value to the pursuit of science, but also strengthened its cognitive operation—it set the stage for an easier and more effective advancement and diffusion of scientific knowledge.

According to Elena Mamchur, a sociocultural historian of science, the aesthetic element produced by the simplicity and logical precision of the structure of theoretical knowledge did not by itself enrich the cognitive side of science, but it did help in making it easier to follow a more rational course in opening new perspectives for its progress.[11] Mamchur cited B. S. Griaznov's remark that an attempt to perfect the organization of a theory should not be viewed as pedantry, crippling the creative impulse in theoretical work, but should be viewed as an efficient tool in the development of both science and philosophy.[12] In the modern emphasis on the harmony and symmetry of the physical structure of the universe, eloquently underscored by Einstein, post-Stalinist Soviet scholars saw both an aesthetic quality and a grand design for a deeper cognitive penetration into the mysteries of the physical world.

Einstein and Dostoevsky

At the end of the 1910's, Einstein, disturbed by the occasional venomous criticism of his work in physics, sought solace in the rich and challenging prose of Feodor Dostoevsky's *Brothers Karamazov*, a work in which he discovered confirmations of his own humanistic philosophy. In his lifetime, he made three comments on Dostoevsky's classic that demonstrate his deep admiration for the great Russian novelist. The first of these came in 1919, in a letter to his close friend Paul Ehrenfest, professor of physics at Leiden University, in which he wrote: "I am reading with enthusiasm *The Brothers Karamazov*. It is the most wonderful book that I've laid my hands on."[13]

The second comment came in 1921, when Alexander Moszkowski, a Polish philosopher, published a biography of Einstein that dealt heavily with questions of philosophy, aesthetic expression, and the ethical dilemmas of modern man. According to Moszkowski, Einstein confessed to him that he had learned more from the great novelist than from the Prince of Mathematics himself, Karl Gauss.[14] In Einstein's words, as recorded by Moszkowski: "If you ask in whom I am most interested at present, I must answer Dostoevsky— Dostoevsky gives me more than any scientist, more than Gauss." As his next words indicate, Einstein's interest in Dostoevsky had a vibrant moral overtone:

> It is the moral impression, the feeling of elevation, that takes hold of me when a work of art is presented. And I was thinking of these ethical factors when I gave preference to Dostoevsky's works. There is no need for me to carry out a literary analysis, nor to enter on a search for psychological subtleties, for all investigations of this kind fail to penetrate the heart of a work such as *The Brothers Karamazoff*. This can be grasped only by means of feelings that find satisfaction in passing through trying and difficult circumstances, and that become intensified to exaltation when the author offers the reader ethical satisfaction. Yes, that is the right expression, "ethical satisfaction"! I can find no other word for it.[15]

The third reference to Dostoevsky came in 1930. In a "German dialogue" with the mathematician J. W. W. Sullivan and the Irish writer James Murphy, Einstein pronounced Dostoevsky "a great religious writer." But that was only in the sense, he said, that Dostoevsky was concerned with presenting a picture of "the mystery of spiritual existence . . . clearly and without comment"; Dostoevsky was not interested in interpreting specific religious problems. Science, preoccupied with the work of nature, Einstein continued, had no power over the depths of spiritual existence—over "the valuation of life and all nobler expressions" of humanity. Built on the foundations of spiritual existence, ethical norms could not be reduced to scientific formulas; they could be felt only through the kind of aesthetic creation Dostoevsky bequeathed.[16]

An unusually large group of Soviet scholars, ranging from physicists to literary critics, harnessed their best skills in searching for a scientific interpretation of Einstein's statements. In a give-and-take between Ivan Karamazov and Aliosha, many detected four ideas that occupied a strategic place in Einstein's scientific theory and worldview: the inadequacy of Euclid's geometry in interpreting cosmic space, the pressing need to extend general scientific vision beyond the traditional framework of a three-dimensional world, Leibnitz's principle of the universe's "predetermined harmony" as the organizing principle of human wisdom; and the view of "truths" as a negation of common sense.[17]

Soviet commentators produced copious literature on the common denominators of Dostoevsky's and Einstein's interpretations of how science related to art. They were particularly eager to show the inner philosophical unity of Dostoevsky's art and Einstein's science. All commentators cast their discussions within a general humanistic framework. They were not interested in adducing arguments in favor of Dostoevsky's influence on Einstein, only in detecting and explaining basic similarities in their general philosophical outlooks.

The search for key similarities in Einstein's and Dostoevsky's creative work produced a great variety of impressionistic statements, supported in many cases less by logical arguments than by liberal doses of imagination. Such statements were more important as expressions of and contributions to a widening scope of the values and regulative norms that were now seen to be shared by art and science than as summaries of verifiable conclusions. The process was facilitated by a de-Stalinization of such key notions as intuition, which went from being a treacherous term, leading directly to decadent mysticism, in the first *Great Soviet Encyclopedia* (1935) to a valuable tool in science no less than in art in the third edition (1972).[18] The new encyclopedia obviously echoed Einstein's statement that scientific discoveries were made by "sins" against reason, and that the laws of nature were discovered not by the ways of logic but by the power of intuition.[19] It was now popular to go along with Descartes' assertion that intuition was not a deceptive judgment of a disordered imagination, but a product of a clear and attentive mind.

Igor Zabelin, a literary critic, thought that Dostoevsky and Einstein were united in their heavy reliance on images in generating and ordering cognitive material.[20] Zabelin may have borrowed that idea from Einstein himself, who, in answer to a questionnaire from the French mathematician Jacques Hadamard, stated that the elements of his scientific thought were not words, whether oral or written, but signs and images.[21] Zabelin was convinced that music, "the most abstract form of thinking in images," was an integral part of Einstein's intellectual makeup. In Zabelin's opinion, the ability to think in images was a primary condition for creative work in modern science. To think in images meant also to bring science closer to art and to endow it with a more satisfying aesthetic quality.[22]

But neither Einstein nor Zabelin was prepared to push the idea of the affinity of science and art too far. In a friendly discussion with the eminent Indian writer Rabindranath Tagore, Einstein was categorical in his claim that beauty, but not truth, depended on the presence of man.[23] Human reason, he said, tried to understand a reality that existed independently and outside of it. In the view of one Soviet commentator, B. M. Runin, Einstein saw science and art as distinct paths to truth, neither of which was privileged. Runin felt, however, that Einstein, in paying homage to Dostoevsky, expressed a personal conviction that art must be considered a powerful source of impulses for scientific work.[24] Runin found it important to emphasize that Ivan Pavlov, in distinguishing two general personality types—the "artistic type" and the "thinking type"—did not intend to consider one superior to the other.[25] The scientific community, in its search for ideological liberation from political imperatives, was now prepared to recognize science as just one road to truth.

Soviet commentators were much attracted to cosmic harmony, a grand idea that Dostoevsky and Einstein shared. M. N. Vol'kenshtein felt that one of the things Einstein greatly admired in Dostoevsky was his search for concrete and symbolic indicators of harmony as the central architectonic principle of the universe. The quest for harmony, Vol'kenshtein wrote, gave Einstein's scientific creativity an admirable aesthetic quality.[26]

To some writers, the search for cosmic unity included, as a special case, the search for a fertile combination of the "paradoxical" and the "normal" in both the physical universe and human behavior. The physicist E. L. Feinberg, for example, a specialist in quantum mechanics and the theory of relativity and a corresponding member of the Academy of Sciences, thought that Einstein was particularly attracted to two unusual features of Dostoevsky's novels: the discordant personal attributes of individual characters and the sudden unexpected and nonlogical turns in plots or in the characters' behavior.[27] Dostoevsky appealed to Einstein because he made all these turns and behavioral mutations most convincing heuristic tools for a deeper understanding of the order of nature. It was often noted, or implied, that Dostoevsky depended on the ambivalent behavior of his characters to reach the depths of human be-havior. Einstein, in his turn, was undaunted by the total discordance of the high claims of his theory with "common sense." In a report to the members of the Berlin Academy in 1921, he made the bold assertion that the more the con-clusions of modern physics disagreed with everyday experience, the higher the degree of their scientific relevance.[28] In its message, that statement did not dif-fer from Dostoevsky's direct claim that "commonplace phenomena and the conventional views of them are not realism, . . . but the very opposite."[29] Dosto-evsky also recognized the creative role of intuition—of unexpected and non-logical paths to the deep secrets of the physical and human universes. More than any other novelist, he prepared modern man to appreciate the paramount role of intuitive flashes in the twentieth-century revolution in science.

E. I. Kiiko agreed with the suggestion that Einstein was especially impressed with Dostoevsky's unceasing search for unity in contradictory views on natural and social reality. Kiiko asserted that the search for causal links between the paradoxical and the normal was sharply manifested not only in *The Brothers Karamazov* but also in *Crime and Punishment* and *The Idiot*. Dostoevsky pre-sented Prince Myshkin, for example, as a man who appeared to be unfit to meet the rigors of the social environment and as a personification of higher moral values.[30] The philosopher R. A. Zobov held close to the same view: he ar-gued that Dostoevsky and Einstein looked at their specific universes of inquiry from unorthodox angles, a method that helped them in their efforts to reach the depths of the structural harmony of the universe.[31]

Boris Kuznetsov, who wrote more extensively and more perceptively about the affinity of Dostoevsky's and Einstein's approaches to reality than any other Soviet writer, referred to the search for harmony in "the infinitely paradoxical existence" as a common feature of the world outlooks of the two giants. Just as Einstein was the first scientist to make use of a mathematical paradox that ac-commodated both Euclidean and non-Euclidean geometries, so Dostoevsky was the first novelist to make the same paradox a literary topic.[32] In a searching discussion with his brother Aliosha, Ivan Karamazov lamented his own inabil-ity to rise above the narrow limits of the Euclidean frame of reference to reach the deeper secrets of the universe. Indeed, according to Kuznetsov, Ivan went even farther: he anticipated the theory of relativity by expressing sorrow for his inability to rise above the three-dimensionality of earth-locked thought about problems of cosmic scope and significance.

In an elaborate argument, Kuznetsov suggested that the essence of Dosto-evsky's creativity lay in combining a "subjective"—or "irrationalist"—world outlook and an "objective"—or "rationalist"—adherence to the canons of po-etics. His statement that in this combination he saw a particular case of Bohr's complementarity principle at work was challenged by another Dostoevsky specialist. M. Gus claimed that Kuznetsov did not make it clear whether the principle of complementarity was a special method of his literary analysis or an internal attribute of Dostoevsky's work. He also thought that the idea of dialectics, rather than the principle of complementarity, offered a more tenable explanation of Dostoevsky's—and Einstein's—creativity, dominated by a search for cosmic unities of "paradoxical phenomena." Complementarity did not mean a synthesis of opposite views; it implied that opposite views of the same reality were not mutually exclusive—that they shared some common ground. On this basis, Gus thought that the main thrust of Kuznetsov's comparison of Einstein and Dostoevsky was groundless.[33]

For Kuznetsov, Einstein needed harmony as the integrative principle of the universe because he assumed that without it there could be no science, no power of mathematics. Dostoevsky suggested a broader view of harmony—a non-Euclidean harmony—to satisfy the ambition of individuals represented by Ivan Karamazov. In evoking non-Euclidean geometry, he was involved in a "thought experiment," not in an affirmation of irrationalism. By adding a poetic touch to his meditations, Kuznetsov found a link between the ethical problems of Dostoevsky's books and the physical conclusions implicit in Einstein's theories.[34]

G. N. Volkov, an expert in the fast-growing sociology of science, was under the spell of Einstein's humanistic reasoning when he asserted in 1968 that Dostoevsky's genius demonstrated that the creative processes in belles-lettres and in science, though different in their modes of expression, were similar in their "research" procedures and dominant impulses. In his "creative laboratory," Volkov wrote, Dostoevsky placed his heroes into situations propelled by emotional stress and by the unsubmissive world of the subconscious. Ivan Karamazov and Raskolnikov looked longingly for elements of cosmic harmony in the staggering disharmonies of human existence. In Volkov's assessment, the novelist and the physicist had sought to unravel the lines of order in a world of apparently endless chaos. In carrying out that task, they showed that there was no sound science without an element of poetry, just as there was no sound poetry without an element of science.[35]

The literary theorist Mikhail Bakhtin, taking note of both the intensity and depth of Dostoevsky's anti-rationalism and the inordinate and broad compass of Einstein's rationalism, was led to disagree with any effort at a "scientific comparison" of the two intricate systems of philosophical thought. Bakhtin did not oppose an "artistic analysis" of distinct modes on which the two eminent figures depended in fashioning their world outlooks. He acknowledged the intellectual richness of Einstein's idea of the inseparability of space and time; but he chose to use "time-space" (or chronotope, as he called it) as a useful metaphor rather than as a point of departure in search of a theoretical structure.[36] Bakhtin's effort must be seen as an invitation to a broader comparative study of Dostoevsky's and Einstein's cosmic philosophies. According to Gary S. Morson and Caryl Emerson, however, "although Bakhtin focuses on the chronotope in literature, he means us to understand that the concept has much broader applicability and does not define a strictly literary phenomenon. To use his own terms we may say that he reveals the canceled 'potential' of the Einsteinian concepts of space-time."[37]

Many critics agreed that Einstein's insistence on the aesthetic attributes of science was deeply set in ethical considerations. In their analysis of Einstein's view of Dostoevsky's creative work, most Soviet writers stayed close to the Moszkowski text: in stating that he had learned more from Dostoevsky than from Gauss, Einstein underscored the importance of aesthetic creativity as "ethical satisfaction." Feinberg, for example noted that Dostoevsky's art chronicled an ethical struggle of global proportions, the same struggle that Einstein detected in the depths of cosmic philosophy grounded in science.[38] Einstein the scientist favored Dostoevsky the novelist because the great writer gave artistic expression to the ethical axis of humanity and because he made aesthetic expression a formidable link between science and morality. All the same, D. Danin, a literary critic and accomplished biographer, was on the right track when he asserted that "the Dostoevsky-Einstein problem" required a multidimensional answer, and that any one answer could be no more than a first approximation of truth.[39]

Universal ethical norms, as Einstein comprehended them, presented a dominant part of the general humanistic aspect of cosmic harmony. In Max Born's observation, Einstein was convinced that "all human actions spring up from the depths of an ethical feeling which is primary and almost completely independent of reason."[40] Einstein made it clear that scientific knowledge and moral norms came from different sources: one from reason and the other from a sentiment deeply rooted in basic humanity. Of different origin, the two had essential features in common: both were validated by means of empirical verification, and both were logically integrated into larger patterns of cultural complexes.[41] Both were also insoluble components of the world of ideas expressing cosmic unity. Soviet commentators, following Einstein, were interested in showing that links between science and the moral code should not obscure their distinctive natures as separate manifestations of cultural activity.

There can be little doubt that Einstein and Dostoevsky shared the idea of the moral essence of humanity. This did not mean, however, that their ethical philosophies and concerns were identical. Whereas Dostoevsky dealt primarily with the emotional origins and complexities of the norms of ethics, Einstein was preoccupied with the logical integration of moral principles.[42] Nevertheless, to both Dostoevsky and Einstein, the aesthetic and ethical denominators of culture were inseparable. Both agreed with Kant that "the beautiful is the symbol of the morally good."[43] To give his view dramatic expression, Dostoevsky asserted that beauty was first an ethical and then an aesthetic category.[44] Einstein came close to Dostoevsky's point when he asserted that *The Brothers Karamazov*, a zenith of achievement in literary art, appealed to him primarily as a source of ethical satisfaction.[45]

To this point, virtually all the Einstein-Dostoevsky studies discussed came from the pens of philosophers, literary theorists, and others unconnected with the hard sciences. To sum up their views briefly, most of them turned their efforts to identifying the many characteristics common to Einstein and Dostoevsky. Both, according to them, were masters in relying on paradoxical situations as openings to new depths of reality; both recognized nonlogical—intuitive—sources of fundamental knowledge; both considered the multiple forms of aesthetic expression a vital link between scientific knowledge and rules of morality; both relied on the notion of cosmic harmony as the most general key for a full understanding of the physical and cultural universe; and both were true masters in making broad use of images as cognitive tools. All these characteristics accounted, in their view, for Einstein's enthusiastic expression of his philosophical proximity to Dostoevsky.

Few members of the Soviet scientific community wrote on this specific subject. They were more interested in Einstein's poignant and oft-repeated argument that though science was in a strong position to contribute to the advancement of the moral code, it played no role in the origin of moral norms.[46] In Einstein's words:

> I do not believe that a moral philosophy can ever be founded on a scientific basis. . . . Every attempt to reduce ethics to scientific formulas must fail. . . . On the other hand, it is undoubtedly true that scientific study of the higher kinds and general interest in scientific theory have great value in leading men toward a worthier evaluation of the things of the spirit. . . . The popular interest in scientific theory brings into play the higher spiritual faculties, and anything that does so must be of high importance to the moral betterment of humanity.[47]

One of the first scientists to sound this note was the mathematician P. S. Aleksandrov. He wrote, in 1967, that science and morality were not only made of different cultural material but also subject to different principles and tempos of development.[48] A. F. Shishkin, a tested hand in questions of ethics, expressed his disappointment in the strong tendency in Soviet philosophy to emphasize "the cognitive unity of science and morality." He also noted that the time had arrived for a radical change in Soviet ethics, especially for a shift in emphasis from the study of narrowly defined problems of "Communist morality" to a direct confrontation with "the moral riddles of the century."[49]

Einstein's ideas helped to create a new picture of the moral basis of science, a picture to which Soviet readers had not been previously exposed. From the leading scientist of the age they received the sobering word that science did not set up standards of morality; that the cultural roots of morality were deeper and more extensive than the roots of science; that science could not only strengthen but also lower moral standards; and that the sharp disparity between the high achievements of science and crumbling moral principles was the dominant feature of contemporary civilization. The discussion of the humanistic domain of Einsteinian thought contributed to a fundamental recasting of the sociological setting of dialectical materialism.

The Einstein-Dostoevsky discussion was part of a growing and unwieldy movement committed to a critical examination of Soviet values. It was one of many ramifications of political dissidence. The physicist Andrei Sakharov, the leader of that movement, did not enter the sprawling debates on Einstein's and Dostoevsky's humanistic thought, but he raised many of the same questions. He was clearly the first Soviet writer to remind his countrymen that the Soviet Union, no less than other industrial nations, faced the critical problem of a concurrence of the accelerated growth of science and rapidly declining moral values.

Post-Stalinist writers recognized that morality, art, and science were based on different human endowments, and that each had its own mode of expression and substantive concerns. Following Einstein, however, they also recognized that they formed parts of a tightly intertwined universal cultural process. Boris Kuznetsov noted some key manifestations of this unity:

> Moral criteria do not determine the content of scientific theories, but they do determine the interests of the scholars. In Einstein's creativity, especially visible is the link of the quest for cosmic harmony with the quest for moral harmony. The social debt of science is dictated by the internal logic of scientific knowledge and by its aesthetic norms. At the present time, this debt calls for a peaceful and constructive strengthening of science in its dealing with the conditions bolstering the authority of reason. It appears that the higher the level of generality of scientific problems, the more indissoluble the intertwining of scientific methods and moral criteria, and the more united the ideals of truth, goodness, and beauty.[50]

Both Einstein and Dostoevsky defended the close ties between science and philosophy. Science, according to Einstein, owed it to itself to maintain close and fruitful ties with philosophy, which not only widened its humanistic base but also provided intellectual guidance in solving knotty theoretical and methodological dilemmas.[51] This contact, he said, was imperative and particularly useful at times of deep crises in science, as happened in the seventeenth century and at the beginning of the twentieth. Dostoevsky made it clear that he favored scientists with "universal minds" like Claude Bernard and Hermann Helmholtz, whose scientific ideas were embodied in carefully articulated philosophies. Dostoevsky was convinced that philosophy provided the main link between science and enlightenment. He was impressed with Bernard's contention that the time was coming when the poet, the philosopher, and the scientist would appreciate cooperating with one another. Dostoevsky argued that science, left to itself, made no impact on enlightenment.[52]

The post-Stalinist thaw encouraged Marxist philosophers to move away from the "naive realism" of Stalinist philosophers, which identified sensory experience as a true mirror of external reality, and to come closer to Einstein's avowal that "the concepts which arise in our thought and in our linguistic expressions are all—when viewed logically—the free creations of thought which cannot be gained inductively from sense experiences."[53]

Unlike Dostoevsky, Einstein believed strongly in the preeminence of science as the intellectual base of modern civilization. He criticized the philosophy of scientism on the ground that it could only weaken science by extending its authority to domains outside its legitimate competence. Whereas Dostoevsky devoted his life to the elaboration of a complex system of philosophical irrationalism, Einstein was first and foremost a spokesman for the philosophy of rationalism.

Dostoevsky recognized the intellectual power and cultural identity of science as the most sublime expression of the triumphs of reason. However, speaking through the unnamed hero of his *Notes from Underground*, he informed his readers that reason accounted for only 20 percent of being human—the rest being made up by other "conscious and unconscious" attributes of "individuality." In "human law," he detected a contradiction to "the laws of logic."[54] Dostoevsky actually recorded his opposition to the Nihilists' claims that science alone provided a panacea for all social ills and the safest guideposts to the general enlightenment. He praised the neurophysiological research of I. M. Sechenov, but objected to its lack of a broader humanistic anchor.[55]

It would not be unreasonable to conclude that in his reference to Gauss and Dostoevsky, Einstein demonstrated his interest in comparing not so much two specific and concrete individuals as two distinct and quite different—yet equally essential—modes of cosmic inquiry that the two giants happened to

represent. He learned less from Gauss only because the illustrious mathematician and astronomer worked within the same—scientific—universe of inquiry. Gauss was one of his own kind, relying on the same general style of work
and method of decoding the mysteries of nature. Dostoevsky, by contrast, introduced him to an intricate arsenal of approaches to the mysteries of the universe that did not coincide with the logic of the scientific method. Nor would it
be unreasonable to count national sentimentality among the leading factors
influencing such a wide and enthusiastic response by Soviet scholars to Einstein's tribute to Dostoevsky. On one occasion, Einstein mentioned Leo Tol-
stoy as one of his favorite writers, but in that instance, he gave no clues for his
choice.

Einstein emphasized the unity of universal culture, an idea that he based on
the cosmic principles of symmetry, harmony, and simplicity. His views on the
subliminal aspects of culture stood clearly above the limiting factor of ethnic,
religious, national, or mythological favoritism. As most clearly manifested in
his *Diaries* and *Letters*, Dostoevsky occupied a completely different position:
his view of an evolving world culture was neither more nor less than a projected triumph of the ethical purity and nascent humanity of Russian culture.
His "Messianic cult," as Joseph Frank aptly named it, was not an isolated outcry of a supercritical mind; on the contrary, it was a resonant echo of "the main
movement of Russian thought and the socio-political reality of his time."[56] He
presented N. Ia Danilevskii's *Russia and Europe* as a careful assemblage of incontrovertible proofs for his claim that the Russian soul and its religious underpinnings were destined to become the prime contributor to the anticipated
emergence of a new "type" of world culture.[57]

Soviet comments on the alleged affinity of Einstein's and Dostoevsky's
world outlooks were consistent in disregarding Dostoevsky's emphasis on the
nonessential place of science in Russian culture. Technology and science, he
wrote, could be learned from Germans without the need for taxing Russia's in-
ventive capacities. He could not be criticized, however, for disavowing recent
(Nihilist) claims that science alone was in a position to play a key role in determining the moral behavior of individuals and nations.[58] Einstein, in his turn,
claimed that Dostoevsky's religious meditations—revealing the essence of his
world outlook—were a philosophical and a psychological commitment rather
than a theological exercise, and that they served as a stimulant for higher intellectual (including scientific) achievements.[59]

Alongside the fundamental differences separating Einstein as a powerful
champion of rationalism and Dostoevsky as the most articulate exponent of
philosophical irrationalism, there were deep-rooted similarities. Both were inspired by the idea that intuition was the safest and most productive path to

sublime wisdom. Both admitted that the real truth reached beyond—and often openly challenged—common sense, and that the road to a comprehension of cosmic unity was full of paradoxical contradictions and obstacles. Both exalted the ethical dimension in the search for truth. And both—each in his own way—made extensive and profitable use of "thought experiments."

Preoccupied with exploring the tangled bonds of affinity in Einstein's and Dostoevsky's world outlooks and expressive mechanisms, post-Stalinist commentators showed almost no interest in differences between the two illustrious men, though they were surely more pronounced than the similarities. The commentators achieved their goal not only by bringing Dostoevsky closer to Einstein, but also by bringing Einstein closer to Dostoevsky. The comparison strengthened the humanistic base of science, which, in turn, served as a powerful weapon of the scientific community in its struggle for the kind of "freedom of reason" that Einstein had defined for them: "the independence of thought from the restriction of authoritarian and social prejudices."[60] Perestroika invited a partial revival of the traditional view of Dostoevsky as a great writer guided by a deeply rooted and painstakingly articulated system of religious beliefs and sentiments unaffected by and unrelated to the world of science.[61]

The efforts to penetrate the depths of—and to discover fundamental similarities between—Einstein's and Dostoevsky's cosmic philosophies and ethical and aesthetic principles were part of a massive involvement in the study of the relations of science to all major domains of humanistic culture. The ultimate aim of those efforts was to prevent science from forming a separate culture without a humanistic base; the goal was to achieve a "humanitarianization" of science. The studies considered in this chapter were supplemented by equally bold and challenging efforts to unravel the humanistic essence of Einstein's thought without taking the routes leading to Dostoevsky comparisons.

The Soviet scholars' great emphasis on the similarities of Einstein's and Dostoevsky's philosophies and general world outlook, despite the weighty and extensive differences between them, undoubtedly represented a sentimental and appreciative response to Einstein's fervent praise of the monumental achievements of one of the most distinguished Russian literary figures. But the major intent of that enterprise, ultimately, was to strengthen the scientific community by bringing it closer to nonscientific universes of inquiry, and to elevate it to the position of final authority in assessing the potentialities and limitations of science and in certifying scientific knowledge. The remarkable feature of the Dostoevsky-Einstein discussion was that it sought little help from dialectical materialism.

9 ∎

Einstein in the Light
of Perestroika

During the long years of the post-Stalinist thaw, the historians and philosophers of science argued that Einstein gradually drifted from Ernst Mach's neopositivist positions to the vantage points of materialistic epistemology. Without denying that argument, perestroika historians preferred to emphasize that
the change was due to contemporary efforts in the Soviet Union to eradicate
the deliberate misrepresentations and unwitting distortions of his ideas by a
long succession of Stalinist philosophers and their allies in the scientific community. The experts agreed, however, that Einstein's philosophical views could
under no circumstances be stereotyped as materialistic in a dialectical sense.
Even if he drifted toward materialism, he landed a long distance from it.

The Reconsideration of Einstein's
Philosophical Thinking

In a book published in 1987, D. P. Gribanov offered a general interpretation of
the philosophical footing of Einstein's contributions to physics as seen by Soviet scholars in the early phase of perestroika.[1] From the very beginning, according to Gribanov, Einstein was not attached to any system of idealistic
thought. Although Einstein took serious note of "healthy kernels" in various
expressions of idealistic thought, he was in fundamental disagreement with the
philosophical ideas of Berkeley and Hume, Kant and Mach, logical positivism,
and various other outcroppings of epistemological subjectivism. As Gribanov
saw it, the perestroika spirit encouraged Soviet scholars to follow Einstein's example in viewing Western idealism as a valuable source of positive thought,
which, if handled carefully, could strengthen Marxist theory and thereby enrich the intellectual culture of the USSR in general. Einstein showed that Berkeley went beyond the limits of "subjective idealism" and "solipsism," that

there was more to Hume than "agnosticism" expressed in a denial of "objective causality," that Kant's thought covered more ground than "transcendental idealism" indicated, and that Mach was a misguided philosopher but a positive contributor to science.[2]

Gribanov was particularly critical of the tendency of earlier Soviet philosophers to place the label subjectivism on Einstein's thought. He wrote:

> The authors who identified Einstein with subjectivism relied . . . on his statements related to certain general questions of mathematics. In this connection, the critics usually referred to selected propositions stated in his "Geometry and Experience," where he claimed specifically that "the propositions of mathematics referred to objects of our mere imagination, and not to objects of reality," and that mathematics is "a product of human thought which is independent of experience." Actually, by relying only on these ten citations it is possible to talk about subjectivism. However, an analysis of this study in all its aspects, as well as of Einstein's other works related to the general methodological problem of mathematics, shows that every assertion that ascribes idealistic views to him is unfounded. Einstein did not deny that the roots of mathematics are in the external world, and that its origins are in the practical needs of society.[3]

Without much substantiation, Gribanov argued that, also from the very beginning, Einstein's thinking had been replete with elements of dialectics, so that he had in fact been much closer to the Soviet mode of elucidating the philosophical underpinnings of modern scientific thought than to the so-called idealists.

Where the Stalinist philosophers had seen Einstein's ideas as an invitation to mysticism and fideism, for modern Marxist scholars like Gribanov, Einstein's epistemology invited a deeper and more versatile reliance on the dialectical mode of viewing the physical universe. As now seen, Einstein's thinking was permeated with methodological designs to establish the "dialectical" unity of such opposite components in the general structure of scientific thought as the absolute and the relative, the objective and the subjective, the continuous and the discrete, and the empirical and the rational.

The perestroika atmosphere engendered a nostalgic attitude toward the 1920's, which were often now idealized as prizing pluralism in all aspects of intellectual inquiry. If the 1920's were labeled "democratic," it was because they tolerated heterodox and "incorrect" views in science and philosophy caught in a whirlwind of revolutionary change.[4] The pre-Stalin era devoted much attention to the vital role of philosophical discourse in the advancement of science. But it was clearly guided by Einstein's suggestions, forcefully defended by V. I. Vernadskii, that philosophy was useful to science, and to itself, only insofar as it provided ample room for controversial thought. Philosophical pluralism, faintly hinted at in the age of the post-Stalinist thaw, became the reigning ideal in the succeeding period.

A. A. Nikiforov, a philosophical spokesman for perestroika, reasoned that the presence of many philosophical systems, ideas, and views was not a sign of crisis or decadence but, on the contrary, proof that philosophers performed their duty in a superb fashion—that they created systems of ideas expressing differences in the world outlooks of individual groups. He noted the broader significance of his argument: "The diversity of ideas and opinions is important for us today when our society is in need of a wide spectrum of different models for its future development in order to be able to select the one that offers the best solution for our problems."[5]

Perestroika created favorable conditions for special emphasis on polymorphism, another expression for pluralism. Diametrically opposed to the monolithic attributes of Stalinist ideological aspirations in the fields of thought, polymorphism stood for giving every theory a "free path" to refurbish or replace old concepts. To emphasize polymorphism meant, to a large extent, to emphasize the inner logic and impulse of the mobility of scientific ideas untrammeled by disruptive influences of external origin.[6] It discouraged any reliance on concepts frozen within a definite ideological or any other framework. The word pluralism acquired an additional meaning: it came to symbolize the defense of the vibrancy and richness of universal cultural values through an unceasing resistance to the encroachments of monolithic philosophies.[7]

In the Stalin era, and in fact well beyond, dialectical materialism had been presented as a science, and its practitioners were officially treated as full members of the scientific community. The liberalizing atmosphere of perestroika encouraged a critical reexamination of that claim. Whether dialectical materialism was actually a science, or a scientific method, became a matter of heated discussion in philosophical seminars organized in most institutions of higher education and research centers. It was a popular topic in published debates.

Ardent discussions worked in favor of treating dialectical materialism (and any other philosophical school for that matter) as a system of ideas completely different from scientific concerns and methods. Philosophy, in its new dress, was seen not as a discipline dealing with objective reality, existing independently of the human mind, but as a "culture-creating" human effort, as a unique expression of personality, and as a mode of contemplation elevating human beings to new heights of achievement.[8] A strong wing of Marxist (or ex-Marxist) philosophers came around to agreeing with Nietzsche that "every great philosophy has so far been the self-confession of its originator, and a kind of unintentional, unawares memoir." Whereas science undertook an objective and generalized study of cultural content as a historical and concrete reality,

philosophy offered a subjective and individualized expression of cultural values. The outcome of these discussions, their direct aim in fact, was to expose
the intellectual sterility of the close association of dialectical materialism with
science—a vote of no confidence in the Marxist philosophy of natural science.
While recommending, implicitly, a divorce of philosophy from science, the
new critics assigned philosophy a vital task: to study and generate the type of
wisdom capable of transcending the rigid authority of logic and the scientific
method to place the primary emphasis on the human personality as a creator
and a product of culture. With its attack on old values and tradition-laden social reality, and with its unquestionable accent on individuality and subjectiv-
ity, perestroika appeared as a modern version of the Nihilism of the 1860's, at
least in a limited sense.

The new philosophers marshaled astute arguments against treating dialectical materialism exclusively as an arm of official Soviet ideology. As K. N. Liubutin and D. V. Pivovarov pointed out in *Philosophical Sciences* in 1989, philosophy could not survive if it only responded to external ideological orders issued by official institutions. It must have its own interests, logic of development, and independently conceived challenges and critical problems.[9] Briefly,
in order to develop a healthy discipline of philosophy, a society must create the
conditions that encouraged a diversity of views and recognized the independence of the internal resources and impulses of philosophical thought. The
authors welcomed the emergence of "counterideologies" challenging the
sanctity of doctrinaire unity and *partiinost'* as the beginning of a new era in
Marxist thought.[10]

A willingness to reexamine Western idealistic philosophy helped Soviet
scholars broaden their intellectual perspectives and discover new avenues
linking the theoretical problems of science with fertile criticism emanating
from nonscientific sources. Hitherto, Marxist critics had shown almost no interest in Edmund Husserl's phenomenology, which they considered an expression of unalloyed idealism diametrically opposed to Marxist theory. Stimulated by the moderating winds of perestroika, many philosophers now found
the Husserlian search for new depths of structured cognition both fascinating
and rich in exploratory thought. They appreciated Husserl's phenomenology
as an elaborate system of philosophical ideas, always within sight of the most
critical problems of modern science. Husserl's penetrating analysis of problems directly related to the foundations of mathematics—for example, his
criticism of the psychological foundations of Gottlob Frege's logicism—
attracted favorable attention.[11] The masters of hermeneutics—Hans-Georg
Gadamer et al.—received a most sympathetic consideration for their effort to
add new and more penetrating tools to the method of "understanding" the role

of cultural tradition as related to different theoretical problems of modern science, both social and natural. The representatives of postmodernism—particularly Thomas S. Kuhn and Imre Lakatos—became the most frequently cited philosophers in the Soviet Union, almost always as leading contributors to the philosophical foundations of the history of science as an academic discipline.[12]

Interest in the new philosophical "internationalism" did not work against the awakening of Russian nationalism in various cultural fields. Perhaps most of that heightened patriotism came as a reaction to the growing waves of nationalism among the non-Russian nations and ethnic groups. A ringing revival of religious philosophy, espoused by V. S. Solov'ev, N. A. Berdiaev, and a long list of religious philosophers and allied metaphysical idealists in the late nineteenth and early twentieth centuries, was the most conspicuous manifestation of this movement. A growing number of its supporters argued that Marxism, in its entirety, was an alien importation without roots in the history of Russian culture. The time had come, they believed, to search for and resuscitate the reigning currents in national thought. Small wonder, then, that the search for the "national character" of Russian philosophy became a topic of considerable interest. This kind of pursuit took on a dual task: to help illumine the uniqueness of Russia's intellectual tradition and culture and to explain Russian contributions to the world pool of philosophical knowledge.[13] The two leading philosophical journals, *Questions of Philosophy* and *Philosophical Sciences*, chose to emphasize the Russian idealistic philosophy of the 1890's and the first two decades of the twentieth century as the true expression of the Russian national character. Although that philosophy was steeped in anti-rationalist metaphysics and mysticism, both journals normally avoided reprinting articles carrying the more abstruse and openly anti-democratic metaphysical messages.

Suggestions for new models of philosophical discourse came not only from Western philosophy and Russian idealistic thought of pre-Soviet vintage, but also from Soviet scholars whose works had been suppressed for years because of their open challenge to the Marxist style and substance of philosophical engagement. Among them were some of the most notable names in various domains of intellectual endeavors, including Pavel Florenskii, A. F. Losev, Mikhail Bakhtin, and Vladimir Vernadskii. Florenskii worked earnestly on building a bridge between science and religion, Bakhtin made an impressive effort to create a science of literary criticism, Losev gave scientific dimension to the study of the history of aesthetic expression, and Vernadskii acclaimed the historical intertwining of science, philosophy, and religion. All viewed science in a broad humanistic context—as both a contributor to and a beneficiary of symbiotic relations with religious, aesthetic, and ethical cultures. All helped reintegrate science into the broad framework of modern culture, the same idea that received eloquent support from Einstein's humanistic thought.

The time had also come to recognize Lenin's negative attitude toward epistemological relativism. Unaware of Einstein's theory, Lenin had cautioned, in *Materialism and Empirio-Criticism,* that "to place relativism at the base of the theory of knowledge means inevitably to become a victim of absolute skepticism, agnosticism and sophistry, or of subjectivism."[14] Stalinist philosophers, quick to relate cultural relativism to Einstein's scientific relativity, condemned the former as a philosophical stance dedicated to a full negation of objective truth and, particularly, to a treatment of moral norms as unstable parts of culture.[15]

Their successors now, for the first time (or at least the first since the Soviet state was founded), began to look favorably on the notion of cultural relativism as elaborated by Oswald Spengler, Arnold Toynbee, Pitirim Sorokin, and Nikolai Danilevskii. According to one writer, to accept cultural relativism meant to make the "irrationalist" method a tool of the philosophy of history. That method, as he saw it, concentrated on the "uniqueness," "nonrepetitiveness," and "individuality" of specific cultures or culture aggregates, rather than on crosscultural uniformities and universalities. He thought that only a combination of interests in unique and universal phenomena could lead to a deeper understanding of social history.[16] His recognition of cultural relativism was a tacit admission that the grand evolution of culture could not be totally reduced to the inexorable laws of rationality. It also confirmed Dostoevsky's contention that to be rational was only 20 percent of being human.

Perestroika, particularly in its later phase, brought another notable change in the use of the word relativism: the scholarly community now showed little hesitation in deploying the word in reference to the most essential part of Einstein's scientific legacy as well. The general softening of attitude toward philosophical idealism and the shifting of emphasis away from the conflict between materialism and idealism helped give "relativism" a broader and more respectable meaning. When the conflict between idealism and materialism ceased to be the fundamental problem of Soviet philosophy, the lines separating the two mainstream orientations became increasingly more fluid and uncertain, allowing such concepts as relativism to find a more comfortable place in the expanding realm of philosophical thought.

Taking a new and more critical look at Marxist thought, many Soviet philosophers reckoned that the time was ripe also for a reexamination of the dialectical process itself. "The time is opportune," observed a philosopher, "for a new interpretation and understanding of fundamental changes in modern science, and for critical comparison of achieved results with the existing princi-ples, laws, and categories of dialectics."[17] There were recurrent calls for a thor-ough revamping of Marxism as a philosophical system. In growing numbers, however, philosophers readily admitted that a modernization of dialectics would not answer all the philosophical needs of post-Newtonian science, and that they needed to look for help to non-Marxist systems of thought as well.

Still, it would not be correct to say that the last residues of Stalinist ideology had been eliminated. In *Science: A Component of the Social System*, published in 1988, V. Zh. Kelle was adamant in his claim that socialist systems alone were on track toward finding a favorable solution to C. P. Snow's "two-culture" syn-drome. He admitted, however, that socialist societies had a distance to cover in order to fully answer Snow's challenge.[18] At a time when more and more Soviet scholars turned to the West for fresh ideas on the relations of science to hu-manistic culture, Kelle's commentary was anticlimactic, to say the least.

In the post-Stalinist era, especially in the age of perestroika, the Marxist philosophers' dream of the full congruence of science with dialectical materi-alism gave way to the widespread conviction that only a rational combination of various orientations could ensure the progress of scientific thought. Bohr's principle of complementarity, much criticized in the Stalin era, was now highly valued as a mode of scientific exploration. Marxist philosophers presented it as incontrovertible proof of the work of dialectics in nature. They now agreed, for example, that the nature of mathematical knowledge could best be explained by combining the mutually exclusive theories in that branch of science.

As the scientific and philosophical communities prepared to enter the post-Soviet period, they showed their readiness to build on two types of activities established during the thaw and perestroika. Some philosophers and scientists continued to liberalize dialectical materialism, a process destined to lead to the full abandonment of philosophical ties with the encrusted dogmatism of Marxist theory and to make it easier to evolve new philosophical positions. Other philosophers and scientists chose to build directly on outlooks that had little in common with Marxism. At any rate, the transition to the new philo-sophical and ideological vistas in science proved to be much less arduous than the transition to new systems of economic relations and political institutions. Most philosophers agreed with Einstein that philosophical discussion was useful to science only insofar as it offered a variety of propositions to choose from.

The New Vision of Relativistic Cosmology:
The Riddle of Mass and Energy

In the 1980's, Soviet scholars largely abandoned the Stalinist ideologues' long cherished dream of developing a unique and unified Marxist theory of the structure and evolution of the universe. Instead, it was clearly admitted that only a combination of different theories could produce a pool of knowledge explaining the basic dynamics of the universe. A. N. Pavlenko, for example, said as much when he asserted that a *maximal* explanation of cosmic processes could be achieved only by depending on more than one model of the universe.[19] Here, too, mutually exclusive theories, or models, were expected to explain different aspects of the same reality. Marxist scholars came around to recognizing the principle of complementarity, not only as a key strategy of quantum-mechanical explanation, but also as a law governing the cosmos.

In the Soviet Union, as in the West, developments in astronomy followed a paradoxical course: the unbelievable snowballing of new information produced by the tools of radio astronomy was taking man much deeper into the hitherto locked mysteries of the cosmos, but also making it more difficult to achieve general theoretical unity, most of all in cosmology. Here again the pluralist tendencies of perestroika made themselves felt as Soviet cosmologists went their separate ways in a search for a general theory of the structure and evolution of the universe.

Some astronomers now spoke of entering a post-Einsteinian phase in the evolution of modern cosmology. That label meant different things to different people. In general, however, the new cosmology had several easily discernible characteristics. It took Einstein's legacy as a starting point, but held that his theories needed much additional work in the way of strengthening their empirical base, making their theoretical underpinnings more precise and consistent, and improving their mathematical apparatus. Zel'manov's suggestion, made in the 1950's, to link the general theory of relativity with an anisotropic model gained a new lease on life and articulate supporters. But there was far from universal agreement. Controversies blossomed in what was now a wide-open system where premises outpaced empirical testing.

A good part of the problem, as the academy scholars openly admitted, was a pronounced lag in the production of modern research equipment. Much of their work required advanced microphysical research in fields related to cosmology. At the end of the 1980's, the academy appointed a so-called Cosmomicrophysical Council to explore ways of tapping the resources of quantum mechanics for cosmological research. "Cosmomicrophysics" was expected to strengthen the search for a comprehensive theory covering all types of interaction in nature and to add to the understanding of the origin and evolution of the universe.[20] Its intent was to press farther afield than Einstein's conceptual framework for a unified field theory allowed.

If not exactly a problem, the astronomers so recently a target of the Stalinist ideologues were assuredly aware that cosmology qua science had not been fully separated from cosmology qua philosophy. Encouraged to build on the scientific knowledge of cosmology, philosophers were discouraged from trying to create or to advance a monolithic Marxist "science of the universe." The persistence on the part of some philosophers to criticize their scientific brethren further fueled the controversies in the cosmological community. G. I. Naan became widely recognized as a model philosopher of cosmology: his "philosophy" rested on solid developments in the theory and methodology of modern astronomy. The new conditions influenced Marxist philosophers to concentrate more on adjusting dialectical materialism to the advances in cosmology than to adjusting cosmology to the philosophical postulates of dialectical materialism.

The culminating point in recognizing A. A. Friedmann's contribution to relativistic cosmology came in 1988. In June of that year, 250 Soviet and foreign scholars gathered in Leningrad to celebrate the centennial of his birth. Although the speakers covered all phases of Friedmann's rich repertory of scientific activity, their attention centered on his crowning achievement, the idea of the expanding universe. Andrei Sakharov added his own glowing praise via the pages of *Priroda*, pronouncing Friedmann's theory one of the two most treasured achievements of modern science. Just as Darwin had blessed the nineteenth century with the idea of the evolution of life on earth, so Friedmann had blessed the twentieth century with the idea of the evolution and general development of the universe.[21]

The ebbing of ideological pressure, the softening of political threats, and the dismantling of critical parts of philosophical orthodoxy helped resolve the riddle of energeticism, a mixture of science and philosophy and the central point in Lenin's attack on physical idealism. After 1955, Marxist philosophers, more than the physicists, had continued to attack energeticism as a general expression of idealistic thought in the natural sciences, but they no longer treated Einstein as one of its representatives or allies. Without demoting matter to a secondary position, they viewed energy as a general measure of the motion and interaction of all kinds of matter and as a link between all natural phenomena.[22] No longer did they disagree with the reasoning that led Einstein to conclude that the conservation of energy and the conservation of mass were one and the same law of nature. They accepted the formula expressing the equivalence of mass and energy, but gave it a specific meaning: it only demonstrated that mass and energy were integral characteristics of all physical objects, and that they were related to each other in a definite mathematical proportion. They at once accepted Einstein's formula and preserved the materialistic ontology of Marxist philosophy. Energy became one of the pillars of the materialistic view of the universe.

Summing up the more recent views, Gribanov concluded in 1987 that most of the misrepresentations of the principle of the equivalence of mass and energy resulted from a chronic tendency to confuse matter with mass and energy. He noted that matter was a philosophical rather than a physical (or a scientific) notion, and that it had nothing to do with the mass-energy equivalence principle. Mass and energy were physical notions—the former referring to inertia and gravitation, and the latter to a type of motion. With matter out of the picture, the idea of the "de-materialization of atoms" ceased to be of scientific and ideological relevance.[23] In Marxist theory, the concept of matter had primarily an epistemological definition: it covered the entire world of reality that had an objective existence, that is, everything existing independently of human consciousness. The official interpreters found a solution to Einstein's claim that "the concept of the material object was gradually replaced as the fundamental concept of physics by that of the field."[24] They simply identified the field as a particular state of matter.

New Interest in the History and Ethos of Science

The ferment of perestroika stimulated a potent and long overdue inquiry into the tributaries and early history of the theory of relativity. Specialists in the history of science worked on tracking down the wide range of influences on the formation of Einstein's ideas. They examined the pertinent developments in post-Maxwellian physics that had made Einstein aware of the need to recast the theoretical foundations of physics and had provided some of the ingredients that found their way into the theory of relativity. In the post-Stalin era, Soviet historians of physics produced significant literature on such developments as the Michelson-Morley experiments on the speed of light, Fizeau's experiment with ether drag, the Lorentz transformation, the FitzGerald-Lorentz "contraction hypothesis," and Poincaré's direct confrontation with the idea of "relativity."

As Soviet scholars shifted from the idea of Einstein's indebtedness to Mach's philosophy, they became interested in Mach's influence on Einstein's scientific—rather than philosophical—creations. V. P. Vizgin, a leading expert on the historical roots of Einstein's contributions to physics, deserves the major credit for pointing out the clear traces of Mach's influence on the general theory of relativity.[25] Although Vizgin subscribed to the consensus view that Einstein was the sole creator of the general theory, he thought it undeniable that Einstein had consulted individual scientists or their works in seeking solutions to specific problems. He believed, in this connection, that Einstein received more help from Mach's physical theories than from any other scientific source. In fact, he noted that Mach's influence was formidable in every phase of the germination of the general theory of relativity, from the formulation of the equivalence principle in 1907 to the construction of the general covariance equations of the gravitational field in 1915. Vizgin brought Mach closer to Einstein by clarifying some of his more obscure ideas and keeping a record of his contacts.

For one historian of science, the question of interest was the extent to which the intellectual climate at the turn of the century might have influenced Einstein's thinking. Attentive to Paul Forman's suggestive and elaborately documented claim that Heisenberg transplanted the idea of "indeterminacy"—or "acausality"—from the intellectual atmosphere of the Weimar Republic to microphysics, Elena Mamchur was prompted to ask whether Einstein had absorbed the idea of relativity from the intellectual atmosphere and cultural impulses of Western Europe and the United States at the turn of the century.

Obviously stimulated by Lewis S. Feuer's discussions of "the social roots of Einstein's theory of relativity," Mamchur identified several suggestive early-twentieth-century prototypes of relativism, floating whimsically but adding up to a formidable change in the spheres of high culture. She recognized strong elements of relativism in the writings of Thornstein Veblen, the musical atonality of Arnold Schoenberg, the cubism of Georges Braque and Pablo Picasso, and the avant-garde poetry of Max Jacob and Guillaume Apollinaire.[26] As Mamchur wisely noted, the mere enumeration and fleeting scrutiny of parallel developments in a society was a far cry from a full explanation of turning points in the history of thought and aesthetic values. She did point out, however, a challenging and fertile area for future historical-sociological research.

By 1987, when Mamchur published her article, Soviet scholars had gone a long way toward sorting out the historical influences on Einstein and his own enormous influence on the development of modern science. They displayed inordinate strength in piecing together the labyrinthine complexities of Einstein's role in enriching the substantive concerns and methodological tools of modern science. They did not make much progress, however, in tackling the ever-interesting and challenging problem of the sociocultural moorings of the principal ideas built into the theory of relativity.

There were many indications that the future would bring a more extensive and systematic effort to give specialists in the philosophy and methodology of science a solid grounding in the sociological aspects of their discipline. A prospectus describing future training plans, issued in 1991, demanded a heavy concentration on the general cultural background of science and on the anatomy and dynamics of the scientific community. In the list of theoretical suggestions for organizing methodical and in-depth studies of the evolution of scientific thought, the prospectus made no reference to the writings of Marxist pioneers or to Soviet studies. Instead, it showed a keen interest in the theoretical models advanced by Robert K. Merton, Thomas S. Kuhn, Karl Popper, Imre Lakatos, Gerald Holton, Stephen Toulmin, and Paul Feyerabend. Merton's conception of the ethos of science underscored the principles of "universality," "collectivity," "objectivity," and "organized skepticism" as the pivotal norms of the moral code of the scientific profession. It also defined the exclusive right of the scientific community to certify scientific knowledge.[27] Merton's conception should prove helpful in the ongoing efforts to gain a deeper understanding of Stalin's violations of each major norm of the ethos of science and of the historical dynamics of the Soviet scientific community.

Spokesmen for the Soviet scientific community needed no reminders about violations of the ethos of science in their country, especially in the era of Stalinism. They agreed with Einstein that selfish interests and the opportunistic actions of individual scientists were common, and that a pure dedication to science of the Max Planck type was not only a rare phenomenon, but also the only guarantee for the advancement of scientific knowledge.[28] They were ready to censure the kind of behavior exhibited by Trofim Lysenko, who had relied on outside authorities to achieve a commanding position in Soviet biology. The moral transgressions of such pseudosciences as Olga Lepeshinskaia's "noncellular cytology" and N. Ia. Marr's "new theory of language" still await a thorough sociological study. All those dubious disciplines had become official parts of Soviet science without having been ratified by the scientific community. They were applauded as original creations of Soviet science and had played a major role in the Stalinist war on cosmopolitanism.

The scholars of the 1980's made another notable contribution to the history of Soviet science: reviving the memory of the Einsteinian scholars in the Soviet Union who had perished in the forced labor camps. That decade saw the publication of detailed biographical material on Semkovskii, Frederiks, Hessen, and Bronshtein—studies that at once gave those early proponents of the theory of relativity their due and brought to full light the mammoth proportions of the Stalinist terror of 1936–38. Without them, it would have been impossible to assess the degree to which pro-Einstein writings were equated with anti-Soviet activities.[29] Other newly released archival material had likewise shown that, in general, to write favorably about Einstein was to become a certain candidate for close secret police scrutiny and severe penalties. Iakov Frenkel, a consistent, brilliant, and gentle supporter of Einstein's scientific principles, was not sent to prison or into forced labor, but his punishment was no less severe. Marxist newspapers and journals had made him the prime personification of the most intolerable attributes of unpatriotic behavior and exposed him to the most vitriolic attacks and ominous threats from 1930 to 1952, the year of his death.

M. P. Bronshtein, dead at the age of thirty-two, was one of the most active and prominent physicists to vanish in Stalinist penal camps. In addition to original research in the challenging area where quantum mechanics and the theory of relativity met, he played a pioneering role in ensuring a constant flow of Western ideas in physics to the USSR. He wrote about experimental studies of the spectral composition of anomalously scattered radiation, which he considered the most promising initial step to more advanced studies in relativistic quantum mechanics.[30] In 1931, Bronshtein was the first Soviet scientist to present a comprehensive, methodical, and original picture of current trends in relativistic cosmology and the first to integrate Friedmann's mathematical ideas into the mainstream of modern cosmological thought.[31] Those views brought him the opprobrium of an unidentified censor, who found it necessary to append critical remarks to the published article. He expressed total disagreement with the notion, deeply embedded in current cosmology, that the future advancement of the science depended primarily on mathematical methods; and fastened on Bronshtein's reference to the "annihilation of matter" as an idea in direct opposition to both the law of the conservation of matter and Soviet ideology. He advised readers to question the scientific validity of the modern cosmological emphasis on the finite nature of space and time, which, he thought, offered a gloomy picture of the future of the universe. Instead, they would be well advised to concentrate on the dialectics of infinity and finitude. As if all this was not enough, the censor was outraged by Bronshtein's favorable reference to Arthur Eddington's ebullient endorsement of the idea of the expanding universe.[32] In Marxist circles at that time, Eddington was considered the major architect of a system of mystical thought built on the theory of relativity. More generally, the publication of studies on the life and work of the four eminent Einsteinian scholars helped to expose how the Stalin reign had shattered the last vestiges of academic autonomy and paralyzed both the moral code and the esprit de corps of the Soviet world of scholarship.

In the years of the thaw, historians of science had tended to view the Stalinists' ideologically motivated attacks on physicists as minor detours on an otherwise straight path to the unity of dialectical materialism and science. Passionate condemnations of the theory of relativity received only a passing men-

tion. The historians consistently overlooked or glossed over the more unpleas-
ant episodes in the fiery ideological crusade against modern developments in
physics. They showered one of the most resolute of Einstein's foes, I. V. Kuznetsov, with praise for his effort to show a fundamental agreement of Bohr's
correspondence principle with the laws of materialistic dialectics. In fact,
Questions of Philosophy remembered him, in 1971, as the founder of a school of
thought in Marxist philosophy. The thaw philosophers tended also to skim
over the more absurd demands of their Stalinist predecessors—like Kuznetsov's demand that Einstein's general theory of relativity be completely eliminated from Soviet scientific thought. An irate physicist noted sardonically in
1990 that Kuznetsov's portrait continued to decorate a wall at the Institute of
the History of Natural Science and Technology.[33]

M. B. Mitin, the chief architect of Stalinist ideology not only escaped sharp
rebuke but made something of a comeback in the Brezhnev era, when he occupied the prestigious position of editor of *Questions of Philosophy*. Working am-
bitiously on keeping the thaw close to the Stalinist tradition in that period of
stagnation (*zastoi*), he claimed, for example, that the evolution of de Broglie's
and Einstein's epistemological orientations had moved in the direction of materialism. The inverse, and more accurate, idea that dialectical materialism
may have moved in the direction of de Broglie's and Einstein's philosophical
positions did not enter his mind.[34]

Only in the age of perestroika were the Marxist philosophers finally identified as "charlatans" and "vulgarizers" of Marxist theory, as men of low moral
standards, as major contributors to the stagnation of philosophical thought in
the USSR, and as purveyors of pseudoscientific ideas. Nor did their scientific
allies get anything less. Even so eminent a man as A. I. Oparin, the creator of an
internationally recognized theory of the origin of life, did not escape severe
criticism for his close association with the Stalinist attacks on science and the
scientific community. Oparin had supported Lysenko as a means for personal
aggrandizement.

Although Stalin's death in 1953 and Khrushchev's exposé of the evils of Stalinism in 1956 had gone some way in setting the scientific community on the
road to recovery, only perestroika would bring it close to full health. To a significant extent, the struggle went on under the banner of the "de-ideologization" of science—of freeing the world of scientific scholarship from total subservience to the ideology of the political party in power. The cultural and psychological forces unleashed by perestroika faced the Gargantuan task of rees-
tablishing the full authority of the scientific community.

Einstein and the New Humanism

Moved by the searching spirit of perestroika, a *Communist* writer asked the rhetorical question: "Can we assume the full unity of the true, the good, and the beautiful?" Although he was not prepared to answer in the affirmative, he did acknowledge that at certain times ethical and aesthetic factors made more "effective" and "real" contributions to society than scientific ideas. The kind of beauty embodied in the paintings on terra-cotta statuettes, temples, and murals, for example, had reigned supreme for more than a millennium of cultural creativity. These statements were part of the stern rebuttals of Stalinist policy that translated art and morality into cognitive subdivisions of science, and all of them into tightly controlled vehicles of ideology. In no small measure, they were also an echo of Einstein's "newly discovered" comments on the humanistic foundations of science.

Until the 1980's, Soviet writers extolled the general scientific achievement of their country as the mainspring of humanistic thought and values. With the advent of perestroika, they no longer concealed the fact that Soviet science, no less than science in capitalist countries, did not always work in full harmony with humanistic ideals. Inspired by the warnings of Einstein and Sakharov, they ascribed the galloping militarization of the world and the threat of nuclear and biological warfare to the channeling of much of scientific effort toward goals contradicting the cardinal norms of humanism. At long last, the Soviet political authorities recognized that the function of the scientific community was not just to exploit nature, but also to protect the world and its resources. Not many Soviet commentators disagreed with Einstein's statement that modern societies, including their own, were caught between rapid advances in science and the erosion of moral bonds.

The widening of the base of humanistic approaches to science, reinforced by an earnest de-ideologization of social and cultural values, invited a fresh analysis of Einstein's views on socialism. It was not until the late 1980's that this problem drew the close attention of Soviet writers. In 1989, *Communist*, the official journal of the rapidly fading Central Committee of the Communist Party, carried a translation of Einstein's "Why Socialism?," originally published in 1949. That was followed, in 1990, by an analysis of Einstein's views on socialism and their relevance to the present state of affairs in the USSR.[35] G. E. Gorelik, the author of the article, recognized Einstein's claim that the success of socialism depended on the solution of an extremely difficult sociopolitical problem: how to prevent the centralized state from becoming all-powerful and overweening, and how to protect the rights of the individual by instituting democratic counterweights to bureaucracy. He did not hesitate to identify Einstein's "solution" with the ultimate goal of perestroika. Nor did he hesitate to mention approvingly Einstein's statement that science was not a creator of social goals, a direct challenge to the ideology of Stalinism. The fate of socialism, including the Soviet variety, depended not on its scientific foundations but on its ethical constitution.[36]

In a particularly thoughtful paper published some years earlier, Gorelik, the leading expert on the social and political history of Soviet physics, offered insightful comments on Einstein's statement that moral predisposition, aesthetic impulse, and religious sentiment encouraged a steady search for wider horizons of positive knowledge. More specifically, he wrote about Einstein's explanation of how the feeling of beauty, the power of moral sentiment, and religious inspiration exercised a favorable influence on the professional work of physicists.[37] In elaborating his argument, Gorelik observed that Einstein's worldview, dominated by the idea of cosmic unity, recognized no rigid boundary separating the world of human morality from the physical universe. As Einstein saw them, the physical and moral worlds obeyed the rigorous authority of determinism. He treated the inexorable power of causality as the basic mechanism of scientific explanation in both the physical and the social order. He could not accept Heisenberg's principle of indeterminacy for the simple reason that it violated the law of cosmic unity. Einstein readily admitted that Spinoza's powerful arguments in favor of ethical determinism played a vital role in shaping his worldview. Spinoza, he wrote, deserved credit for discovering the universality of causal relations at a time when that idea was only beginning to ignite the imagination of naturalists.[38] By stretching the logical links in his reasoning, Gorelik presented Einstein's cosmic unity as a synthesis of humanistic motivations and physical determinism. He thought the synthesis worked because it had no room for either anthropocentrism in religion or free will in the world of moral prescriptions.

Gorelik's essay reflected a major change in Soviet views on the relations of science to ethical values. Unlike the Stalinist philosophers, for whom moral law rested exclusively on scientific foundations, the latter-day philosophers found the notion of the moral foundations of science both eminently attractive and essential to any search for universal values. Whether they depended on philosophical meditation or on sociological analysis, they were encouraged and assisted by the ideas built into Einstein's humanistic predisposition.

The continued study of analogies in Einstein's and Dostoevsky's humanistic views added a new dimension to the burgeoning inquiry into the literary and philosophical legacy of the great Russian novelist. It gave more depth to the professional interest in Dostoevsky's general views on the place of science in modern culture. To be sure, the research in that area was slow to come. R. K.

Balandin's thoughtful and provocative essay "Dostoevsky and Natural Science," published in 1992, represented the first systematic analysis of the great writer's position.[39] Although Dostoevsky did not make a habit of offering scientific explanations of natural and social reality, he made extensive use of scientific metaphors and expressed strong views on the turning points in the development of modern science—on the promises of non-Euclidean geometry, Darwin's evolutionary theory, and geological historicism, among others. Balandin did not mention the crucial fact that Dostoevsky had a favorite scientist, Claude Bernard, who, in his opinion, blended a powerful and accomplished scientific intellect with a deep and refined artistic appreciation, a combination Einstein had hailed on many occasions. Always the visionary, Dostoevsky was the first to use the word "sputnik" to designate an artificial object orbiting the earth.

Perestroika writers, much impressed by Einstein's glowing references to Poincaré's humanistic views, began to take a fresh look at the French scientist. They found notable similarities between their views on the aesthetic aspects of science. Einstein gave full support to Poincaré's praise of science as a source of deep appreciation of the artistic attributes of nature; in mathematical methods, he saw the mechanisms for a most perfect portrayal of the harmony of nature.[40] According to A. B. Migdal, a member of the Academy of Sciences, the aesthetic quality of science was two-pronged: it revealed the harmony of the material world, and it presented the beauty of logical constructions codified in mathematical expressions.[41] "The search for beauty, that is, for the unity and symmetry of the laws of nature," he added, "is a characteristic of twentieth-century physics, particularly during the most recent decades."[42]Another writer observed that "in the final analysis, the aesthetics of science is essentially a knowledge about the harmony of the world, which is deduced from our experience of chaos and complexity."[43] Still another writer noted that "simplicity, elegance, and harmony are characteristics of all leading scientific theories of our age. . . . As a principle of knowledge, symmetry is closely linked with harmony—with beauty."[44] Soviet writers were lavish in giving Einstein the main credit for identifying harmony, simplicity, and symmetry as the pivotal attributes of the beauty of both nature and successful scientific theories.

The study of the ties between science and morality received as much attention as the study of the ties between science and art. Here again Einstein's influence played a vigorous part. Soviet writers showed a strong inclination to accept Einstein's rule that it was much more farsighted to explore the moral foundation of science than the scientific foundation of morality. Einstein gave a forthright description of the combined contributions of the major nonscientific modes of inquiry to the advancement of scientific knowledge: "Our moral leanings and tastes, our sense of beauty and religious instincts are all tributary forces in helping the reasoning faculty toward its highest achievement."[45] In an article on socialism, Einstein delivered the same message: "Socialism is directed toward a social-ethical end. Science, however, cannot create ends and, even less, instill them in human beings; science, at most, can supply the means by which to attain certain ends. . . . For this reason, we should be on our guard not to overestimate science and scientific methods when it is the question of human problems."[46]

Post-Stalinist scholars also detected a basic similarity in Poincaré's and Einstein's views on the relations of science to morality. When Poincaré stated that "there is not now and there will never be scientific ethics in the strict sense of the world, but science can be an aid in an indirect manner," he anticipated both a central idea of Einstein's humanistic thought and a cardinal interest of perestroika supporters.[47]

The searching temper of perestroika favored a closer examination of the cultural role of religion as a dynamic part of "planetary humanism," a marshaling force for the protection of universal cultural values from the threat of nuclear warfare and other massive problems of modern civilization. It was an echo of both Einstein's dedicated efforts to mobilize international opinion for world peace and "the new internationalism" based on the most fundamental principles of humanism. As two contributors to the journal *Philosophical Sciences* wrote in 1990: "Modern thought, based on a recognition of the primacy of universal human values, opened new perspectives for closer cooperation between Marxists and Christians."[48] The new dialogue concentrated not on doctrinal differences and ideological preferences, but on common interests in strengthening the universal bonds of mankind. For the first time, Soviet writers found something positive to say about the role of the Russian Church in modern society; for example, one praised the church for sponsoring a series of international conferences devoted to the grave issues of world peace.[49] They admitted that a successful—and most desirable—dialogue with religious spokesmen must transcend the narrow and blinding domain of ideological issues.

As part of the new religious inquiry, Soviet scholars looked afresh at the Protestant Reformation and its impact on scientific thought. Their predecessors had produced respectable work on the Renaissance—on both its creations in the arts and its scientific legacy—but they had tended to dismiss its central development as a sideshow in the grand course of Western culture. In the view of scholars like V. V. Lazarev, who wrote on the subject in 1987, the earlier historians had got it all wrong. Not the Renaissance, but Protestantism, had set philosophy and science free to search for secular wisdom. It had strengthened the cultural forces that replaced the Bible by the book of nature, God by reason, religion and the church by politics and the state, and the sky by the earth. Lazarev found the beginnings of modern philosophy and science not in the structure of socioeconomic relations, the mainstay of Marxist theory, but in the cultural values and outlooks brought forth by a religious revolution. In basic agreement with the interpretations advanced by Max Weber and Robert K. Merton, Lazarev viewed the Protestant emphasis on the rationality of action, the practical orientation of knowledge, the worship of work, and the individual interpretation of experience and intellectual stimuli as cultural values that created a healthy atmosphere encouraging involvement in science as a moral duty.[50]

Stalinist historians and philosophers of science had similarly disparaged Newton's world outlook as a combination of theology, science, and alchemy, a muddle of ideas that reflected the confused and contradiction-ridden state of seventeenth-century thought and social fermentation. Under the thaw and particularly under perestroika, Soviet scholars came up with a radically different interpretation of Newton's world outlook. It was now held to be an organic blend of consistent mechanistic orientation in "natural philosophy" and carefully articulated theological thought. Swayed by Einstein's ideas, scholars found common ground for Newton's religion and science: a firm belief that Biblical writing was not simply "a book of revelation," beyond the power of human understanding, but a special kind of historical testimony encouraging a rational study of the universe.[51]

Under Stalinism, all discussions and publications about religion had served only one purpose: to strengthen atheism as a major pillar of official Soviet ideology. In the age of perestroika, the cause of atheism continued to be very much alive, but it was no longer forbidden—or indeed particularly unusual—for writers to deal with all sorts of religious topics without mentioning, let alone promoting, atheism. They wrote, for example, on such previously untapped topics as the harmonious relations between science and religion, modern efforts in Moslem countries to bring Islam closer to science, and the religious connotations of the "holographic paradigm," combining the "new physics" and the "eternal philosophy" of such classical belief systems as Taoism, Vedantism, Hermetism, and mystical Christianity.[52] The basic intent of these and similar studies was simply to inform readers about religious ideas to which they had never been exposed, not to demonstrate their ideological errors.

In searching for a smooth transition to new interests in the cultural role of religion in modern society, Soviet scholars received much help from Einstein's suggestive ideas. Einstein offered them models of religious beliefs that stopped midway between traditional religion and atheism. His views represented a combination of Spinozistic pantheism and an original idea of "religiosity" that generated enthusiasm for science by favoring a picture of a universe dominated by harmony and rationality.[53] Suggestions like Einstein's found a direct expression in the declaration that the combination of scientific and religious meditations of "Descartes, Galileo, Newton, Huygens, and Leibnitz" led to the "formation of the mechanistic view of classical science."[54]

The Mainstream View of Einstein's General Theory

In the long and dramatic flow of history from the October Revolution in 1917 to the dismemberment of the Soviet Union in 1991, no criticism of Einstein's scholarly legacy stood above challenge by opposing views. Even during the harshest years of Stalinist efforts to create a Soviet version of the general theory of relativity, opposition was clearly present and sufficiently crystallized to be readily identified. A steady weakening of ideological controls and philosophical restrictions during the post-Stalinist thaw produced nearly universal acceptance of Einstein's physical theories in the age of perestroika. Simultaneously, and perhaps not surprisingly, philosophers on the whole showed a declining interest in Einstein. Some did continue, under his influence, to probe more deeply into the humanistic aspects of science, but many more, fired up by the critical spirit of the times, set themselves to the arduous task of detecting and repairing the crevices in the structure of Soviet society. What was of most concern now was not the ideological enemies of the state, but the critical problems of a rudderless society.

The most outspoken supporters of Einstein's general theory in the perestroika period were V. L. Ginzburg, Ia. B. Zel'dovich, I. D. Novikov, and L. P. Grishchuk, all astrophysicists. According to Ginzburg, there was no room for questioning the scientific authenticity of the application of the general theory of relativity to "a wide spectrum of phenomena," particularly of a cosmological nature. He noted that the principle of the equivalence of inertial and gravitational masses was at the top of the list of experimentally confirmed Einsteinian principles, and that the verification of isotropy was near the top. He concluded that the general theory of relativity was "well substantiated by experimental proofs."[55] The continuing search for experimental proofs for Einstein's theory, he wrote, added streams of new ideas to physical theory and of new research designs to modern technology. Ginzburg did not deny that Einstein's key ideas required refinement and amplification to realize their full potential in both theoretical and experimental research. He protested against the indiscriminate use of cosmology in the ideological war against idealistic philosophy. Cosmology, he noted, could neither prove nor disprove the existence of God.[56]

Ginzburg's unalloyed optimism about the scientific strength of the general theory of relativity was best expressed in the following statement:

> Today this theory is in a more secure position than ever before. No errors mar
> the explanations of its relations to weak gravitational fields; all tested predic-
> tions of the theory have produced positive results. Experimental research in
> weak fields continues and promises to accumulate much richer material in the
> near future. At the same time, the central attention has shifted to strong gravi-
> tational fields. Established as a classical theory of gravitation, the general theory
> of relativity is presently concerned with quantum generalizations and with the
> problems close to classical singularities. Here cosmology, including the full
> range of the general theory of relativity, meets microphysics. In the domain of
> classical (nonquantum) interests, the problem of the discovery and study of
> black holes has been the central concern of the general theory of relativity. Until
> black holes are discovered, it will be difficult, if not impossible, to ascertain the
> correctness of applying the general theory of relativity to strong fields. A dis-
> covery of black holes would prove to be one of the greatest triumphs of the gen-
> eral story of relativity.[57]

Zel'dovich thought that, in contrast to the theory of elementary particles,
which was still very much in the formative stage, the theory of gravitation, qua
the general theory of relativity, was fully developed in its cardinal principles.
He opposed the popular opinion in the Soviet scientific community that the
future development of the general theory of relativity depended mainly on astronomical data. Nor did he think that astronomy, relativistic or not, could
offer much help to future advances in the study of nuclear relations and the
discovery of new particles. Zel'dovich agreed with Ginzburg that there was no
need for major readjustments in the general principle of relativity. In an article
coauthored with Grishchuk, he stated that the general theory of relativity was
"a completely satisfactory theory of gravitation," and that there were neither
internal causes nor experimental contradictions that would warrant any
changes.[58]

"Einstein's general theory of relativity," according to Novikov, was "the
only perfectly self-consistent relativistic theory of gravity, free of internal contradictions and conceptually very beautiful. Not only direct experimental tests,
but also its internal structure and interrelations with other branches of physics,
have convinced most physicists that the general theory of relativity is above
challenge."[59]

Ginzburg and Zel'dovich worked on two levels. They made a deliberate and
successful effort to reduce epistemological involvement to a minimum. That
strategy helped them avoid paths leading directly and inexorably to ideological
issues of the most inflammatory nature. By deliberate design, they avoided

philosophical metaphors. At the same time, as their scientific work clearly indicates, they managed with remarkable skill to keep abreast of both the rich and often confusing streams of theoretical thought and the latest achievements in experimental physics related to the theory of relativity and relativistic cosmology. In general, the two astrophysicists reminded their colleagues that they found it much more profitable to take advantage of the rich potential of Einstein's ideas than to undertake a premature building of new theoretical structures. They also pointed out the need to take a careful look at theories presented as substitutes for Einstein's ideas or as alternate routes to the pivotal ideas built into the general theory of relativity.

Grishchuk gave a fairly accurate picture of the typical attitude of his fellow scientists when he asserted that the general theory of relativity provided an explanation of gravitation based on the incontrovertible fact of "the curved space-time continuum," perfectly upheld by all experiments so far conducted. Einstein's theory, in his view, presented an ideal combination of intellectual depth and mathematical harmony. Like many other leading scientists, from Landau to Zel'dovich, he saw in the general theory an achievement of great aesthetic value.[60]

Grischuk and the others thought that every branch of modern physics was seriously affected by Einstein's ideas. That view, however, was occasionally challenged by other scientists. Especially in the age of perestroika, some effort was made to show that powerful twentieth-century currents of Newtonian science had grown and advanced independently of Einstein's theoretical views. There were claims that the expansion and revitalization of Newtonian science were much stronger in the second half of the twentieth century than in the first. Soviet writers pointed out that Newtonian science was particularly strong in the study of unstable and nonlinear mechanical systems, and agreed with Ilya Prigogine's statement that classical dynamics continued to be a blossoming science and had a very promising future.[61] Most Soviet physicists honored Einstein by claiming that his contributions to science were as productive and epoch-making as those of Newton. They agreed with Einstein that it was Newton who prepared the way for the field theory and enriched science with "the law of causation" and "the differential law."[62]

A total acceptance of the fundamental principles of the general theory of relativity did not imply a uniform interpretation of their messages. The fact that almost all components of modern cosmology were anchored in it did not prevent the existence of many cosmological models. In no other science was the work of intuition invited on a larger scale. The anthropic cosmological principle, the idea of the plurality of universes, and the reliance on topological foundations in constructing cosmological models were only a few theoretical schemes finding their niches in the continuous flow of suggestive thought under the umbrella of the general theory of relativity.

Moreover, there were intermittent rumblings from dissident quarters. Ginzburg and Grishchuk, in particular, were occasionally accused of treating Einstein's theories as closed systems of thought growing on their self-generated momentum and internal impulse. The critics alleged that Ginzburg and his group ignored or fought suggestions not readily reducible to Einstein's theoretical legacy. The more zealous critics suggested that Ginzburg and Zel'dovich were inclined to think that the theory of relativity should be integrated with quantum mechanics only on the condition that the unified system fully preserved all Einstein's principles. True to their limitless faith in the superior intellectual power and promises of Einstein's scientific creations, Ginzburg and his colleagues dedicated much of their scholarly effort to popularizing the achievements of relativistic cosmology.

Many scholars chose to express their commitment to the general theory by emphasizing specific features. E. L. Feinberg, for example, gave Einstein much credit for helping create an "intellectual revolution" that brought together C. P. Snow's "two cultures": one based on the logical rigor of thinking traditionally attributed to the natural sciences, and the other on the "translogical"—intuitive—sources of humanistic explanation.[63] A. T. Grigor'ian noted that the future belonged to Einstein's unique way of applying the principle of correspondence to the history of the natural sciences.[64]

The Competing Positions

Although the Fock-Aleksandrov orientation had all but dropped completely out of the picture by the 1980's, individual questions posed by the two eminent scientists continued to be alive and to influence the effort to eliminate the contradictions between the general theory of relativity and dialectical materialism. The role of the observer in defining physical reality, the need for privileged coordinate systems, and the links between the general covariance and equivalence principles as formal arrangements of mathematical procedures continued to preoccupy the experts in relativistic physics.

Thinking in Fock's terms, Gribanov, who emerged as one of perestroika's principal philosophical spokesmen, ended his *Philosophical Views of Albert Einstein* (1987) with a reminder that, despite Einstein's epoch-making achievements, he showed imprecision and inconsistency in several segments of his thought. He, for example, isolated entire sections of his elaborate mathematical operations from physical content, failed to advance a consistent conception of "relativity," did not avoid contradictions in his interpretations of the "objective world," and was lax in his portrayal of the role of observer in physical analysis.[65] These objections, however, did not stand in the way of Gribanov's firm conviction that the theory of relativity gave a deeper understanding of the structure of the universe than classical physics, and "revolutionized the notions of time, space, and motion—the entire field of cosmology and cosmogony."[66] In his opinion, scientists should focus on refining and advancing Einstein's general theory as both a primary theoretical source of cosmology and a true springboard in the search for a unified field theory. He ruled out any need for substitute theories.

The views of Fock and Aleksandrov were dismissed in part because their effort to negate the physical relevance of a portion of Einstein's mathematical apparatus was suspected of an ideological bias, even though no charges were openly stated. T. A. Artykov claimed that much of Fock's criticism of the mathematical apparatus of the general principle of relativity was based on a misinterpretation of Einstein's ideas. He thought that Fock erred in transferring J.-L. Lagrange's formal statements on covariance to Einstein's theory. In his opinion, Fock considered the general principle of relativity an "auxiliary" tool rather than a physical concept, and he failed to relate that principle to the concepts of Galilean-Newtonian physics, where it had real content and methodological meaning.[67] Artykov thought that Fock was not correct in claiming that Einstein considered the general theory of relativity a generalized version of the special theory of relativity, overlooking the fact that the two theories referred to different types of physical reality. In any case, already in the thaw years, as we saw, a rapidly increasing number of physicists and philosophers were unwilling to part with the idea—and the name—of relativity, in popular use since the early years of the twentieth century.

As a work of science alone, Fock's *Theory of Space, Time and Gravitation* continued to stand out as one of the Soviet Union's most notable contributions to the discussion centered on the general theory of relativity, but it steadily lost ground from the 1970's on because it tried to reconcile two irreconcilables: an open scientific theory and a closed philosophical stance. Its most lasting influence was to help Soviet Einsteinian studies focus on the physical phenomenon of the space-time manifold, particularly in its relevance to the theory of gravitation. A literature of monumental proportions treated the changing conceptualization of space and time, and of space-time, as the most reliable path to the understanding of fundamental differences between the mechanistic and relativistic revolutions in science. As seen by Soviet writers, the changes in interpretations of space-time had the potential to affect all other regions of knowledge. Physics, we were told, became the leading science because it gave a fuller picture of the characteristics of space-time than any other branch of objective knowledge.[68] The concept of space-time found its way to literary criticism (Bakhtin's chronotope), social psychology, sociology, philosophy, anthropology, and biology. It harnessed the limitless power of imagination in treating the possible multiplicity of physically distinct cosmic worlds. The growing preoccupation with the plurality of worlds, based on distinct space-time configurations, was a direct challenge, perhaps only a transitional one, to Aleksandrov's notion of absolute space-time.

Physicists and historians of science remembered Fock sentimentally for his eloquent endorsement of Lobachevskii's non-Euclidean geometry as a direct contribution to the general theory of relativity. "Every new step in the advancement of Einstein's theory of relativity," he had declared in 1950, was "a step forward in the elaboration and refinement of Lobachevskii's ideas."[69] More to the immediate point, in this period when the full horrors of the Stalin era were exposed, they gave full marks to this man so widely known as a strict adherent of dialectical materialism for his courageous campaigns against recurrent attacks on Soviet physicists by "ignorant" Marxist philosophers. He was praised, in particular, for writing letters to key representatives of the central government pleading for lenience toward scientists accused of crimes against the state.[70]

Ivanenko's enormous contributions, most of which were made before perestroika, were recognized in an international festschrift in 1991. That work, *Modern Problems of Theoretical Physics*, showed Ivanenko's unusually broad span of scientific interests—he had delved into quantum gravity, quantum cosmology, gravity and new forces (including the problem of torsion), and stochastic quantum mechanics.[71] But his attention was never far from a dedicated search for a unified field theory. Nothing in the intricate theory of gravitation—and its moorings in both quantum mechanics and the general theory of relativity—had escaped his attention.

Stated before 1980, Ivanenko's ideas became particularly attractive in the free atmosphere of perestroika. Streams of new ideas gushing from the West brought forth a profusion of fresh views on relativistic cosmology springing from advances in physics and mathematics outside the Einsteinian framework, and revitalized the search for a unified field theory. Interest in the subject received a powerful boost in 1985, when Vladimir Vizgin, a leading historian of modern physics, published a pioneering study on the developments in the first third of the twentieth century. The book showed the relevance of various views on the unity of the physical universe to the theoretical foundations of modern physics. It dissected the up-and-down search for common theoretical grounds of the theory of relativity and quantum mechanics.[72] Vizgin devoted a special study to controversies over Newton's study of universal gravitation.[73]

Although Vizgin did not say so explicitly, he undoubtedly agreed with most Soviet historians of physics that it would take the emergence of a new genius to resolve the problem of the unity and symmetry of the universe. As they saw it, Einstein's special theory of relativity and Max Planck's quantum theory were major forward steps in the search for cosmic harmony started by Newton's mechanics and Maxwell's electrodynamics. In their firm opinion, the new physics created by Einstein and Planck represented not a rejection, but an expansion and enrichment, of classical physics. The views of men like Thomas S. Kuhn and Gaston Bachelard that the Newtonian tradition had no place in modern physics did not have any supporters in the Soviet Union. Soviet scholars recognized both the revolutionary sweep of the new physics and its continuity with the old.

"Post-Einsteinianism," a label that made its appearance at this time, but was seldom used, did not point to a parting from the rich legacy of Einsteinian thought. Without a specific and clearly stated meaning, it usually designated nothing more than an effort to add another level to the Einsteinian physical edifice and to achieve a synthesis of the most promising achievements of modern physics. As G. M. Idlis pointed out, in concentrating on the study of elementary particles as bridges between quasi-closed macroworlds, post-Einsteinianism was effectively pursuing an idea suggested by Einstein and Rosen in "The Particle Problem in the General Theory of Relativity" (1935).[74] The term, in whatever context, merely expressed the hope of Soviet physicists to elevate Einstein's legacy to momentous new heights, to link it more firmly to more recent theories like the vacuum theory and the theory of supergravitation.

No matter what the differences between the groups of physicists we have been discussing, they all aspired, to one degree or another, to achieve an accommodation between the new physics and dialectical materialism, which, perestroika or not, was still the official philosophy of the state. Only a few diehard physicists resisted that move, notably A. A. Logunov, rector of Moscow University and a member of the Party's Central Committee, and his handful of supporters. Logunov made a last-ditch attempt to create a unique Soviet (Marxist) theory of relativity. Following Stalinist philosophers, he argued that Einstein erred in overloading his theoretical constructions with mathematical operations unrelated to physical reality, in depending heavily on such subjective categories as observer, frame of reference, and coordinate systems, and in showing a total disregard for conservation laws. Logunov constructed a substitute for Einstein's general theory of relativity, which he named "the relativistic theory of gravitation," and which in his opinion fully conformed to the Faraday-Maxwell notion of the electromagnetic field as a physical entity.[75] In 1984, Logunov published reprints of two of Henri Poincaré's papers on the theory of relativity, together with his own extensive commentary, aimed at showing that Poincaré had formulated all the basic ideas Einstein had presented in his special theory of relativity.[76]

As the USSR teetered ever nearer to complete collapse, more and more physicists came to speak out publicly against Marxism as the only acceptable philosophy of science. Despite their repeated calls for philosophical pluralism, no effort was made to produce a systematic and comprehensive non-Marxist study of the theory of relativity. By then, had those pluralists but known it, the day of dialectical materialism as a monopoly philosophy was about to pass.

In the twilight of the Soviet reign, relatively little was written about Einstein, particularly by philosophers. Where *Questions of Philosophy*, since its founding in 1946, had fastened time and again on the theory of relativity and its founder, it now showed no serious interest in either the one or the other. The country's leading philosophical periodical carried only a few articles on Einstein's science from 1985 to 1991. During the 1980's no important conference, usually a rich source for publications, dwelt on topics devoted exclusively or primarily to Einstein's work and life. That development did not mark a general decline in philosophical concern with Einsteinian science, merely a transitional transfer of the journal's interest away from the epistemological intricacies of modern science. It was at this time that the journal made a concerted attempt to revive the idealistic-metaphysical thought of V. S. Solov'ev, N. A. Berdiaev, and other religious philosophers who had previously been treated with utmost scorn.

There was one man, though, whose interest did not flag: the Uzbek philosopher of science T. A. Artykov. In 1990, he published, at his own expense, *A Philosophical Analysis of the Principle of Relativity*. He could obviously not have anticipated the historical importance of his small and unpretentious-looking volume. It was the last book on the theory of relativity to be published in the Soviet Union.

Conclusion

In the pre-perestroika age, Soviet philosopher-ideologues saw the universe as a manifestation of perfect order, a universe based on the regularities of the laws of nature and on the predictability of cosmic processes. The laws of conservation were the basic glue that held the universe together and explained both its structure and its dynamics. Those laws accounted for the unity of nature—the unity of mass, matter, energy, and electrical charge—which, in turn, was closely related to the principle of causality and to the universal determinism that explained the bewildering complexity and evolution of the interacting parts of nature. All these laws and fundamental conditions revealed the lavishly emphasized and exalted symmetry of the world.

That picture of the world had a distinct ideological bearing. It represented a cosmic projection of the orthodox Marxist notion of a perfect society—a society ruled by predictable and inexorable laws; dominated by the unity of cultural values and social attitudes; sustained by harmony in class relations; and open to predictable and planned evolution. Soviet philosophers made an earnest attempt to demonstrate a fundamental consonance of Einstein's and Dostoevsky's views on the cosmic unity of moral values with official Soviet ideology. In the age of perestroika, Soviet ideology became too loose and unsettled to provide working guides for cosmic views.

Each of the four periods we have studied represented a unique and clearly separated historical reality. Each made a distinct use of dialectical materialism in clarifying and codifying the norms for judging the consonance—or dissonance—between the cardinal elements of Einstein's theories and the dictates of Marxist-Leninist ideology. Each served as a significant reflection of the national temper at a given period of time.

During the first phase—the 1920's before the emergence of Stalinism—Bolshevik ideology drew nourishment directly or indirectly from the original writings of Marx, Engels, and Lenin, which were published, lectured on, and written about on a mammoth scale. The raft of new Marxist associations designed to reach every segment of the population spent most of their time trying to suffuse the Bolshevik dreams of a future society with the ideological norms prescribed by Marxist classics. But where Bolshevik ideology was simple and forthright, Marxist philosophy was a morass of tangled and contradictory ideas. Dialectical materialism—the Marxist philosophy of nature—boiled in controversy compounded by personal animosities. In the Communist Academy of the Social Sciences, founded in 1918 for the purpose of applying Marxist theory to new developments in society and culture, complete philosophical anarchy prevailed.

Underdeveloped and amorphous, dialectical materialism could not serve as an authoritative clearing-house for scientific ideas. Split to the core, the community of Marxist philosophers, which included a segment of unusually active revisionists, produced a full arc of attitudes toward the theory of relativity, from categorical rejection to total and enthusiastic acceptance.

Most physicists took advantage of the philosophical confusion to dwell exclusively on the scientific principles offered by the theory of relativity and to interpret them as a culminating achievement of modern science. Ia. I. Frenkel, V. K. Frederiks, A. F. Ioffe, S. I. Vavilov, and most other members of the scientific community were engaged primarily in presenting the theory of relativity to the general reading public and in explaining why they held optimistic views on the future role of Einstein's ideas in uncovering the mysteries of the universe.

A. A. Friedmann was the most notable exception. He undertook not only to give a popular account of the key principles of the theory of relativity, but also to enrich Einstein's legacy: he introduced the idea of the expanding universe, one of the cornerstones of relativistic cosmology. Friedmann was the only Soviet scientist with whom Einstein had a crucial dialogue. Totally detached from Einstein's views on the epistemological aspects of scientific knowledge, he showed a sophisticated familiarity with and a profound interest in the mathematical apparatus of the theory of relativity. Friedmann's cosmological idea was so novel that both Soviet physicists and Marxist philosophers took a relatively long time to integrate it into their thinking. When he died in 1925, the scientific community remembered him primarily for his contributions to meteorology.

Despite their tolerance of pluralism in the interpretation of Einstein's theories, Soviet Marxist scholars and ideologues aimed ultimately to bring scientific views into absolute unity with their philosophical tenets. Most Marxist philosophers favored granting dialectical materialism an unrestricted monopoly in handling questions of scientific methodology and in acting as a clearing-house for scientific ideas. Marxist theorists found themselves involved in seething intramural squabbles, but they generally agreed that dialectical materialism should be vested with supreme authority in evaluating scientific knowledge and in formulating substitutions for concepts, hypotheses, and theories rejected on ideological grounds.

Most of these philosophers viewed new scientific ideas with skepticism. They were ideologically conditioned to treat all non-Marxist ideas emanating from the West with much suspicion. In any event, criticizing such ideas was a safe and easy way to show their political loyalty, certainly safer than endorsing them. In their custody, Marxist theory became a conservative intellectual force. Lunacharskii's rancorous attack on genetics as a decadent creation of the bourgeois mind signaled the coming of a frontal attack on one of the most advanced branches of biology. Marxist philosophers showed a readiness to reject the theory of sets *in toto* because Georg Cantor, the founder of that branch of mathematics, wrote that his mathematical constructs were products of intuition and pure reasoning, rather than logical results of accumulated human experience. Many criticized Einstein for making a similar statement in respect philosophical relativism, which flew in the face of Lenin's observation that relativism not only denied the objective nature of knowledge but also negated the existence of absolute and universal guides for the development of science.[2]

For a long time, Soviet Marxists were suspicious of quantum theory because the so-called discontinuities, uncertainties, and complementarities introduced by Planck, Heisenberg, and Bohr, respectively, appeared to be extremely difficult to reconcile with such guiding principles of dialectical materialism as absolute causality and universal determinism. Attacks on the achievements of science in the West served as automatic parts of a general war against Western culture and its bourgeois impulses. At the very end of the 1920's, with Stalin's quick and carefully calculated moves to purify and consolidate the ideological front, the following disciplines were either outlawed or seriously impaired: psychology, demography, sociology, soil science, linguistics, and economics.

Soviet writers differed widely on the kind of reception Einstein's ideas got in their country before Stalinism took hold. At one extreme was O. Iu. Shmidt, a leading Communist academic, who stated at an important conference of Marxist scholars held in 1929 that in the past Einstein's ideas had encountered overwhelming opposition in Marxist circles. He suggested that A. K. Timiriazev's "strong" influence and the popularity of the "idealistic" interpretations of the theory of relativity in the West were the primary reasons for that opposition.[3] At the other extreme was the historian G. E. Gorelik, who stated many years later that the Soviet Union had received Einstein's ideas most favorably. He claimed that a majority of scholars thought the revolutionary restructuring of physics precipitated by the theory of relativity harmonized with the social revolution brought about by the Bolsheviks. And he added: "Small wonder that during the 1920's and 1930's Soviet scholars were in first place in advancing Einstein's ideas."[4]

In fact, the statements of both writers are obvious exaggerations; the truth was somewhere between the two extremes. Although most Marxist philosophers criticized the ideological implications of Einstein's scientific ideas, a clear majority of physicists both recognized the theory of relativity as a gigantic step forward in the growth of modern physics and stayed away from philosophical-ideological involvement. With the exception of Friedmann and a few other scholars, the scientists made little effort to enrich Einstein's contributions. The handful of physicists who criticized Einstein belonged to the older generation, loyal exclusively to the Faraday-Maxwell tradition. In no country in the 1920's was the underlying philosophy of the theory of relativity looked at more thoroughly, openly, and resolutely through an ideological prism than in the Soviet Union.

Nevertheless, even so devout a Marxist-Leninist as M. B. Mitin maintained a relatively open-minded attitude as the country entered the second—or Stalinist—period. Marxists, he wrote, should not overlook the important scientific and philosophical points of mutually beneficial contacts between dialectical materialism and the theory of relativity. While recognizing the existence of broad disagreements, he was particularly eager to point out a promising future for friendly and cooperative relations between the two disparate sets of theoretical thought. Mitin obviously had no grasp of the principles of Einstein's theory and was the last person who could have been expected to push for closer ties between Einsteinian and Marxist views. At the end of the decade, he became one of the chief champions of Lysenko's crusade against genetics.

As Stalin's terror began to unfold at a galloping pace after 1935, Marxist philosophers became less patient with Einstein. By then, any of their fellows who had seen the theory of relativity as a confirmation of dialectical materialism were gone, most of them victims of Stalinist terror. The new warriors, dominated by A. A. Maksimov and Ernest Kol'man, were eager to intensify the attack on Einstein's idealism. They recognized the grandeur of Einstein's scientific achievement, but they did not dwell on it. Their main guns were aimed at those of his pronouncements that allegedly gave comfort to mystical philosophers and the ideologues of "dying capitalism."[5]

Aided by the atmosphere of fear brought about by the Stalinist purges, most philosophers launched blistering personal attacks on the leaders of Soviet physics, accused of subversive efforts to sacrifice science to physical and mathematical idealism. Tamm, Fock, Frenkel, Ioffe, Landau, and Vavilov became regular targets of public chastisement for unpatriotic behavior and "pseudoscientific" work. Stalinist philosophers accused them of erecting barriers to a Soviet theory of relativity and a Soviet quantum mechanics. The Stalinist terror in 1936–38 took a heavy toll on other leaders in Einsteinian studies. Seven top experts in the theory of relativity—two philosophers (S. Semkovskii and B. M. Hessen) and five physicists (M. Bronshtein, V. K. Frederiks, V. A. Fock, Iu. B. Rumer, and L. D. Landau)—were sent to Stalinist prisons, from which four did not return.[6]

The wartime alliance with the Western democracies helped create a more conciliatory attitude toward physical idealism in general and the theory of relativity in particular. On the occasion of the celebration of the twenty-fifth anniversary of the October Revolution, Mitin told an assembly of the Academy of Sciences that the theory of relativity, despite its idealistic aspects, was generally in accord with the philosophical postulates of dialectical materialism. As late as 1947, M. E. Omel'ianovskii, in his *Lenin and Physics*, essentially reaffirmed Mitin's views and spoke of looking forward to further scientific and philosophical explorations of the rich world of Einsteinian thought.

Despite Mitin's prestige as an architect of Stalinist Marxism, many of the Stalinists continued their attacks on physical idealism, though they shifted their attention to a different set of Western scientists. As one of Vavilov's articles showed clearly, the main targets now were Eddington, Jeans, Milne, and Jordan, who, in their philosophical treatises, waged a direct and open war on materialism, determinism, and objective epistemology. In his attack on physical idealism, Vavilov chose not to mention, or to mention sparingly, Einstein, Bohr, and Heisenberg, obviously a quiet effort to move the attack away from the top echelon of modern physicists.

The hope for a more conciliatory attitude toward Einstein was quickly shattered. In the same year that Omel'ianovskii's book came out, A. A. Zhdanov, a member of the Politburo, told a hastily assembled group of Moscow scholars and writers that the time had come for a reinforced war on ideological impurities in the arts and sciences. He called for a frontal assault on all ideological deviations from Soviet norms and on cosmopolitanism, defined as a servile attitude toward Western cultural values and achievements. Zhdanov's clarion call resulted in an all-out war on physical idealism and its residues in Soviet science, in an emphasis on documenting the originality and intellectual power of the Russian philosophical tradition, and in a quick removal of all elements of "fideism" and "mysticism" from relativistic cosmology. The ideological house-cleaning was accompanied by an officially stimulated and supervised search for unique cultural qualities and expressions of Russian nationalism. Zhdanov called for a critical and thorough reexamination of Einstein's affinity with epistemological subjectivism and creationist metaphysics. He ordered a return to the 1930's: to a rigorous and uncompromising attack, both scientific and philosophical, on the idealistic moorings of Einstein's principles.

Despite Zhdanov's call for a merciless war against ideological impurities in physics, neither the physicists nor the philosophers sprang into immediate action. Like Omel'ianovskii, most felt that Einstein's idealistic lapses were much overshadowed by his great scientific achievements. But Zhdanov's policy statement could not be ignored, and in 1949, the two groups' main journals, *Advances in the Physical Sciences* and *Questions of Philosophy*, responded by launching a mammoth attack on both Einstein's epistemology and his physics. The criticism reached a peak in 1952, when I. V. Kuznetsov—in the symposium *The Philosophical Problems of Modern Physics*—announced that Einstein's scientific ideas were so absurd as to be beyond repair. He wrote that Einstein's "wrong" philosophical ideas had caused him to formulate faulty scientific principles. As clearly as anything else, that collection exposed the deep ideological rift between the Marxist philosophers and the leading physicists, who now found themselves subjected to even more acrimonious and threatening attacks than in the 1930's.

How did the leading physicists respond to Kuznetsov's condemnation of the theory of relativity? Although the nervous atmosphere he had helped create and the increased ideological pressure of government authorities compelled certain philosophically attuned physicists to intensify their criticism of Einstein's principles, not a single leading physicist endorsed the recommendation for a total rejection of Einstein's theories. Two scientists, the physicist V. A. Fock and the mathematician A. D. Aleksandrov, themselves dedicated Marxists, went so far as to attack Kuznetsov, Maksimov, and other representatives of the anti-Einsteinian campaign by exposing their unfamiliarity with modern physical theories, their flawed interpretations of dialectics, and their loose play with the laws of logic. For tactical reasons, they concentrated much more on the weaknesses of the Kuznetsov group's arguments than on the strengths of Einstein's contributions. They objected to Einstein's primary emphasis on the principle of relativity but otherwise stayed within the framework of Einstein's thought. Their intent was not to weaken but to strengthen the basic configuration of Einsteinian principles. Implicit in their arguments was the suggestion that scientific—in contrast to philosophical—criticism of the theory of relativity should be the task of physicists exclusively. They never criticized specific aspects of the theory of relativity without reminding their readers of Einstein's vast contributions to science. It is important to note, however, that in these exceedingly hard times, only a few other scientists and even fewer philosophers chose to join Fock and Aleksandrov in challenging the officially encouraged onslaught on Einstein's legacy. Nor should it be overlooked that there was no scarcity of physicists who chose to support Kuznetsov's campaign.

All this leads to the question: why was there no Lysenko in physics? How did physics, the science that evoked more extensive and explosive philosophical-ideological deliberations than any other discipline, escape falling prey to an administrative system modeled on Lysenkoist formulas? Although there is no one precise answer to that question, the unity of the community of physicists clearly played a large role.

Biology, at the time of the rise of Lysenko, was so severely split that the legitimacy of entire disciplines was disputed. Leo Berg, who was by no means alone in his position, challenged the scientific validity of the entire Darwinian tradition in evolutionary biology. A. N. Severtsov, a man of considerable influence, did not go along with the effort of I. I. Shmal'gauzen to advance a synthesis of Darwin's evolutionary views and the basic principles of genetics, insisting that the time for such a synthesis had not yet arrived. A blossoming branch of neo-Lamarckism, strong in Marxist circles, challenged the scientific legitimacy of genetics rooted in the Mendelian tradition.

Physics, by contrast, was virtually free of any fundamental splits. Only a handful of physicists considered the legacy of Faraday and Maxwell the last word in their science and refused to ally themselves with relativity and quantum-mechanical theories. In the words of a Soviet commentator, they did not recognize twentieth-century physics, nor were they moved by the enormous scope of its revolution. Most leading physicists recognized the immense potential of quantum mechanics and the theory of relativity. Most acknowledged the rapid decline of Newtonian supremacy in science. Biology, plagued by inner disunity, became easy prey to outside interference. Physics, dominated by an overarching intellectual unity, was in a relatively good position to resist it.

Particularly in 1936–38, numerous physicists became inmates of Stalinist political prisons, from which many did not return. The physicists, however, did not have as hard a time as as the biologists, because, in the words of Andrei Sakharov, their assailants were less aggressive than Lysenko and his official supporters.[7] There were nevertheless clear signs even then that the authorities were becoming increasingly alarmed about the tolerant attitude of a large number of leading physicists toward the "idealistic" tenets of the theory of relativity and were giving much thought to a Lysenkoization of physics.

The first effort to subject physics to Lysenkoist methods of ideological purification took place in 1938, when V. F. Mitkevich, a full member of the Academy of Sciences and a partisan of the old physics, came up with the idea of organizing a special session of leading Soviet physicists and philosophers to discuss the burning problems of the "corrupting" influence of idealistic philosophy on their science. His goal was to turn the projected debate into a public condemnation of the theory of relativity and quantum mechanics as scientific developments totally incompatible with the spirit and the letter of Soviet ideology and dialectical materialism. The conference was not held, however, for the simple reason that no "idealist" of high academic standing was willing to participate. Obviously, government authorities did not give Mitkevich's venture the support he expected. Perhaps they were too deeply involved in the Lysenko affair to open yet another ideological front in the difficult world of science.

A few months after the full Lysenkoization of biology had been achieved in 1948, the government ordered the academy to make a new effort to use the Lysenko method of "public debate" to expose the deep inroads of idealism and cosmopolitanism into Soviet physics. After meeting more than forty times, the organizing committee, made up mainly of selected members of the academy and University of Moscow physicists, discontinued its work without having reached an agreement. In this case, the leading physicists relied on delaying tactics to create one deadlock after another. That project, too, came to a sudden halt, obviously on orders from Stalin, who was fully aware of the world-wide criticism of Lysenkoist attacks on the integrity of science and was no doubt determined to avoid, for the time being, taking on the community of physicists. The explosion of the first Soviet atomic bomb may have played some part in that decision.

It would not be unreasonable to assume that in making physical idealism a target of attack, Stalin and other members of the Politburo were not interested so much in Lysenkoizing physics as in slowing down the more radical efforts of Soviet scientists to achieve an accommodation with Western physics. After all, Stalin and his aides allowed M. A. Markov to publish a virtual endorsement of the epistemological positions of the Copenhagen school in quantum mechanics, and they did not prevent Ioffe, Landau, Lifshitz, and Naan from disregarding at least some ideological restrictions in recognizing the revolutionary sweep of the theory of relativity. "Thaw" is an apt word for the period after Stalin's death, for the process of de-Stalinization in some instances moved at glacial speed. In the sciences, for example, biology continued to suffer under the oppressive reign of Lysenkoism for eleven years after 1953. In physics, by contrast, the thaw had immediate and lasting effects. Once released from the ideological controls of the Stalin years, the community of physicists managed to go their own way even under Brezhnev's calculated policy of "stagnation" (*zastoi*) or re-Stalinization.

Einsteinian studies, both technical and popular, benefited enormously from the decision of political authorities to deprive the philosophers of the right, used excessively under Stalin, to harass scientists suspected of ideological deviation. After 1956, no philosopher attacked an Einsteinian scholar because of detected idealistic aberrations in his thinking. No philosopher challenged Einstein on scientific grounds. It was not that the philosophers were no longer concerned about the elements of idealism in Einsteinian thought, only that they concentrated almost exclusively now on the affinity of Einstein's theoretical principles with materialistic dialectics. Most now took the position that Einstein's science and Einstein's philosophy were separate provinces of thought. Predictably, there were some recalcitrants who refused to part with the idea of the close connection between science and philosophy. In 1965, S. G. Suvorov continued to argue that "the positions Einstein took in physics [could] be understood only in the light of his general philosophical views."[8] Nor was Suvorov particularly eager to join in the widespread effort then under way to bring materialistic dialectics and the broader messages of Einstein's views closer to each other.

Just as the philosophers of the thaw turned their energies to accommodating the principles of dialectical materialism to Einstein's ideas, so they tried hard to accommodate Einstein's ideas to dialectical materialism. Stalinist philosophers had complained bitterly about Einstein's recognition of the primacy of "formal" geometry over "empirical" physics (they thought that Einstein considered physics a branch of applied mathematics). In a grand reversal, post-Stalinist philosophers came to the opposite conclusion: that Einstein—particularly in "Geometry and Experience," presented at the Prussian Academy of Sciences in 1921—considered geometry a tool of physics, and an empirical science at that. Soviet scholars thought that they came closer to Einstein's philosophical views, not only by attaching more amenable meanings to the principles of dialectical materialism, but also by detecting what they thought to have been Einstein's gradual retreats from idealistic positions in the theory of knowledge. They were particularly interested in documenting the intellectual odyssey that took Einstein from positivism in the direction of materialism. In their firm opinion, however, he stayed a good distance from dialectical materialism.[9] They continued to recognize strong differences between Einstein's views and Marxist-Leninist philosophy, but they now showed a stronger interest in exploring similarities.

The philosophers of the thaw, of course, did not work in a scientific vacuum. Whether they moved from dialectical materialism to Einstein's ideas, or from Einstein's ideas to dialectical materialism, Soviet physicists, particularly those with a flair for philosophy, contributed a large share to making the Marxist philosophy of science much more flexible and responsive to the onrushing streams of new ideas. They also helped make Marxist philosophy much more eclectic and intent on finding a middle ground in the whirlpool of philosophical thought. Dialectical materialism lost much of its identity as an integrated and internally coherent system of philosophical principles, but it gained in general flexibility and in adaptability to modern developments in intellectual culture.

Perestroika, introduced in 1985, put the finishing touches on the post-Stalinist thaw. It saw the elimination of numerous branches of intellectual pursuits—notably scientific communism, dialectical logic, historical materialism, and Marxist philosophy—from the domain of science. It created a new political atmosphere that helped provide institutional safeguards for protecting the world of scholarship from ideological abuses by authorities external to the pursuit of science. And it brought an end to a monolithic philosophy shielded by state authorities and strengthened by the official suppression of heterodox thought. These revolutionary developments produced a phenomenon previously known only during the early 1920's: a proliferation of competitive orientations in all universes of inquiry. As in the 1920's, the attitudes toward the theory of relativity covered the full gamut of positions, from full rejection at the one extreme to full acceptance at the other. In contrast to the 1920's, however, by far the vast majority of interested scholars favored the theory of relativity.

Through its leading representatives, the scientific community recognized the influence of Einstein's humanistic ideas on political attitudes and cultural values, matters of cardinal interest in this new searching age. M. A. Markov, the much-respected physicist who in 1947 had warned the Communist leaders about the dangerous political-ideological moves that threatened to separate Soviet physics from the most fundamental aspects of modern science, went so far as to suggest that Gorbachev's "new orientation in international relations" bore clear signs of Einstein's influence. Markov referred particularly to changes in Soviet polices that helped inaugurate the age of nuclear disarmament.[10]

To put all that has gone before in proper context, this final word. Soviet experts in the wide and elusive field known as the "science of science" wrote much about the inner workings and contributions of the various schools that had come to be formed around eminent scientists. They published numerous studies on the schools founded by Bohr, Rutherford, Born, and Heisenberg, among others. Many Soviet scientists, including Ioffe, Landau, Vavilov, and Vernadskii, were also treated as leaders of recognized schools. But not one of those writers ever spoke of an "Einstein school." In the prevalent opinion in the Soviet world of scholarship, the theory of relativity was so deeply ingrained in modern physics and its related sciences that the Einstein school encompassed essentially the entire scientific community.

Notes

For complete authors' names, titles, and publication data on the works cited in short form in these Notes, see the References Cited, pp. 257–84.

Introduction

1. *Dnevnik XII S"ezda*, no. 5, pp. 170–72.
2. Pauli, pp. 11, 133. See also Klein, *Paul Ehrenfest*, pp. 152–54.
3. Vizgin and Gorelik, p. 278. One of the official discussants ("opponents") of the Epstein paper read at the Twelfth Congress of Russian Naturalists and Physicians was N. N. Shiller, a Kiev University professor of physics known for his experimental studies in electrodynamics and thermodynamics and his elaborate and perceptive philosophical writings at the end of the 1890's on the anticipated doom of the mechanistic view of nature. Einstein published a brief communication in *Physikalische Blätter* in 1905 on one of Shiller's original contributions to the theory of entropy (Einstein, *Collected Works*, vol. 2, pp. 122–24).
4. Borgman, "Elektrichestvo i svet."
5. Eikhenval'd, "Materiia i energiia."
6. Gol'dgammer, p. 188.
7. Vizgin and Gorelik, p. 279.
8. Umov, "Znachenie Dekarta," p. 519.
9. Gulo, p. 284.
10. Umov, *Izbrannye sochineniia*, pp. 492–502.
11. As cited in Pisarzhevskii, p. 122. See also Umov, "Evoliutsiia fizicheskikh nauk," p. 27. For a Soviet interpretation of Umov's views on relativity, see Gulo, pp. 240–46.
12. Umov, "Evoliutsiia atoma," p. 24; Umov, "Evoliutsiia fizicheskikh nauk," p. 27.
13. Khvol'son, *Populiarnye stat'i*, p. 325.
14. Khvol'son, "Printsip otnositel'nosti," p. 1279.
15. Khvol'son, "Otnositel'nost' (printsip)," p. 922.
16. Khvol'son, *Printsip otnositel'nosti*.
17. P. N. Lebedev, *Sobranie sochinenii* (Moscow, 1963), p. 366.
18. Čapek, p. 253; Vizgin and Gorelik, p. 268.
19. *Novye idei v fizike: teoriia otnositel'nosti*, vol. 3. See also *Novye idei v matematike: printsip otnositel'nosti v matematike*, vol. 5.

20. Iushkevich, "Review."
21. Grdina, "Fizicheskii ili ogranichennyi printsip." See also G. R., "Review."
22. Orlov, "Osnovnye formuly," pp. 163–64, 175.
23. Gurevich, part 3 (no. 7), pp. 112–13.
24. Iushkevich, *Materializm*, p. 131.
25. Iushkevich, "Printsip otnositel'nosti," part 1 (no. 7), p. 206.
26. Ibid., part 2 (no. 8), p. 210.
27. Many of Iushkevich's translations are listed in Koialovich.

Chapter 1: The Early Soviet Reception of Einstein's Theories

1. Ioffe, Lazarev, and Steklov, p. 42.
2. Einstein, "Predislovie avtora," p. 5.
3. Maksimov, "Eshche raz o populiarno-nauchnoi literature," p. 123.
4. Bogoraz-Tan, pp. 3–4, 116. For comments on Bogoraz-Tan's views on the similarities between Einstein's theory of relativity and mythology, see Akhundov, *Kontseptsii prostranstva*, pp. 55–56.
5. Bekhterev, p. 314.
6. Florenskii, *Mnimosti v geometrii*. For critical comments, see Akhundov and Bazhenov, "Estestvoznanie i religiia," p. 53.
7. Jakobson, p. 718.
8. Vasil'ev, *Space*, pp. xi–xxiii.
9. The text analysis of Vasil'ev's views draws on ibid., pp. 80–91.
10. Ibid., p. 90. John Bernard Stallo was an American Hegelian of the Cincinnati school.
11. Gol'tsman, "Nastuplenie," pp. 80–81. For Vasil'ev's earlier favorable comment on Mach's philosophy of science, see Vasil'ev, "Vzgliady Ogiusta Konta," p. 553.
12. Vasil'ev, *Space*, p. 81.
13. Ibid., p. 227.
14. Gol'tsman, "Nastuplenie," p. 80.
15. Nevskii, "Marksizm," p. 205.
16. Vasil'ev, *Space*, p. 228.
17. Schlick et al., *Teoriia otnositel'nosti*.
18. Bogdanov, "Ob'ektivnoe ponimanie," pp. 332–34, 347.
19. Bogdanov, "Uchenie ob analogiiakh," pp. 78–79.
20. Ibid., p. 97.
21. For a comment on Machian elements in Bogdanov's thought, see Cohen, p. 159.
22. Polikarov, p. 421.
23. Lenin, *Materialism and Empirio-Criticism*, chap. 3.
24. Nevskii, "Marksizm," p. 204.
25. Predvoditelev et al., eds., vol. 2, p. 13. See also Kudriavtsev, *Kurs istorii fiziki*, pp. 265–67.
26. Maksimov, "Review," pp. 303–4.
27. Timiriazev, "Novaia neudachnaia popytka," p. 252.
28. Kudriavtsev, "A. K. Timiriazev," pp. 252–53.

29. Maksimov, "Review," p. 306.

30. Timiriazev, "Teoriia otnositel'nosti Einshteina i dialekticheskii materialism," part 1 (no. 8–9), p. 142; Timiriazev, "Teoriia otnositel'nosti Einshteina i makhizm," no. 7, p. 378.

31. Timiriazev, "Novaia neudachnaia popytka," pp. 243–52.

32. Einstein, *Ideas*, pp. 232–33.

33. Timiriazev, "Printsip otnositel'nosti," p. 159.

34. In urging Marxist philosophers to wage a war against idealistic distortions of modern physical theories (in "Importance of Militant Militarism"), Lenin made a direct reference to Timiriazev's review of Einstein's *Relativity: The Special and General Theory* (Lenin, *Polnoe sobranie sochinenii*, vol. 45, p. 29). See Timiriazev, "Review," pp. 70–73.

35. Timiriazev, "Teoriia otnositel'nosti Einshteina i dialekticheskii materializm," part 1 (no. 8–9), p. 146.

36. Ibid., part 2 (no. 10–11), pp. 108–9.

37. Timiriazev, "Review" (1922), p. 73.

38. For discussions of the "anti-dialectical" features of the philosophy of the mechanists, as articulated especially by Timiriazev, see Deborin, pp. 189–241; Mitin, "Mekhanisty"; Egorshin, "Review," pp. 272–79; P. Alekseev; "Ob odnom filosofskom dispute"; L. Suvorov, pp. 38–39; and Delokarov, "Bor'ba," pp. 129–30. For a perceptive Western critique of Timiriazev's general views in the philosophy of science, see Joravsky, pp. 78–80, 107–9, 116–19.

39. See particularly Timiriazev, "Volna idealizma." "Idealistic propaganda," according to Timiriazev, concentrated on five themes: the substitution of energy for matter; the transformation of the action-at-a-distance theory into a new kind of mysticism; the advocacy of the finitude of cosmic space and time and of "thermal death"; the loss of faith in the power of science; and the rejection of the law of causality (pp. 99–100).

40. Sonin, "Trevozhnye desiatiletiia sovetskoi fiziki (1920–1940)," p. 76.

41. Timiriazev, *Estestvoznanie*, pp. 317–29.

42. Orlov, "Klassicheskaia fizika," pp. 75–76.

43. Nevskii, "Restavratsiia idealizma," pp. 123–24.

44. Orlov, "Klassicheskaia fizika," p. 75.

45. Gol'tsman, "Nastuplenie," pp. 100–101; Maksimov, "Teoriia otnositel'nosti," pp. 155–56.

46. Gol'tsman, "Einshtein," p. 126.

47. Gol'tsman, "Nastuplenie," p. 101.

48. Maksimov, "Sovremennoe sostoianie diskussii," pp. 118–19.

49. Maksimov, "O printsipe otnositel'nosti."

50. Ibid., p. 204. Maximov's general criticisms are presented in his comments on Max Born's book on Einstein (Maksimov, "Eshche raz o populiarno-nauchnoi literature," pp. 124–29).

51. Tseitlin, "Teoriia otnositel'nosti," part 2 (no. 4–5), p. 115.

52. Tseitlin, "Review of Timiriazev," p. 227.

53. Tseitlin, "Neskol'ko vozrazhenii," p. 167.

54. Ibid.

55. Tseitlin, "O 'misticheskoi' prirode," p. 101.

56. Dyshlevyi and Petrov, p. 87.

57. Semkovskii, "K sporu v marksizme," p. 155.

58. Semkovskii, *Dialekticheskii materializm*, p. 202.

59. Joravsky, p. 284.

60. Semkovskii, "K sporu v marksizme," p. 126.

61. Ibid., p. 159.

62. This view was held by the philosopher L. Rudash as well: "Modern science (Einstein, Bohr, etc.) takes on a decidedly *materialistic* stance. Space and time, 'the transcendental forms of [Kant's] pure reason,' . . . have become physical objects, while mathematics has found it obligatory to recognize their properties. . . . Now mathematics has been removed from the pedestal. As we see it, it has become an auxiliary tool helping us improve our understanding of space and time, or, more correctly, spaces and times, communicated by experience. Space does not exist without matter. It has become a physical notion. This idea represents the *materialistic* side of Einstein's theory, even though it has not come close to dialectical materialism" (as cited by Gott and Delokarov, p. 8).

63. Semkovskii, "K sporu v marksizme," p. 155.

64. Ibid., p. 156.

65. Semkovskii, *Teoriia otnositel'nosti*, p. 58.

66. Semkovskii, *Dialekticheskii materializm*, p. 75.

67. Iurinets, p. 131.

68. "Pervoe Vsesoiuznoe soveshchanie," pp. 137, 151.

69. Tseitlin, "Review of Semkovskii, " p. 224.

70. Ibid., p. 228.

71. Dyshlevyi, *V. I. Lenin*, p. 28; Delokarov, "Bor'ba," p. 128.

72. Shmidt et al., p. 42.

73. Khvol'son, *Teoriia otnositel'nosti*, p. 92.

74. Hessen, *Osnovnye idei teorii otnositel'nosti*, p. 107.

75. Marx and Engels', p. 1.

76. Delokarov, "B. M. Gessen," p. 73.

77. Ksenofontov, *Dialekticheskii materializm*, pp. 63–64. For comments on the feuds and bitterness between the mechanists and the dialecticians, see Ogurtsov, pp. 357–61. See also Gott and Delokarov, pp. 5–7; and Yakhot, "Podavlenie filosofii," part 1 (no. 9), pp. 55–66.

78. Hessen, *Osnovnye idei teorii otnositel'nosti*, p. 64. See also Delokarov, "Materialisticheskaia dialektika," p. 78; and Delokarov, "B. M. Gessen," p. 80.

79. Gorelik and Frenkel, p. 79.

80. See Graham, "Socio-Political Roots," pp. 713–15.

81. Kedrov, p. 173.

82. Deborin, *Filosofiia*, p. 174.

83. For details on Deborin's criticism of Timiriazev's mechanist orientation, see Deborin, *Dialektika*, pp. 218–21.

84. Ksenofontov, *Leninskie idei*, p. 16.

85. Joravsky, p. 240.
86. Shmidt, "Doklad," p. 13.
87. Hessen, *Osnovnye idei*, p. 68. (This is the work that Shmidt referred to in "Doklad.") See also Delokarov, *Filosofskie problemy*, pp. 182–83; Shmidt et al., pp. 40–41; and Gott and Delokarov, pp. 11–12.
88. Ioffe, "O polozhenii na filosofskom fronte," p. 134; Delokarov, "Bor'ba," p. 132.
89. "O raznoglasiiakh," pp. 58–69.
90. Joravsky, p. 294.
91. P. Alekseev, p. 54; "Peredovaia."

Chapter 2: The Scientific Community and the Theory of Relativity

1. Kasterin, "Sur une contradiction," pp. 90–98.
2. Ibid., pp. 89–91.
3. "Protokol Obshchego sobraniia," p. 368.
4. Khvol'son, *Teoriia otnositel'nosti*, p. 6.
5. Khvol'son, *Die Evolution des Geistes*, p. 173.
6. Khvol'son, *Teoriia otnositel'nosti A. Einshteina i novoe miroponimanie*.
7. Maksimov, "Eshche raz o populiarno-nauchnoi literature," p. 133.
8. Ioffe, *O fizike i fizikakh*, pp. 18–20.
9. Frederiks, "Obshchii printsip otnositel'nosti."
10. Arnol'd, *Teoriia otnositel'nosti*.
11. I. Frenkel, *Teoriia otnositel'nosti*.
12. I. Frenkel, *Na zare*, pp. 136–46.
13. I. Frenkel, *Lehrbuch der Elektrodynamik*.
14. Dorfman, "Iakov Il'ich Frenkel'," p. 6.
15. Io. Frenkel, *Na zare*, p. 148.
16. Ioffe, "Chto govoriat opyty o teorii otnositel'nosti."
17. Vavilov, *Sobranie sochinenii*, vol. 4, p. 11.
18. Ibid., p. 17.
19. Ibid., p. 106.
20. Vizgin, "O n'iutonovskikh epigrafakh."
21. Bronshtein, "Vsemirnoe tiagotenie," p. 252.
22. Vizgin and Gorelik, p. 298.
23. V. Frenkel, "Novye materialy," pp. 13–14.
24. Tropp et al., pp. 203–4; V. Frenkel, "Iurii A. Krutkov," p. 216.
25. Lorentz et al., *Principle of Relativity*, pp. 183–85, 188.
26. Friedmann, "Über die Krümmung des Raumes." See also Friedmann, "Über die Möglichkeit einer Welt mit konstanter negativer Krümmung des Raumes." For comments on Friedmann's theory, see Fock, "Raboty A. A. Fridmana"; and Zel'dovich, "Teoriia rasshiriaiushcheisia Vselennoi." The best short biography is V. Frenkel, "Aleksandr Aleksandrovich Fridman." For a detailed biography, see Tropp et al.
27. Friedmann, *Izbrannye trudy*, p. 398.
28. Gamow, p. 320.
29. For the full text of Friedmann's letter to Einstein, see V. Frenkel, "Novye mate-

rialy," pp. 8–11. For a short but cogent survey of the prehistory and early history of Friedmann's cosmological ideas, see V. Frenkel, "Iurii A. Krutkov," pp. 218–20; and Tatarinov.

30. Einstein, "Zum kosmologischen Problem."

31. Einstein, *Meaning of Relativity*, p. 112. See also Einstein, *Relativity*, pp. 133–34.

32. Weinberg, p. 34.

33. Friedmann, *Mir*, p. 106.

34. Fock, "Raboty A. A. Fridmana," p. 399.

35. Orlov, "Review."

36. Friedmann, *Mir*, pp. 106–7. See also Kaziutinskii, "Sistema mira," p. 132.

37. Sonin, "Grustnaia sud'ba," p. 95

38. For a short biography of Vasil'ev, see Bazhanov and Iushkevich, pp. 221–28.

39. Vasil'ev, "N. I. Lobachevskii," pp. 26–27. See also Einstein, *Ideas*, p. 233.

40. Clifford, p. 298.

41. Vasil'ev, *Space*, p. 110.

42. Ibid., p. 88.

43. Einstein, "Autobiographical Notes," p. 21.

44. Vasil'ev, *Space*, pp. 138–39.

45. *Born-Einstein Letters*, p. 104.

46. Timiriazev, "Teoriia otnositel'nosti Einshteina i dialekticheskii materializm," part 1 (no. 8–9), pp. 150–52.

47. Vernadskii, *Mysli*, p. 11.

48. Einstein, *Sidelights*, p. 23.

49. For details, see Gorelik and Frenkel, pp. 78–83, 254–65; and Vavilov, "Novye opytnye podtverzhdeniia," pp. 457–60.

50. Einstein, *Sidelights*, p. 19; Gorelik and Frenkel, p. 264.

51. Eddington, p. 31; Gorelik and Frenkel, p. 264. For more details on Soviet controversies, see V. Frenkel, *Iakov Il'ich Frenkel*, pp. 363–72.

52. Ia. Frenkel, *Na zare*, pp. 136–46. In his textbook on electrodynamics, published in 1926 in both Russian and German editions, Frenkel stated that ether had completed its role in physics and should be assigned a memorable place in the annals of science. For comments, see V. Frenkel, *Iakov Il'ich Frenkel*, pp. 183–84.

53. There is an element of truth in the following assessment of the state of affairs in Marxist philosophy offered by two later commentators: "The complexity of the general situation in the country did not prevent the philosophy of the 1920s from following a sufficiently democratic course, and from allowing room for realistic inquiries and errors induced by the unsettled nature of the many problems science and the new society faced. Discussion took on a democratic character" (Gott and Delokarov, p. 7). Philosophical "democracy," however, was not the general direction in which Soviet society and politics were moving.

54. Vavilov, "Shestoi s"ezd."

55. See, for example, Pais, *Inward Bound*, one of the most comprehensive studies in the history of modern microphysics, especially pp. 303, 409–11, 418, 426.

Chapter 3: Early Stalinism and Einstein's Theory

1. For details, see "O raznoglasiiakh," pp. 69–89; and Mitin, *Boevye voprosy*, pp. 17–39.

2. "Sovremennye zadachi," p. 18.

3. Mitin, "Ocherednye zadachi," p. 22; "Bor'ba za leninskuiu partiinost'," p. 8; "O raznoglasiiakh," p. 81.

4. Mitin, *Boevye voprosy*, p. 400.

5. Ibid. For comments on the rise of Stalinist orthodoxy in Marxist philosophy, see Ogurtsov, pp. 364–69.

6. Kol'man, "Za marksistsko-leninskuiu nauku," p. 22; "Sovremennye zadachi," pp. 17–18; "Bor'ba za leninskuiu partiinost'," p. 7; Takser, pp. 89–91. For a critical and well-documented discussion of the philosophical orientations of, and attacks on, the mechanists and dialecticians, see Joravsky, chaps. 8–14. See also Yakhot, "Podavlenie filosofii," part 1 (no. 9), pp. 55–64.

7. Sitkovskii, "O antimarksistskoi sushchnosti mekhanizma," p. 32.

8. To give the new philosophy an official stamp, the academy published a two-volume study of Marxist theory by a team of younger philosophers: Mitin and Razumovskii, eds., *Dialekticheskii i istoricheskii materializm.*

9. "O polozhenii," p. 26.

10. Kol'man, "Khod zadom filosofii," p. 11, as cited by Gorelik, "U istokov novogo politicheskogo myshleniia," p. 24.

11. Lenin, *Materialism and Empirio-Criticism*, p. 241.

12. Ibid.

13. For an example, see Maksimov, "O fizicheskom idealizme," pp. 177–83.

14. Einstein, *Ideas*, p. 233.

15. Selinov, p. 27.

16. Einstein, *Ideas*, p. 274.

17. Kuznetsov, "Printsip sootvestviia," p. 64.

18. Maksimov, "Sovremennoe fizicheskoe uchenie," p. 109.

19. Maksimov, "Klassovaia bor'ba," p. 26.

20. Kol'man, "Uzlovye problemy," pp. 30–34.

21. Ioffe, "O polozhenii na filosofskom fronte," p. 143.

22. Ogurtsov, p. 366.

23. Mitin, *Boevye voprosy*, pp. 372–73, 379.

24. Mitin, "Istoricheskoe znachenie," pp. 78–79.

25. "O raznoglasiiakh," p. 59.

26. For more details on the ideas and strategies of Timiriazev, Mitkevich, and Kasterin, see Gorelik, "Obsuzhdenie," pp. 19–25. See also Delokarov, *Filosofskie problemy*, pp. 167–73; and Vizgin, "Martovskaia (1936 g.) sessiia," p. 78.

27. The following discussion is drawn from Semkovskii, " 'Dialektika prirody' Engel'sa," pp. 12–13.

28. "O polozhenii," p. 26.

29. For more on Hessen's life and work, see Gorelik, "Moskva," pp. 16–31; Graham, "Socio-Political Roots"; and Delokarov, "B. M. Gessen."

30. Hessen, "Einshtein, Albert: filosofskie vzgliady," p. 154.

31. Hessen, "Efir," pp. 17–18.

32. See particularly Gorelik, "Moskva," pp. 16–31.

33. Gorelik, "Vikhri," p. 109.

34. In 1931, Kol'man published an article in *Pod znamenem marksizma* in which he issued a stern warning to Hessen to adopt a Bolshevik style of work in "science" ("Boevye voprosy," p. 77). Kol'man made no effort to give concrete examples of Hessen's deviations from the Bolshevik line.

35. Gorelik, "Moskva," p. 18.

36. See, for example, "O polozhenii," p. 26.

37. Tamm, "O rabote filosofov-marksistov," p. 222. See also Yakhot, "Theory," pp. 110–11.

38. "Sessiia Instituta filosofii," pp. 90–91. Both had published slightly different versions of those papers a year earlier in the academy's volume commemorating the 50th anniversary of the death of Karl Marx: Bukharin and Deborin, eds., *Pamiati Karla Marksa.*

39. Ioffe, "Razvitie atomisticheskikh vozzrenii," pp. 467–68.

40. Vavilov, "Dialektika svetovykh iavlenii," p. 479.

41. "Sessiia Instituta filosofii," p. 96.

42. Mandel'shtam, p. 303.

43. Ibid., pp. 304–5.

44. Ibid., p. 93.

45. I. Frenkel, "Teoriia otnositel'nosti Einshteina," p. 38. Minkowski was the first to suggest that the theory of relativity could be called the theory of the absolute nature of scientific laws (Minkowski, "Time and Space," p. 83).

46. Ia. Frenkel, "Einshtein," p. 148.

47. Vavilov, "V. I. Lenin," citing Einstein, *Essays in Science,* pp. 17–18.

48. Vavilov, "Staraia i novaia fizika," pp. 218–19.

49. Vavilov, "V. I. Lenin," p. 38.

50. Vavilov, "Novaia fizika," pp. 30–31.

51. Vernadskii, *Filosofskie mysli naturalista* (1988).

52. Vernadskii, "Izuchenie iavlenii zhizni," "Problema vremeni," "Po povodu kriticheskikh zamechanii," and "Zapiska."

53. Vernadskii, *Khimicheskoe stroenie,* p. 190.

54. Makedonov, pp. 141–42.

55. Kapitsa, *Experiment,* p. 349.

56. In a 1924 article, written jointly with H. A. Kramers and J. C. Slater, Bohr proposed new ideas for the theory of the interaction of light and matter, but he found no place for the photon. In Pais's apt description, that article was "a postscript to the theory of photon and a prelude to the Bohr-Einstein dialogue" (*"Subtle Is the Lord,"* p. 416).

57. Einstein, Podolsky, and Rosen, p. 777.

58. Ibid.

59. Bohr, p. 7.

60. Fock et al., "Mozhno li schitat', chto kvantovo-mekhanicheskoe opisanie fizicheskoi real'nosti iavliaetsia pol'nym?"

61. Nikol'skii, "Otvet V. A. Foku."

62. Nikol'skii, "O putiakh," p. 171.

63. Delokarov, *Metodologicheskie problemy*, pp. 318–19.

64. Nikol'skii, "O putiakh," p. 160.

65. Fock, "K stat'e Nikol'skogo." For more details on Nikol'skii's views, see Delokarov, *Metodologicheskie problemy*, pp. 318–26. See also Jammer, *Philosophy of Quantum Mechanics*, pp. 248, 444.

66. Ksenofontov, *Dialekticheskii materializm*, p. 58.

67. Maksimov, "O filosofskikh vozzreniiakh," p. 25.

68. Ibid., p. 52.

69. See, for example, "K stat'e A. Einshteina."

70. Kol'man, *Noveishie otkrytiia*, pp. 71–72.

71. Einstein, "Remarks," p. 674.

72. Hoffmann, p. 186.

73. Pippard, "Picturing the Atom," p. 3.

74. Maksimov, "O filosofskikh vozzreniiakh," p. 47.

75. Mitkevich, "Osnovnye vozzreniia," pp. 242–43.

76. For more details on Mitkevich's project, see Gorelik, "Naturfilosofskie problemy," pp. 97–98.

77. Ioffe, "O polozhenii na filosofskom fronte," pp. 136–37.

78. Ibid. For pertinent comments on Ioffe's personality and activity, see Josephson, p. 274.

79. Planck, *Universe*, pp. 106–7.

80. Fock, "K diskussii po voprosam fiziki," p. 156.

81. Einstein, *Out of My Later Years*, p. 93.

82. "K stat'e Einshteina," pp. 110–13.

83. Maksimov, "Dialekticheskii materializm i sovremennaia fizika," pp. 77–78.

84. Ioffe, "Sovetskaia fizika." Shpol'skii was likewise careful to avoid the subject of Soviet work on Einstein's theories in his "Fizika v SSSR."

85. Maksimov, "O protivorechiiakh sovremennoi fiziki," p. 120.

86. Kol'man, "Teoriia kvant," p. 135. See also Kol'man, "'Materializm i empiriokrititsizm' V. I. Lenina," pp. 66–67.

87. Kol'man, "Teoriia otnositel'nosti i dialekticheskii materializm," p. 120.

88. Fock, "Al'bert Einshtein," pp. 96–97. Readily recognizing that the special and general theories of relativity were the most powerful tools available to modern physicists, Fock was convinced that Einstein's work on the unified field theory—on the unity of the theory of electromagnetism and the theory of gravitation—was poorly designed and executed: it ignored the research avenues opened by quantum mechanics (p. 97).

89. Gorelik, "Zakony OTO," p. 31.

90. Einstein and Infeld, p. 31.

91. Maksimov, "'Materializm i empiriokrititsizm'—materialisticheskoe obobshchenie," pp. 60–61.

92. Rainov, "Review."

93. Vavilov, *Sobranie sochinenii*, vol. 3, p. 35.

94. Blokhintsev, "Uspekhi," p. 1207.

95. Kompaneets, p. 161. The two Fock papers Vavil'ov cited were "O dvizhenii konechnykh mass"and "Sistema Kopernika."

96. Hoffmann, pp. 228–29.

97. Mitin, "Filosofskaia nauka v SSSR," p. 133.

98. Ibid., p. 135. See also G. Frank, pp. 102–9.

99. Mitin, "Filosofskaia nauka v SSSR," pp. 134–35.

100. Vavilov, *Sobranie sochinenii*, vol. 3, pp. 75–76.

101. Fock, "Sistema Kopernika," pp. 180–81.

102. Einstein and Infeld, p. 212.

103. Fock, *Theory of Space*, p. 351.

104. Infeld, *Why I Left Canada*, pp. 70–71.

105. Kudriavtsev, "Kopernik," pp. 80–81.

Chapter 4: Stalinism After the War

1. Omel'ianovskii, *V. I. Lenin i fizika*, pp. 114–15.

2. Zhdanov, p. 43. For details on the Zhdanov speech, see Sonin, "Razgrom," pp. 104–13.

3. Chernov, pp. 31–34.

4. Sviderskii, "Bor'ba," p. 90.

5. Akhundov, "Spasla li atomnaia bomba sovetskuiu fiziku," pp. 93–94.

6. *Einstein on Peace*, pp. 440, 454; Vallentin, pp. 280–87. For more details and general comments, see Gorelik, "U istokov novogo politicheskogo myshleniia," pp. 9–23.

7. Since Einstein's idea in favor of an international authority was not a detailed proposal but merely an invitation for serious discussion, its categorical rejection by the four Soviet physicists was, according to Gorelik, rash and uncalled for ("U istokov novogo politicheskogo myshleniia," p. 20).

8. Sonin, *"Fizicheskii idealizm,"* p. 100.

9. Vavilov, "'Materializm i Empiriokrititsizm,'" p. 34.

10. Landau and Lifshitz, *Teoriia polia*. My quoted material is from the English translation.

11. Landau and Lifshitz, *Classical Theory*, p. 247.

12. Ioffe, *Osnovnye predstavleniia*, p. 42.

13. Ibid., pp. 335–36.

14. Ibid., p. 357

15. Holloway, "Science and Power," p. 147.

16. Ioffe, "K voprosu o filosofskikh oshibkakh," p. 595.

17. Ibid., p. 598.

18. *Sbornik posviashchennyi semidesiatiletiiu akademika A. F. Ioffe* (Moscow, 1950).

19. Sominskii, p. 126.

20. Fock, "Problema," p. 31. See also Fock, "Sovremennaia teoriia," p. 25.

21. Fock, "Problema," p. 32.

22. Karasov and Nozder, pp. 340–42.

23. Maksimov, "Marksistskii filosofskii materializm," pp. 117–18.

24. Ibid., p. 123. For more details on Maksimov's elaboration of Zhdanov's arguments in their application to physics, see Sonin, "Trevozhnye desiatiletiia: 1947–1953," p. 81.

25. Leonov, pp. 362, 427, 450–51.

26. I. Kuznetsov, "Printsip sootvestviia," pp. 16–24.

27. Maksimov, "Obsuzhdenie," p. 387.

28. Ibid., p. 386.

29. Naan, "K voprosu o printsipe otnositel'nosti," pp. 57–58.

30. Graham, *Science, Philosophy, and Human Behavior*," p. 361.

31. For comments on Naan's views, see Akhundov and Bazhenov, *Filosofiia i fizika* (Moscow, 1989), pp. 50–51.

32. Shteinman, "O reaktsionnoi roli idealizma"; Maksimov, "Marksistskii filosofskii materializm"; S. Suvorov, "Leninskaia teoriia poznaniia"; Omel'ianovskii, "Falsifikatory nauki"; Karpov, "O filosofskikh vzgliadakh."

33. Omel'ianovskii, "Falsifikatory nauki," p. 156.

34. As cited in Maksimov, "Marksistskii filosofskii materializm," p. 124.

35. Prokof'eva, "Konferentsiia," pp. 71–72.

36. Ibid., p. 75.

37. Ibid., p. 76.

38. Omel'ianovskii, "Bor'ba materializma," p. 159.

39. Blokhintsev and Frank, p. 12.

40. Sonin, "Soveshchanie," part 1 (no. 3), p. 100.

41. Frish, p. 343.

42. Akhundov, *Aktual'nost' istorii*, p. 55.

43. Sonin, "Soveshchanie," part 2 (no. 4), p. 92.

44. Akhundov, "Spasla li atomnaia bomba sovetskuiu fiziku?," p. 96.

45. Gorelik, "Fizika universitetskaia," p. 36.

46. Ibid., pp. 35–36.

47. Timiriazev, "Eshche raz o volne idealizma," p. 147. For details on Hessen, see Gorelik, "Moskva."

48. Kasterin, *Obobshchenie osnovnykh uravnenii*.

49. Blokhintsev, Leontovich et al.; Timiriazev, "Po povodu kritiki raboty Kasterina."

50. Tamm, "O rabotakh N. P. Kasterina," p. 448.

51. "Rezoliutsiia Ob"edinennogo zasedaniia," pp. 600–601.

52. Gorelik, "Fizika," p. 33.

53. Ibid., pp. 32, 44.

54. Vasetskii, pp. 91–96.

55. Tepliakov; Solov'ev.

56. Lenin, *Materialism and Empirio-Criticism*, p. 281.

57. Einstein and Infeld, p. 142.

58. Einstein, *Essays in Science*, p. 104.

59. Russell, *Human Knowledge*, p. 291.

60. A. P. Aleksandrov, "Kak delali bombu?" (1988). See also Vizgin, "Martovskaia (1936 g.) sessiia," p. 79.

61. I. Kuznetsov, "Protiv idealisticheskikh izvrashchenii," p. 261.

62. "Ot redaktsii" (*UFN*), pp. 145–46; Kravets, pp. 353–55.

63. Einstein, *Meaning of Relativity*, p. 47; "Ether and the Principle of Relativity," in Einstein, *Sidelights*, p. 13.

64. Fock, "Massa i energiia," p. 162.

65. Einstein and Infeld, p. 197.

66. V. N. Veselovskii, *Filosofskoe znachenie zakonov sokhraneniia materii* (Moscow, 1964), pp. 141–43.

67. Kol'man, *Filosofskie problemy*, p. 18.

68. The demise of the law of the conservation of matter is described in Müller-Markus, "Diamat," pp. 74–75.

69. Vavilov, "Filosofskie problemy," pp. 25–30.

70. Ibid., p. 18.

71. I. Kuznetsov, "Sovetskaia fizika," p. 47.

72. Ibid., p. 52.

73. Ibid., pp. 52–53, citing Barnett, *Universe*, p. 11.

74. The following material is based on I. Kuznetsov, "Sovetskaia fizika," pp. 53–63.

75. Ibid., p. 72. Kuznetsov issued a stern warning: "Physics is much in need of a thorough criticism and a full unmasking of all the systems of Einstein's theoretical views, as well as of those of his followers, in the domain of physics, no less than in individual philosophical pronouncements. The idealistic views of Einstein and the Einsteinians have led physics into a blind alley. Denouncing reactionary Einsteinianism in the field of physical science is one of the most important duties of Soviet physicists and philosophers" (ibid., p. 47).

76. Shteinman, "Za materialisticheskuiu teoriiu bystrikh dvizhenii," p. 298.

77. I. Kuznetsov, "Sovetskaia fizika," p. 47. See also Müller-Markus, "Diamat," p. 73.

78. According to David Comey, Kuznetsov and Shteinman "drew heavily on the pre-relativity interpretation of Lorentz's theory of the electrodynamic nature of mass, according to which the mass of an electrically charged body was said to increase because of real internal changes in the fields in which bodies move. Thus at any given point in time there is only one real value for a body's rest mass and length" (p. 191).

79. The text discussion is based on Karpov, "Kritika," pp. 218–33.

80. Not to be outdone in the anti-Einstein campaign, Maksimov wrote in the newspaper *Red Fleet* (June 14, 1952): "There is no doubt that Einstein's theory of relativity propagates antiscientific views on the basic questions of modern physics and on science in general. Einstein's ideas have led physics not in a forward but in a backward direction, in theory no less than in methodology. Many physicists have learned that Einstein's theory of relativity is a blind alley of modern physics" (cited in Sonin, "Trevozhnye desiatiletiia: 1947–1953," p. 82).

81. Hoffmann, pp. 245–46.

82. Maksimov et al., eds., pp. 24–25.

83. "Razvertyvat' kritiku i bor'bu mnenii v nauke."

84. Vavilov, "Filosofskie problemy," pp. 22–23.

85. Terletskii, "Ob odnoi iz knig akademika L. D. Landau," p. 191.

86. See particularly I. Kuznetsov and Ovchinnikov; Omel'ianovskii, "O knige akademika A. F. Ioffe"; and Omel'ianovskii,"Protiv idealizma."

87. Omel'ianovskii, "O knige akademika A. F. Ioffe," pp. 203–7. See also I. Kuznetsov and Ovchinnikov, pp. 117–40.

88. I. Kuznetsov and Ovchinnikov, pp. 111–12. See also Omel'ianovskii, "O knige akademika A. F. Ioffe," pp. 203–7.

89. Mandel'shtam, p. 303.

90. Mandel'shtam, *Polnoe sobranie sochinenii.*

91. A. Semenov, "Ob itogakh," p. 199.

92. Ibid., p. 201.

93. For additional details on the Mandel'shtam affair, see Sonin, "Trevozhnye desiatiletiia: 1947–1953," pp. 82–83.

94. Terletskii, "Ob izlozhenii osnov spetsial'noi teorii otnositel'nosti," pp. 207, 211–12.

95. Maksimov, "Bor'ba za materializm," p. 194.

96. Bazarov, pp. 182–85.

97. The model for this approach was suggested by Terletskii, "O soderzhanii sovremennoi fizicheskoi teorii otnositel'nosti."

98. Fock, "Sovremennaia teoriia," p. 25.

99. A. D. Aleksandrov, "Po povodu nekotorykh vzgliadov," p. 229.

100. Ibid., p. 245.

Chapter 5: Turning Points

1. Kol'man, "K sporam," pp. 183, 185–88.

2. Ibid., p. 188

3. I. Kuznetsov, "Ob osnovnykh voprosakh," pp. 181–82.

4. Ibid., pp. 189–90.

5. Gorelik, "Naturfilosofskie problemy," p. 102.

6. The discussion of the *Voprosy filosofii* report of January 1955 is based on "K itogam diskussii," pp. 134–37. For the full text of the report, see Müller-Markus, *Einstein,* vol. 1, pp. 351–59.

7. Shpol'skii, "Al'bert Einshtein," p. 186.

8. Ioffe, "Pamiati Al'berta Einshteina," p. 187.

9. Tamm, "A. Einshtein," pp. 6–7.

10. Fock, "Zamechaniia," p. 117.

11. Fock, "Poniatiia odnorodnosti," p. 135.

12. Ivanenko and Kuznetsov, p. 29.

13. See particularly ibid., p. 31.

14. Blokhintsev, "Nekotorye voprosy razvitiia sovremennoi fiziki," p. 31.

15. Infeld, "Moi vospominaniia," p. 168.

16. Lazukin, p. 110.

17. Smorodinskii, p. 32.

18. A. D. Aleksandrov, "Dialektika i nauka," pp. 12, 16.

19. Fedoseev, "Zakliuchitel'noe slovo," p. 592.

20. Ambartsumian, "Nekotorye metodologicheskie voprosy," p. 290.

21. A. D. Aleksandrov, "Filosofskoe soderzhanie."

22. Ibid., p. 97.

23. Kursanov, p. 394.

24. Einstein, *Ideas*, p. 233; Kursanov, p. 401.

25. See I. Kuznetsov et al., *Philosophical Problems of Elementary Particle Physics*; and Omel'ianovskii and Kuznetsov, eds., *Filosofskie voprosy*. See also Fedoseev et al., eds.; and "Obsuzhdenie filosofskikh voprosov teorii otnositel'nosti."

26. Fedoseev, "Zakliuchitel'noe slovo," p. 590.

27. Kapitsa, p. 317.

28. Konstantinov et al., eds., pp. 329, 333.

29. I. Alekseev, "Obsuzhdenie knigi A. K. Maneeva."

30. Maneev, pp. 9–10, 80–83.

31. I. Alekseev, "Obsuzhdenie," pp. 139–40.

32. "Istoricheskii dokument," p. 5.

33. Il'ichev, "Doklad," p. 20.

34. N. Semenov, *Nauka i obshchestvo*, p. 218.

35. In 1933, for example, L'vov chastised O. D. Khvol'son for taking David Hume's claim that causality cannot be either proved or disproved too seriously (L'vov, "Voprosy ideologii," p. 117).

36. L'vov, *Zhizn' Al'berta Einshteina*, chap. 13.

37. B. Kuznetsov, *Einshtein*, pp. 88–93.

38. B. Kuznetsov, *Osnovy teorii otnositel'nosti*.

39. Müller-Markus, *Einstein*, vol. 1, p. 115. For a brief survey of the Soviet historical literature on the theory of relativity, see Vizgin, *Reliativistskaia teoriia*, pp. 11–12.

40. Pogodin, "Al'bert Einshtein."

41. Frankfurt, *Spetsial'naia i obshchaia teoriia*. See also Frankfurt, *Ocherki*.

42. M. Mostepanenko, p. 15.

43. Ibid., p. 224.

44. Ibid., p. 19.

45. Omel'ianovskii, *Dialektika i sovremennaia fizika*, p. 108.

46. Einstein, *Sobranie nauchnykh trudov*; *Einshteinovskii sbornik*.

47. Kopnin, *Dialektika*, pp. 360–61.

48. Chudinov, *Teoriia otnositel'nosti*, pp. 22–24.

49. Ibid., p. 26.

50. Gerald Holton wrote: "In the end, Einstein came to embrace the view which many, and perhaps he himself, thought earlier he had eliminated from physics in his basic 1905 paper on relativity theory that there exists an external objective physical reality which we may hope to grasp—not directly, empirically, or logically, or with fuller certainty, but at least by an intuitive leap, one that is guided by experience of the totality of sensible 'facts'" (*Thematic Origins*, p. 245).

51. See particularly Ivanenko, "Ob aktual'nosti trudov Einshteina"; and Vizgin, "Odin iz aspektov metodologii Einshteina."

52. Lavrova, "Trudy Einshteina."

53. Arshinov and Pakhomov, p. 183.

54. Dyshlevyi and Luk'ianets, eds., *Nauchnaia kartina*. See especially "Predislovie," pp. 3–4.

55. Kobzarev et al., *Einshtein i sovremennaia fizika*.

56. V. Frenkel and Iavelov, *Einshtein-izobretatel'*.

57. Balashov, "'Antropnye argumenty'"; Sakharov, *Memoirs*, p. 546.

58. Balashov, "Antropnyi kosmologicheskii printsip," pp. 33–40.

59. Omel'ianovskii, "Teoriia otnositel'nosti," p. 83. See also Omel'ianovskii, *Razvitie osnovanii fiziki*, p. 89.

60. Gribanov, *Filosofskie osnovaniia*, p. 209.

61. Gribanov, *Filosofskie vzgliady*, p. 119.

62. Gribanov, *Filosofskie osnovaniia*, pp. 134–35.

63. Delokarov, *Filosofskie problemy*; Vizgin and Gorelik; Delokarov, "Bor'ba za materialisticheskoe istolkovanie"; Delokarov, "Puti stanovleniia soiuza filosofii i fiziki"; Delokarov, " Theory of Relativity"; Gribanov, *Filosofskie vzgliady*, pp. 5–55; L. Suvorov, "Rol' filosofskikh diskussii."

64. Delokarov, "Theory of Relativity," pp. 484–85.

Chapter 6: Accommodations to Einstein's Theory of Knowledge

1. See, for example, Einstein, *Ideas*, pp. 272–74.

2. Vavilov, *Sobranie sochinenii*, vol. 3, pp. 27–28.

3. Keldysh, "Estestvennye nauki," p. 46; Einstein, *Out of My Later Years*, p. 93.

4. Einstein, *Ideas*, p. 272.

5. Gott, *Filosofskie voprosy*, p. 15.

6. Einstein, "Prologue," pp. 10–11.

7. German, p. 39.

8. "Intuitsiia."

9. Konstantinov et al., eds., pp. 300–301.

10. Spirkin, p. 344.

11. Einstein, *Ideas*, pp. 292–93; Karmin, p. 249.

12. Vol'kenshtein, *Perekrestki nauki*, p. 3.

13. Kopnin, "Teoreticheskie postroeniia," pp. 6–7; Kopnin, *Dialektika*, p. 356.

14. Novikova, pp. 78–79.

15. Panov, p. 196.

16. Einstein, "Autobiographical Notes," p. 5.

17. Lenin, *Materialism and Empirio-Criticism*, p. 260.

18. Bacon, p. 19.

19. Lenin, *Materialism and Empirio-Criticism*, p. 261.

20. Einstein, *Ideas*, p. 233.

21. Markov, "O prirode fizicheskogo znaniia," pp. 140–47.

22. Heisenberg, *Physics and Philosophy*, p. 81.

23. For a general Marxist assessment of the dialectics of the objective and subjective components of physical reality, see Omel'ianovskii, "Filosofskaia bor'ba."

24. Dyshlevyi, "Filosofskoe soderzhanie," p. 64.

25. A. Mostepanenko, "Problema sushchestvovaniia," p. 182.

26. Lakatos, p. 104.

27. Omel'ianovskii, "Filosofskaia bor'ba," p. 132.

28. Dyshlevyi and Kanak, p. 176.

29. Illarionov and Polovinkin, p. 164; Chudinov, "Ratsionalizm Einshteina," p. 34.

30. Einstein, *Letters to Solovine*, p. 85. See also Ovchinnikov, p. 25.

31. Einstein, "Autobiographical Notes," p. 21. For comments on the respective influences of Mach and Boltzmann on the theory of relativity, see Stadler, pp. 63–73.

32. Khramova, pp. 178–79.

33. Dyshlevyi and Kanak, p. 154.

34. I. Kuznetsov, "Sovetskaia fizika," pp. 47, 49.

35. Il'in and Gapich, p. 89.

36. Dyshlevyi and Kanak, p. 165.

37. Kalantar, p. 106.

38. Mitin, "V. I. Lenin," p. 12; Narskii, pp. 245–46.

39. Heisenberg, *Across the Frontiers*, p. 6.

40. Molchanov, p. 235.

41. Einstein, "Remarks," p. 667.

42. Dyshlevyi, "A. Einshtein," p. 97.

43. Ibid., pp. 96–97. See also Aronov and Pakhomov, pp. 62–73.

44. Chudinov, "Ratsionalizm Einshteina," p. 34.

45. Sidorov, p. 37.

46. Konopleva, p. 125.

47. Zahar, p. 301.

48. Bachelard, *New Scientific Spirit*, p. 47.

49. Logunov, *Lektsii*; Logunov, "Novaia teoriia gravitatsii"; Logunov, "Reliativist-skaia teoriia gravitatsii" (*Pr*); Logunov, "Reliativistskaia teoriia gravitatsii i novye pred-stavleniia o prostranstve-vremeni."

50. Chudinov, "Einshteinovskaia 'kontseptsiia nabliudaemykh,'" p. 218.

51. Einstein, *Ideas*, p. 274.

52. Kas'ian, pp. 50–51.

53. Bazhenov et al., eds., *Printsip dopolnitel'nosti*. See also I. Alekseev, *Kontseptsiia dopolnitel'nosti*.

54. Bazhenov et al., eds., *Printsip dopolnitel'nosti*, part 3.

55. I. Alekseev, "Nekotorye soobrazheniia," p. 102.

56. Engels, *Ludwig Feuerbach*, p. 73.

57. I. Alekseev, "Nekotorye soobrazheniia," p. 102.

58. Bolotovskii, p. 111.

59. Einstein, "Quanten-Mechanik."

60. I. Alekseev, *Kontseptsiia dopolnitel'nosti*, p. 64.

61. Kobzarev, "A. Einshtein"; El'iashevich; I. Alekseev, "Nekotorye soobrazheniia"; Aronov and Pakhomov, "Filosofiia i fizika."

62. Aronov and Pakhomov, p. 73. For a more general analysis of the Einstein-Bohr debate, see Markov, "O edinstve i mnogoobrazii form materii." For a detailed com-

parative analysis of their ideas, see I. Alekseev, *Kontseptsiia dopolnitel'nosti*, pp. 64–71. See also Panchenko, pp. 55–60, 181–89.

Chapter 7: Approaches to the General Theory of Relativity

1. Fock, *Teoriia prostranstvo*. All text references are to the second edition of the English translation.

2. Fock, *Theory of Space*, p. 376.

3. For a summary of Fock's studies in quantum mechanics and the theory of relativity, see Veselov.

4. Pais, *Inward Bound*, pp. 339, 357, 379–80, 394.

5. Dirac, *Principles*, pp. 137, 139.

6. Hoffmann, p. 228.

7. Ibid., p. 229; Pais, *'Subtle Is the Lord,'* pp. 290–91, 333; B. Kuznetsov, *Einstein*, p. 329. In the words of A. Z. Petrov, "Whereas in the classical linear field theories the equations of motion are not induced by field equations, but introduced from outside, in the general theory of relativity, in which field equations are nonlinear, the equations of motion are consequences of field equations" ("Obshchaia teoriia otnositel'nosti," p. 66).

8. I. Frenkel, "Teoreticheskaia fizika," p. 317.

9. Fock, "O dvizhenii konechnykh mass," p. 410.

10. Fock, "Sistema Kopernika," pp. 183–84. For more details, see Dyshlevyi, *V. I. Lenin*, pp. 146–59.

11. Chudinov, "Einshteinovskaia 'kontseptsiia nabliudaemikh,'" p. 216.

12. Molchanov, p. 235

13. Terletskii, "O soderzhanii sovremennoi fizicheskoi teorii," p. 196.

14. Fataliev, pp. 129–30.

15. A. Z. Petrov offered an invariant classification of all possible gravitational fields based on an analysis of the structure of the tensor of the space-time curvature in. "Obshchaia teoriia otnositel'nosti," p. 67. See also A. Petrov, *Novye metody*.

16. A. Petrov, *Prostranstva Einshteina*, p. 95.

17. A. Petrov, "Gravitatsiia i prostranstvo," pp. 177–78.

18. Fock, "Fizicheskie printsipy," p. 16.

19. Fock, "Physical Principles," pp. 212–13.

20. Fock, "Otnositel'nosti teoriia," p. 408.

21. Infeld, *Why I Left Canada?*, p. 88.

22. Graves, pp. 140–42.

23. Sviderskii, "Nekotorye filosofskie problemy," pp. 135–37.

24. Sviderskii, *Prostranstvo*, pp. 27–36.

25. A. D. Aleksandrov, "Prostranstvo," pp. 212, 219; A. D. Aleksandrov, "On the Philosophical Content," pp. 113–14.

26. A. D. Aleksandrov, "Filosofskoe soderzhanie," p. 130.

27. Synge, p. ix.

28. A. D. Aleksandrov, "Filosofskoe soderzhanie," pp. 126–30.

29. A. D. Aleksandrov, "Teoriia otnositel'nosti," p. 322.

30. For comments on the terms, see Avaliani, *Absolutnoe i otnositel'noe*, pp. 155–58.

31. Perederii, pp. 246–47.

32. A. D. Aleksandrov, "Prostranstvo," pp. 225–29; A. D. Aleksandrov, "Filosofskoe soderzhanie," p. 136.

33. A. D. Aleksandrov, "Teoriia otnositel'nosti," p. 271.

34. Fock, "Otnositel'nosti teoriia"; A. D. Aleksandrov, "Otnositel'nosti teoriia: teo-retiko-poznavatel'noe znachenie." See also A. D. Aleksandrov, "Otnositel'nosti teor-iia" (*FE*).

35. Fock, "Otnositel'nosti teoriia," p. 410.

36. Ibid., p. 411.

37. For a summary of Omel'ianovskii's arguments, see his "Teoriia otnositel'nosti."

38. Sviderskii, "O nekotorykh metodologicheskikh printsipakh," pp. 139–44.

39. A. Petrov, *Prostranstvo-vremia i materiia*.

40. See, for example, Fock, "O roli printsipa otnositel'nosti; and Fock, "Fizicheskie printsipy" (in Omel'ianovskii, ed., *Lenin*).

41. A. D. Aleksandrov, "Otnositel'nosti teoriia" (*FE*).

42. Dyshlevyi and Petrov, eds., *Filosofskie problemy*; A. Petrov and Dyshlevyi, eds., *Prostranstvo i vremia*. According to the two editors, *Filosofskie problemy* was "the first Soviet study devoted specifically to an analysis of the philosophical problems of Einstein's theory of gravitation and relativistic cosmology" (p. 4).

43. A. Petrov, "Obshchaia teoriia otnositel'nosti," pp. 65–71.

44. Avaliani, *Ocherki*, p. 5.

45. Konstantinov et al., eds., pp. 132–44; Sheptulin, ed., pp. 70–76.

46. Bazhenov, "Problema prostranstva-vremeni," p. 181.

47. Landau and Lifshitz, "Tiagotenie."

48. Graham, "Reception," p. 127.

49. Ivanenko, "O edinoi fizicheskoi kartine mira," pp. 77–87.

50. Ivanenko, "Osnovnye idei," pp. 319–21.

51. Ivanenko, "Bemerkungen," pp. 355–69.

52. Ivanenko, "O edinoi fizicheskoi kartine mira," p. 87.

53. Ivanenko, "Osnovnye idei," p. 322; Ivanenko, "O edinoi fizicheskoi kartine mira."

54. Ivanenko, "K stoletiiu so dnia rozhdeniia," p. 14.

55. For a comment, see Gribanov, *Filosofskie vzgliady*, p. 65.

56. Ivanenko, "Perennial Modernity," pp. 318–19.

57. Ibid., p. 318.

58. Zel'manov, "Kosmologiia," p. 321. For a cogent analysis of the relations of dia-lectical materialism to relativist cosmology, see Mikulak, pp. 48–50.

59. Tursunov, *Gorizonty*, p. 15.

60. Ginzburg, *Key Problems*, p. 132.

61. Feinberg, "Deviat' rubtsov," p. 38; Perfyl'ev.

62. Eigenson, p. 13.

63. V. Petrov, "Ob astronomicheskikh sledstviiakh."

64. "Kosmologiia," pp. 111–13.

65. "Predislovie," in A. A. Mikhailov et al., eds., *Astronomiia* (Moscow, 1967), p. 8.

66. Vorontsov-Vel'iaminov, *Ocherki po istorii astronomii v SSSR.*

67. According to P. C. W. Davies, "In spite of the qualifications concerning the very early stages of the expansion, the Friedmann models remain the basic working model universes for most cosmologists" (*Space and Time*, p. 166).

68. Gorelik, "Rasshireniiu Vselennoi," p. 104. By 1962, the philosopher M. V. Mostepanenko was ready to admit that "the idea of the expanding universe found an excellent explanation within the framework of the general theory of relativity, free of all fideistic ideas about the creation of the universe" (*Materialisticheskaia sushchnost' teorii otnositel'nosti*, p. 215). He also linked the general theory of relativity with the idea of time infinity.

69. Friedmann, *Izbrannye trudy*, pp. 397–409.

70. Zel'dovich and Novikov, *Reliativistskaia astrofizika*, p. 387.

71. A. Petrov, "Obshchaia teoriia otnositel'nosti," p. 67.

72. Naan, "Kosmologiia," p. 256.

73. Sviderskii, "O nekotorykh metodologicheskikh printsipakh," p. 143.

74. Zel'manov, "K postanovke kosmologicheskoi problemy," p. 80.

75. Umarov and Mostepanenko, p. 204; Chudinov, "Obshchaia teoriia otnositel'nosti," pp. 72–74.

76. Zel'dovich and Novikov, *Stroenie*, pp. 438–43.

77. Ambartsumian, *Filosofskie voprosy*, pp. 90–91.

78. Ambartsumian and Kaziutinskii, "Dialektika v sovremennoi astronomii," p. 242.

79. Ambartsumian, *Filosofskie voprosy*, p. 106. See also Dyshlevyi and Kanak, p. 204

80. Ambartsumian and Kaziutinskii, "Problema razvitiia v astronomii," p. 63.

81. Ibid., p. 59.

82. Omel'ianovskii, "Teoriia otnositel'nosti," pp. 85–86. See also Evgrafov et al., eds., p. 614.

83. Dyshlevyi, "Filosofiia i astronomia," p. 121.

84. See, for example, "Obsuzhdenie doklada [L. F. Il'icheva]," p. 47.

85. Ginzburg, "Zamechaniia o metodologii," p. 86.

86. Ginzburg, *O teorii otnositel'nosti*, p. 107.

87. Ginzburg, "Novye fizicheskie zakony," p. 19.

88. Tursunov, *Gorizonty*, p. 41.

89. Idlis, p. 54.

90. For a detailed comment on the philosophical and scientific complexities of the notion of the infinity of the universe, see Naan, "Gravitatsiia," pp. 268–85.

91. Ginzburg, *Kak ustroena Vselennaia*, p. 9.

92. M. Mostepanenko, p. 216.

93. Hawking, pp. 48–49, 105–6. See also A. Petrov, "Obshchaia teoriia otnositel'nosti," p. 68; and Novikov, p. 151.

Chapter 8: Einstein's Humanistic Influence

1. L'vov, *Zhizn' Alberta Einshteina.*

2. Nietzsche, p. 120.

3. For interesting thoughts on the basic unity of the "two cultures" inspired by Einstein's suggestive ideas, see Feinberg, "Intellektual'naia revoliutsiia." Feinberg saw the growing role of "extralogical, intuitive, and synthetic judgment" in mathematics—the key science in the modern world—as a clearly perceptible path to the humanization of the natural sciences" (pp. 44–45).

4. Heisenberg, "Smysl."

5. Zanchenko, p. 56.

6. Einstein, Murphy, and Sullivan, p. 374.

7. Migunov, p. 92. For a comment on the new role of intuition in mathematics, see Cherniak.

8. Kotina, p. 115; Danin, p. 106.

9. Gomilko, p. 42.

10. Einstein, *Ideas*, pp. 224–25; Rozov, pp. 16–17.

11. Mamchur, p. 115.

12. Ibid.; Griaznov, p. 106.

13. Seelig, p. 265: "Ich lese mit Begeisterung 'Die Brüder Karamazoff.' Es ist das wunderbarste Buch das ich je in der Hand gehabt habe."

14. Moszkowski, p. 185. For comments, see Balandin, p. 94; and Kis'unko, p. 283.

15. Moszkowski, pp. 186–87.

16. Einstein, Murphy, and Sullivan, p. 374.

17. B. Kuznetsov, "Einstein and Dostoevsky"; Feinberg, "Vzaimosviaz'"; Kiiko; Volkov; Danin.

18. German; Spirkin.

19. Einstein, "Prologue," pp. 7–12.

20. Zabelin, p. 165. For a critical comment, see Runin, "Put tvorchestva," pp. 115–17.

21. Hadamard, pp. 142–43.

22. A similar view is presented in Miller, p. 96.

23. Einstein, "Autobiographical Notes," p. 23; Runin, "Lichnost'," p. 143.

24. Runin, "Lichnost'," p. 122.

25. Runin, "Logika nauki," p. 134.

26. Vol'kenshtein, "Krasota nauki," p. 16.

27. Feinberg, *Art*, p. 88.

28. Einstein, *Ideas*, p. 233.

29. Dostoevsky, *Pis'ma*, vol. 1, p. 169.

30. Kiiko, p. 128.

31. Zobov, "Strukturnyi podkhod," p. 202.

32. B. Kuznetsov, "Einstein and Dostoevsky," p. 13. See also Gunshenskaia, pp. 158–60.

33. Gus, *Idei*, pp. 571–73.

34. B. Kuznetsov, *Einstein and Dostoevsky*, p. 72.

35. Volkov, pp. 310–11.

36. Bakhtin, p. 18. For a comment, see Catteau, p. 382.

37. Morson and Emerson, pp. 368–69.

38. Feinberg, "Vzaimosviaz' nauki," p. 32. See also Runin, "Logika nauki," p. 134.

39. Danin, p. 95.

40. *Born-Einstein Letters*, p. 155. See also Gorelik, "O roli gumanitarnykh predstavlenii," p. 87.

41. Einstein, *Out of My Later Years*, pp. 114–15.

42. Ibid.

43. Kant, p. 244.

44. Tiun'kin and Stakhanova, p. 31.

45. Moszkowski, p. 187.

46. Antipov, p. 44.

47. Einstein, Murphy, and Sullivan, pp. 374–75.

48. P. S. Aleksandrov, p. 12.

49. Shishkin, pp. 17–18, 26.

50. B. Kuznetsov, "Ob esteticheskikh kriteriiakh," p. 90.

51. Einstein, "Remarks," pp. 683–84. Einstein acknowledged his debt to the philosophers David Hume and Ernst Mach in tackling the central issues of his physical theory ("Autobiographical Notes," p. 52).

52. For Dostoevsky's views on the place of science in Russian culture, see Balandin.

53. Einstein, *Ideas*, p. 22.

54. Dostoevsky, *Notes from Underground*, pp. 24–27.

55. Dostoevsky, *Pis'ma*, vol. 2, p. 259

56. J. Frank, *Through the Russian Prism*, p. 168.

57. Dostoevsky, *Pis'ma*, vol. 2, p. 181.

58. Dostoevsky, *Vozvrashchenie cheloveka*, p. 313.

59. Einstein, Murphy, and Sullivan, pp. 374–75.

60. Einstein, *Ideas*, p. 32; Kisun'ko, p. 283.

61. Tiun'kin and Strakhanova. "Opyt cheloveka."

Chapter 9: Einstein in the Light of Perestroika

1. Gribanov, *Filosofskie vzgliady*. This is actually a revised edition of *Filosofskie osnovaniia*.

2. Gribanov, *Filosofskie vzgliady*, pp. 20–22.

3. Ibid., p. 127.

4. Gott and Delokarov, p. 7.

5. Nikiforov, p. 61.

6. Ilyin and Kalinkin, pp. 215–16.

7. Nyasanbaev and Suleimanov, p. 48.

8. Skorobogatskii, p. 20.

9. Liubutin and Pivovarov, p. 71.

10. Ibid., p. 72.

11. Biriukov, pp. 109–14.

12. For comments on neopositivism and post-positivism, see Zanchenko.

13. Vlasenko, pp. 51–56.

14. Lenin, *Polnoe sobranie sochinenii*, vol. 18, p. 139.

15. See, for example, "Reliativizm," in *Filosofskii entsiklopedicheskii slovar'* (Moscow, 1983), p. 578.

16. Shinkarenko, pp. 110–15.

17. Ruzavin, p. 3.

18. Kelle, p. 186.

19. Pavlenko, p. 54.

20. "Koordinatsiia issledovanii po kosmomikrofizike," p. 40; Sakharov, "Kosmo-mikrofizika."

21. "Goriachie tochki," p. 5. See also Sakharov, *Memoirs*, pp. 542–43.

22. Miakishev, "Energiia," p. 191.

23. Gribanov, *Albert Einstein's Philosophical Views*, pp. 236–37.

24. Einstein, "Foreword," p. xvi.

25. Vizgin, "Rol' idei."

26. Mamchur, p. 38.

27. Merton, chaps. 11, 12. For a theoretical discussion of Merton's "ethos of science," see Semenova, pp. 105–6.

28. Rozov, p. 20; Einstein, "Max Planck," pp. 1077–1078.

29. See, in particular, Delokarov, "B. M. Gessen"; and Gorelik, "Moskva fizika."

30. Bronshtein, "Pogloshchenie," p. 669.

31. Bronshtein, "Sovremennoe sostoianie," pp. 172–75, 177.

32. Ibid., pp. 124–26, 183–84.

33. Feinberg, "Deviat rubtsov."

34. Mitin, "Velikoe ideinoe oruzhie," pp. 24–25.

35. Einstein, "Pochemu sotsializm"; Gorelik, "O 'stalinizme' Einshteina," pp. 77–78.

36. Gorelik, "O 'stalinizme' Einshteina," p. 81; Einstein; *Ideas*, p. 158.

37. Gorelik, "O chuvstve prekrasnogo," part 1 (no. 9), pp. 42–44.

38. Einstein, "Introduction," p. xi.

39. Balandin, "Dostoevsky i estestvoznanie."

40. Planck, Einstein, and Murphy, p. 210. For details on Einstein-Poincaré relations, see Pais, '*Subtle Is the Lord,*' pp. 169–72.

41. Migdal, p. 8.

42. Ibid., p. 9.

43. Vol'kenshtein, "Nauka i estetika," p. 82.

44. Osmanov, p. 197.

45. Einstein, Murphy, and Sullivan, p. 375.

46. Einstein, *Ideas*, p. 152.

47. Poincaré, p. 104.

48. Deriugin and Osipovskii, p. 5.

49. Nikitina, p. 13.

50. Lazarev, pp. 122–23.

51. Dmitriev, p. 67.

52. See, for example, Kaziutinskii, "Antropnyi printsip"; Tursunov, "Kosmologiia"; and Gorelov and Kurbanov. "Ekologiia i religiia."

53. The intricacies of Einstein's attitude toward religion are presented in Delokarov, "Bog," pp. 233–44.

54. Arshinov, Kurdiumov, and Svirskii, p. 99.

55. Ginzburg, O *teorii otnositel'nosti*, p. 184.

56. Ibid., p. 106.
57. Ibid., p. 184.
58. Zel'dovich and Grishchuk, "Tiagotenie," p. 698.
59. Novikov, p. 165.
60. Grishchuk, pp. 147–49.
61. Akhundov, "Mekhanika N'iutona," p. 17.
62. Einstein, *Ideas*, pp. 260–61.
63. Feinberg, "Intellektual'naia revoliutsiia," p. 37.
64. Grigor'ian, "Al'bert Einshtein."
65. Gribanov, *Filosofskie vzgliady*, pp. 266–69.
66. Ibid., pp. 261–63.
67. Artykov, *Filosofskii analiz*, pp. 123–24.
68. Abasov, p. 74.
69. Fock, "Nekotorye primeneniia," p. 86.
70. For a detailed and balanced historical analysis of Fock's academic life, see Gorelik, "V. A. Fok."
71. Pronin and Obukhov, eds., p. vi.
72. Vizgin, *Unified Field Theories*.
73. Vizgin, "On the Road."
74. Idlis, *Revoliutsii v astronomii*. The article by Einstein and Rosen was published in *Physical Review*, 2d series, 1935, vol. 48, pp. 73–77.
75. Logunov set out his ideas most clearly in four works: *Lektsii*; "Osnovnye printsipy reliativistskoi teorii gravitatsii"; "Reliativistskaia teoriia gravitatsii i novye predstavleniia o prostranstve-vremeni"; and "Reliativistskaia teoriia gravitatsii." For criticisms of his "relativistic theory," see Ginzburg, "Obshchaia teoriia otnositel'nosti; Zel'dovich and Grishchuk, "Obshchaia teoriia verna!," pp. 524–26; Zel'dovich and Grishchuk, "Tiagotenie," p. 698; and Grishchuk, p. 159. See also Akhundov, *Aktual'nost' istorii*, p. 51.
76. Logunov, *K rabotam*, pp. 94–96.

Conclusion

1. Cantor, p. 85.
2. Lenin, *Materialism and Empirio-Criticism*, pp. 288–89.
3. Shmidt, "Doklad," p. 13.
4. Gorelik, "U istokov novogo politicheskogo myshleniia," p. 24.
5. See, for example, Kol'man, "Materializm i empiriokrititsizm"; and Maksimov, "Dialekticheskii materializm i sovremennaia fizika."
6. Fock's imprisonment is described in Frish, pp. 224–26.
7. Sakharov, *Memoirs*, p. 46.
8. S. Suvorov, "Filosofskie vozzreniia," p. 546.
9. For comments, see Ivanov, p. 51.
10. Markov, "Otvet," p. 115.

Glossary

Axiomatics: the reduction of a science or part of science to a system of mathematically formulated and integrated principles

Bernoulli's Theorem: the principle of energy conservation for fluids in steady flow. It explains the change of a fluid's velocity with a decrease in fluid pressure.

Brownian motion: the first successful effort to establish a statistical law of molecular motion

Complementarity Principle: the rule that wave and particle theories combine to explain the dual nature of light

Conventionalism: a philosophical orientation articulated by Henri Poincaré claiming that the principles of scientific theory are products of the consensus of the scientific community

Copenhagen school of quantum mechanics: a strong emphasis on the principles of complementarity, correspondence, and uncertainty in the study of the microphysical world. (See the separate entries on each of these principles.)

Correspondence Principle: the proposition that under certain conditions revolutionary changes in modern physics do not imply a full rejection of the classical theories they claim to have replaced

Energeticism: Wilhelm Ostwald's philosophical orientation centered on energy as the substratum of all natural phenomena

Eötvös experiments: the work of the Hungarian physicist Roland Eötvös's on the earth's gravitational field that resulted in proof that inertial mass and gravitational mass are equivalent

Ernst Mach's epistemology: the claim that the roots of both scientific facts and scientific laws are in "sensations"—the "elements" of experience that are both natural and mental. A unique form of empiricism. It rejects the notion of causality as a principle of scientific explanation. Lenin labeled it empirio-criticism, a philosophical theory opposed to materialism.

Ether: the medium postulated by nineteenth-century physicists as the carrier of electromagnetic waves

FitzGerald-Lorentz, see *Lorentz Transformation*

General Theory of Relativity: the culminating point of Einstein's achievement concentrating on accelerated and curved motion and elaborating the general principle of relativity offered by the special theory of relativity. It formulates the principle of the equivalence of inertial and gravitational motion. It is the basis of the relativistic theory of gravitation.

Laplacian Law of Causality: an emphasis on the universality and primacy of causal explanations of natural phenomena

Logicism: the claim that the foundations of mathematics are built on logic

Lorentz Transformation: the hypothesis, confirmed by the special theory of relativity, that electrons become slightly condensed when they are in motion

"Mathematical idealism": a Marxist label for the "excessive" mathematization of science and for denying mathematics empirical footing

Michelson-Morley experiments: the studies of Albert Michelson and Edward Morley on the speed of light that led Einstein to eliminate ether as a physical reality

Neo-Pythagoreanism: a philosophical orientation injecting mystical elements into the Pythagorean mathematical tradition

Non-Euclidean geometry: various new geometries based on the rejection of Euclid's fifth (parallel) postulate. Einstein's relativistic cosmology is based on Bernhard Riemann's non-Euclidean geometry.

Phenomenology: Edmund Husserl's philosophical orientation representing an effort to extend the domain of science beyond the limits of reality accessible to the traditional scientific method. It is known as "transcendental phenomenology."

Photoelectric Effect, Theory of: an explanation of the ejection of electrons from a substance affected by electromagnetic radiation, such as light

"Physical idealism": an orthodox Marxist label for philosophical theories of physical reality as creations of the human mind rather than as objective reflections of the external world

Red Shift: Edwin Hubble's discovery that the light of galaxies moving away from the earth with increasing velocity shifts toward the red. This phenomenon proves the idea of the expanding universe.

Relativistic Theory of Gravitation: a reigning orientation in modern cosmology based on the geometry of curved cosmic space-time and recognizing the dependence of gravitation on the distribution of cosmic masses. It combines physics and geometry.

Space-time continuum: a pivotal principle in physics positing the unity of the three dimensions of space and time

Special Theory of Relativity: Einstein's creation based on three principles: the recognition of the unity of space and time (clarified by Hermann Minkowski); the statement that all inertial systems confirm the absolute authority of the laws of nature; and the assertion that in empty space light is always propagated with a definite velocity that is independent of the motion of the emanating body. It is concerned exclusively with uniform (non-accelerating) and rectilinear motion. The rejection of Newtonian absolute space and time is a major contribution of this theory.

Uncertainty Principle: the claim, formulated by Werner Heisenberg, that it is impossible to determine simultaneously the position and the velocity of an atomic particle. It rejects causality as a reigning principle of scientific explanation of physical phenomena on the nuclear level.

Unified field theory: a mathematical framework that would unite electrodynamics and gravitation.

References Cited

Russian figures whose works are available in English (e.g., V. A. Fock) are listed under the English spelling of their names, usually with the Russian transliteration in brackets. The following abbreviations are used in this list.

BSE   *Bol'shaia sovetskaia entsiklopediia*
FN   *Filosofskie nauki*
OMEN   *Izvestiia Akademii nauk SSSR: Otdelenie matematicheskikh i estestvennykh nauk*
Pr   *Priroda*
PZM   *Pod znamenem marksizma*
UFN   *Uspekhi fizicheskikh nauk*
VAN   *Vestnik Akademii nauk SSSR*
VF   *Voprosy filosofii*
VIET   *Voprosy istorii estestvoznaniia i tekhniki*
VKA   *Vestnik Kommunisticheskoi akademii*

Abasov, A. S. "Prostranstvo i vremia, prostranstvenno-vremennaia organizatsiia," *VF*, 1985, no. 11, pp. 71–82.

Agapov, B. N., et al. *Khudozhnik i nauka.* Moscow, 1966.

Akhundov, M. D. *Aktual'nost' istorii: uchenyi i ideologiia.* Moscow, 1990.

———. *Kontseptsii prostranstva i vremeni: istoki, evoliutsiia, perspektivy.* Moscow, 1982.

———. "Mekhanika N'iutona i formirovanie issledovatel'skoi programmy," in Akhundov and Illarionov, eds., *N'iuton*, pp. 16–34.

———. "Spasla li atomnaia bomba sovetskuiu fiziku?," *Pr*, 1991, no. 1, pp. 90–97.

———. "Vzaimodeistvie teoreticheskogo i empiriokriticheskogo aspekta v razvitii fiziki," in Sachkov, ed., *Teoriia poznaniia*, pp. 127–48.

Akhundov, M. D., and L. B. Bazhenov. "Estestvoznanie i religiia v sisteme kul'tury," *VF*, 1992, no. 12, pp. 42–53.

———. "U istokov ideologizirovannoi nauki," *Pr*, 1989, no. 2, pp. 1, 90–99.

———, eds. *Estestvoznanie v bor'be s religioznym mirovozzreniem.* Moscow, 1988.

Akhundov, M. D., and S. V. Illarionov. "Preemstvennost' issledovatel'skikh program v razvitii fiziki," *VF*, 1986, no. 6, pp. 56–65.

———, eds. *N'iuton i filosofskie problemy fiziki XX veka.* Moscow, 1990.

Aksenov, G. P. "Prostranstvo-vremia zhivogo v biosfere," in Sokolov and Ianshin, eds., *V. I. Vernadskii*, pp. 129–39.

"Al'bert Einshtein, 1879–1979," *Pr* 1979, no. 3, pp. 3–61.

Aleksandrov, A. D. "Dialektika i nauka," *VAN*, 1957, no. 6, pp. 3–17.

———. "Filosofskoe soderzhanie teorii otnositel'nosti," in Fedoseev et al., *Filosofskie problemy*, pp. 93–136.

———. "On the Philosophical Content of the Relativity Theory," in *Einstein and Philosophical Problems*, pp. 102–21.

———. "Otnositel'nosti teoriia," *Filosofskaia entsiklopediia*, 1967, vol. 4, pp. 177–81.

[———]. "Otnositel'nosti teoriia: teoretiko-poznavatel'noe znachenie," *BSE*, 2d ed., 1955, vol. 31, pp. 411–13.

———. "Po povodu nekotorykh vzgliadov na teoriiu otnositel'nosti," *VF*, 1953, no. 5, pp. 225–45.

———. *Problemy nauki i pozitsiia uchenogo*. Moscow, 1988.

———. "Prostranstvo i vremia v sovremennoi fizike v svete filosofskikh idei Lenina," in Omel'ianovskii, ed., *Lenin i sovremennoe estestvoznanie*, pp. 202–29.

———. "Teoriia otnositel'nosti kak teoriia absolutnogo prostranstva-vremeni," in Omel'ianovskii and Kuznetsov, eds., *Filosofskie voprosy*, pp. 269–323.

Aleksandrov, A. D., and G. M. Idlis. "Vklad v reliativistskuiu teoriiu prostranstva, vremeni i tiagoteniia," in Grigor'ian, ed., *Issledovaniia*, pp. 106–12.

Aleksandrov, A. P. "Kak delali bombu?," *Izvestiia*, July 22, 1988.

Aleksandrov, P. S. "Neobkhodimo vdokhnovenie," *Literaturnaia gazeta*, January 25, 1967, p. 12.

Alekseev, I. A. *Kontseptsiia dopolnitel'nosti: istoriko-matematicheskii analiz*. Moscow, 1978.

———. "Nekotorye soobrazheniia po povodu diskussii Einshteina i Bora," *VF*, 1979, no. 1, pp. 96–103.

———. "Obsuzhdenie knigi A. K. Maneeva, 'K kritike obosnovanii teorii otnositel'nosti,'" *VF*, 1961, no. 6, pp. 139–44.

Alekseev, P. V. "Diskussiia s mekhanistami po probleme vzaimosviazi filosofii i estestvoznaniia," *VF*, 1966, no. 4, pp. 44–54.

Ambartsumian, V. A. *Filosofskie voprosy nauki o Vselennoi*. Erevan, 1973.

———. "Nekotorye metodologicheskie voprosy kosmogonii," in Fedoseev et al., eds., *Filosofskie problemy*, pp. 268–90.

Ambartsumian, V. A., and V. V. Kaziutinskii. "Dialektika v sovremennoi astronomii," in Omel'ianovskii, ed., *Lenin i sovremennoe estestvoznanie*, pp. 230–56.

———. "Problema razvitiia v astronomii," in Sachkov et al., eds., *Materialisticheskaia dialektika*, pp. 40–63.

———. "Revoliutsiia v astronomii XX v. i ee filosofskoe znachenie," in Omel'ianovskii, ed., *Sovremennoe estestvoznanie*, pp. 246–68.

Ambartsumian, V. A., et al., eds. *Fizicheskaia teoriia (filosofsko-metodologicheskii analiz)*. Moscow, 1980.

———. *Razvitie astronomii v SSSR*. Moscow, 1967.

Antipov, G. A. "Prisushche li nauke nravstvennoe nachalo?," in A. N. Kochergin, ed., *Nauka i tsennosti*, pp. 43–57.

Arkhiptsev, F. T., et al. *Leninskii etap v razvitii filosofii marksizma*. Moscow, 1972.

Arnol'd, V. A. *Teoriia otnositel'nosti Einshteina: obshchie osnovaniia*. Moscow, 1924.

Aronov, R. A., and B. Ia. Pakhomov. "Filosofiia i fizika v diskussiiakh N. Bora i A. Einshteina," *VF*, 1965, no. 10, pp. 59–73.

Aronov, R. A., et al., eds. *Filosofskie voprosy sovremennogo estestvoznaniia (Fizika, matematika, biologiia)*. Moscow, 1977.

Arshinov, V. I. "Novoe kachestvo sovremennoi nauki," *Pr*, 1991, no. 2, pp. 13–17.

Arshinov, V. I., and B. Ia. Pakhomov. "Review of 'Einshtein i filosofskie voprosy fiziki XX veka,'" *FN*, 1980, no. 4, pp. 183–85.

Arshinov, V. I., S. P. Kurdiumov, and Ia. I. Svirskii. "Klassicheskaia mekhanika N'iutona i problema samoorganizatsii v sovremennom nauchnom poznanii," in Akhundov and Illarionov, eds., *N'iuton*, pp. 98–116.

Artsimovich, L. A., et al., eds. *Razvitie fiziki v SSSR*. 2 vols. Moscow, 1967.

Artykov, T. A. "Evoliutsiia fizicheskogo printsipa otnositel'nosti," *FN*, 1989, no. 9, pp. 36–43.

———. *Filosofskii analiz printsipa otnositel'nosti*. Tashkent, 1990.

Avaliani, S. Sh. *Absolutnoe i otnositel'noe*. Tbilisi, 1980.

———. *Ocherki filosofii estestvoznaniia*. Tbilisi, 1968.

Bachelard, Gaston. *The New Scientific Spirit*, tr. Arthur Goldhammer. Boston, 1984.

Bacon, Francis. *Advancement of Learning*. New York, 1901.

Bakhtin, Mikhail. *Problems of Dostoevsky's Poetics*, tr. Caryl Emerson. Minneapolis, 1984.

Balandin, R. K. "Dostoevsky i estestvoznanie," *VIET*, 1992, no. 1, pp. 93–103.

Balashov, Iu. V. "'Antropnye argumenty' v sovremennoi kosmologii," *VF*, 1988, no. 7, pp. 117–27.

———. "Antropnyi kosmologicheskii printsip v zerkale kritiki," *FN*, 1990, no. 9, pp. 31–40.

Barashenkov, V. "Einshtein i sovremennoe estestvenauchnoe myshlenie," *Kommunist*, 1980, no. 4, pp. 62–73.

———. "Prostranstvo i vremia bez materii?," *VF*, 1977, no. 9, pp. 77–83.

———. "Za predelami teorii Einshteina—supersimmetriia i supergravitatsiia," *Znanie-sila*, 1988, no. 7, pp. 29–37.

Barnett, L. *The Universe and Dr. Einstein*. New York, 1946.

Bazarov, I. P. "Za dialektiko-materialisticheskoe ponimanie i razvitie teorii otnositel'nosti," *VF*, 1952, no. 6, pp. 175–85.

Bazhanov, V. A., and A. P. Iushkevich, "A. V. Vasil'ev kak uchenyi i obshchestvennyi deiatel'," in A. V. Vasilev, *Nikolai Ivanovich Lobachevskii*, pp. 221–28.

Bazhenov, L. B. "Problema prostranstva-vremeni," in Bazhenov et al., eds., *Filosofiia*, vol. 1, chap. 4.

Bazhenov, L. B., et al., eds. *Evoliutsiia materii i ee strukturnye urovni*. Moscow, 1983.

———. *Filosofiia estestvoznaiia*, vol. 1. Moscow, 1966.

———. *Printsip dopolnitel'nosti i materialisticheskaia dialektika*. Moscow, 1976.

Beer, F. *Teoriia otnositel'nosti i eia istoricheskiia osnovy*. Berlin, 1921.

Bekhterev, V. M. *Kollektivnaia refleksologiia*. Petrograd, 1921.

Beletskaia, Vanda. "Vzyvaiushchii," *Ogonek*, 1939, no. 8, pp. 6–7, 30–31.

Biriukov, B. V. "Fenomenologiia v kontekste filosofii matematiki: Gusserl'-Frege-Bekker-Veil, " *FN*, 1969, no. 2, pp. 108–14.

Blokhintsev, D. I. "Nekotorye voprosy razvitiia sovremennoi fiziki," *VF*, 1959, no. 10, pp. 31–34.

———. "Puti razvitiia teoreticheskoi fiziki v SSSR," *UFN*, 1947, vol. 33, no. 2, pp. 285–93.

———. "Uspekhi teoreticheskoi fiziki v Sovetskom Soiuze za 20 let," *Zhurnal eksperimental'noi i teoreticheskoi fiziki*, 1937, vol. 7, no. 11, pp. 1203–1208.

Blokhintsev, D. I., and I. M. Frank. "Predislovie," in *S. I. Vavilov i sovremennaia fizika* (Moscow, 1970), pp. 1–14.

Blokhintsev, D. I., and F. Gal'perin, "Efir," *Front nauki i tekhniki*, 1936, no. 6, pp. 49–55.

Blokhintsev, D. I., M. A. Leontovich et al. "O stat'e N. P. Kasterina, Obobshchenie osnovnykh uravnenii aerodinamiki i elektrodinamiki," *OMEN*, 1937, no. 3, pp. 425–36.

Bogdanov, A. A. "Ob'ektivnoe ponimanie printsipa otnositel'nosti," *VKA*, 1924, no. 8, pp. 332–47.

———. "Uchenie ob analogiiakh," *VKA*, 1923, no. 2, pp. 78–97.

Bogoraz-Tan, V. G. *Einshtein i religiia*. Petrograd, 1923.

Bohr, Niels. *Atomic Physics and Human Knowledge*. New York, 1961.

Bolotovskii, B. M. "V etom spore pobezhdennykh ne bylo," *VF*, 1979, no. 1, pp. 109–11.

"Bor'ba za leninskuiu partiinost' v nauke i zadachi Komakademii," *VKA*, 1931, no. 12, pp. 3–18.

Borgman, I. "Elektrichestvo i svet," *Dnevnik XII S"ezda*, no. 1, pp. 93–107.

Borisovskii, G. B., et al., eds. *Iskusstvo i tochnye nauki*. Moscow, 1979.

The Born-Einstein Letters, tr. Irene Born. London, 1971.

Broglie, Louis de. "Anri Puankare i fizicheskie teorii," in Poincaré, *Izbrannye trudy*, vol. 3, pp. 703–11.

———. "Einstein and Physics," in de Broglie et al., *Einstein*, pp. 31–44.

———. *Savants et découvertes*. Paris, 1951.

———. "The Scientific Work of Albert Einstein," in Schilpp, ed., *Albert Einstein*, vol. 1, pp. 107–27.

Broglie, Louis de, et al. *Einstein*. New York, 1979.

Bronshtein, M. P. "Pogloshenie i rasseianie γ-luchei," *UFN*, 1932, vol. 12, no. 5–6, pp. 649–70.

———. "Sovremennoe sostoianie reliativistskoi kosmologii," *UFN*, 1931, vol. 11, no. 1, pp. 124–84.

———. "Vsemirnoe tiagotenie i elektrichestvo (Novaia teoriia Einshteina)," in Gorelik and Frenkel, pp. 245–52.

Bukharin, N. I., and A. M. Deborin, eds. *Pamiati Karla Marksa: Sbornik statei, k piatidesiatiletiiu smerty: 1833–1933*. Leningrad, 1933.

———. *Pamiati V. I. Lenina: Sbornik statei k desiatiletiiu so dnia smerti—1924–1934*. Moscow-Leningrad, 1934.

Cantor, Georg. *Contributions to the Founding of the Theory of Transfinite Numbers*, tr. Philip E. B. Jourdain. New York, 1955.

Čapek, Milič. *The Philosophical Impact on Contemporary Physics*. Princeton, N.J., 1961.

Catteau, Jacques. *Dostoevsky and the Process of Literary Creation*, tr. A. Littlewood. Cambridge, Eng., 1989.

Cherniak, V. S. "Iskusstvo i matematicheskaia struktura," *Vestnik Moskovskogo universiteta*, 1969, no. 3, pp. 44–52.

Chernov, F. "Burzhuaznyi kosmopolitizm i ego reaktsionnaia rol'," *Bol'shevik*, 1949, no. 5, pp. 30–41.

Chudinov, E. M. "Einshteinovskaia 'kontseptsiia nabliudaemikh' i operatsionalizm," in Dyshlevyi, Kedrov et al., eds., *Filosofskie osnovaniia*, pp. 207–22.

———. "Obshchaia teoriia otnositel'nosti i prostranstvenno-vremennaia struktura Vselennoi," *VF*, 1966, no. 3, pp. 65–75.

———. *Priroda nauchnoi istiny*. Moscow, 1977.

———. "Ratsionalizm Einshteina i sovremennaia fizika," in Kobzarev et al., *Einshtein i sovremennaia fizika*, pp. 22–36.

———. *Teoriia otnositel'nosti i filosofiia*. Moscow, 1974.

Chudinov, E. M., ed. *Leninskoe filosofskoe nasledie i sovremennaia fizika*. Moscow, 1981.

Clifford, William Kingdon. *Lectures and Essays*, vol. 1. London, 1879.

Cohen, Robert S. "Ernst Mach: Physics, Perception and the Philosophy of Science," in Robert S. Cohen and R. J. Seeger, eds., *Ernst Mach: Physicist and Philosopher* (Dordrecht, Netherlands, 1978), pp. 126–64.

Comey, David D. "Soviet Controversies Over Relativity," in *The State of Soviet Science* (Cambridge, Mass., 1965), pp. 186–99.

Danin, D. "Vozmozhnye resheniia (Iz dnevnika literatora)," *Voprosy literatury*, 1964, no. 8, pp. 88–111.

Davies, P. C. W. *Space and Time in the Modern Universe*. Cambridge, Eng., 1977.

Deborin, A. M. *Dialektika i estestvoznanie*. Moscow, 1929.

———. *Filosofiia i politika*. Moscow, 1961.

———. "Lenin i krizis noveishei fiziki," *Otchet o deiatel'nosti Akademii nauk SSSR za 1929 god*, Supplement, vol. 1. Leningrad, 1929.

Delokarov, K. Kh. "B. M. Gessen i filosofskie problemy estestvoznaniia," *VAN*, 1978, no. 12, pp. 75–84.

———. "Bog, religiia i mirovozzrenie A. Einshteina," in Akhundov and Bazhenov, eds., *Estestvoznanie*, pp. 233–43.

———. "Bor'ba za materialisticheskoe istolkovanie teorii otnositel'nosti v sovetskoi nauke (1920–1930 gody)," *FN*, 1967, no. 4, pp. 125–32.

———. *Filosofskie problemy teorii otnositel'nosti (Na materiale filosofskikh diskussii v SSSR v 20–30e gody)*. Moscow, 1973.

———. "Formirovanie soiuza filosofov i estestvoispytatelei v SSSR (Istoricheskii analiz)," in Chudinov, ed., *Leninskoe filosofskoe nasledie*, pp. 62–119.

———. "Materialisticheskaia dialektika i teoriia otnositel'nosti," *VF*, 1979, no. 8, pp. 72–82.

———. *Metodologicheskie problemy kvantovoi mekhaniki v sovetskoi filosofskoi nauke: istoricheskii analiz*. Moscow, 1962.

———. "Puti stanovleniia soiuza filosofii i fiziki v SSSR (20–30e gody)," *FN*, 1971, no. 1, pp. 111–20.

————. "Review of D. P. Gribanov, 'Filosofskie vzgliady Einshteina,'" *FN*, 1989, no. 9, pp. 137–40.

————. "The Theory of Relativity and Soviet Science (A Historical-Methodological Analysis)," in Gribanov et al., *Einstein*, pp. 450–94.

————, ed. *Einshtein i filosofskie problemy fiziki XX veka.* Moscow, 1979.

Deriugin, S. V., and E. G. Osipovskii. "Novyi etap v razvitii otnoshenii mezhdu marksistami i khristianami," *FN*, 1990, no. 2, pp. 3–9.

Dialekticheskii materializm i sovremennoe estestvoznanie: sbornik statei. Moscow, 1957.

Dirac, P. A. M. *The Principles of Quantum Mechanics.* 4th ed. Oxford, 1958.

Dmitriev, I . S. "Religioznye iskaniia Isaaka N'iutona," *VF*, 1991, no. 6, pp. 58–67.

Dnevnik XII S"ezda Russkikh estestvoispytatelei i vrachei. Moscow, 1910.

Dorfman, Ia. G. "Iakov Il'ich Frenkel," in Ia. Frenkel, *Sobranie*, vol. 2, pp. 3–16.

————. *Vsemirnaia istoriia fiziki s nachala XIX do serediny XX vv.* Moscow, 1979.

Dostoevsky, F. M. *'Notes from Underground' and 'The Grand Inquisitor.'* New York, 1960.

————. *Pis'ma*, vols. 1, 2. Moscow, 1929–30.

————. *Vozvrashchenie cheloveka*, ed. M. M. Strakhanova. Moscow, 1989.

Duhem, Pierre. *The Aim and Structure of Physical Theory*, tr. P. E. Wiener. New York, 1962.

Dyshlevyi, P. S. "A. Einshtein o probleme real'nosti v fizike XX v.," *FN*, 1979, no. 4, pp. 87–97.

————. "Filosofiia i astronomiia," *Kommunist*, 1975, no. 4, pp. 120–23.

————. "Filosofskoe soderzhanie i znachenie obshchei teorii otnositel'nosti i metodov Einshteina," in A. Petrov and Dyshlevyi, eds., *Prostranstvo*, pp. 44–69.

————. "Kontseptsiia otnositel'nosti kak vazhneishii element metodologii fiziki XX stoletiia," *FN*, 1973, no. 2, pp. 406–10.

————. *V. I. Lenin i filosofskie problemy reliativistskoi fiziki.* Kiev, 1969.

Dyshlevyi, P. S., and F. M. Kanak. *Materialisticheskaia filosofiia i razvitie estestvoznaniia.* Kiev, 1977.

Dyshlevyi, P. S., and A. Z. Petrov. "Filosofskie voprosy obshchei teorii otnositel'nosti," *VF*, 1968, no. 1, pp. 87–97.

Dyshlevyi, P. S., and V. S. Luk'ianets, eds. *Nauchnaia kartina mira: logiko-gnoseologicheskii aspekt.* Kiev, 1983.

Dyshlevyi, P. S., and A. Z. Petrov, eds. *Filosofskie problemy teorii tiagoteniia Einshteina i reliativistskoi kosmologii.* Kiev, 1963.

Dyshlevyi, P. S., B. M. Kedrov et al., eds. *Filosofskie osnovaniia estestvennykh nauk.* Moscow, 1976.

Eddington, A. S. *The Nature of the Physical World.* Cambridge, Eng., 1930.

Egorshin, V. "Review of 'Dialektika i priroda,'" *VKA*, 1928, no. 27, pp. 269–82.

Eigenson, M. S. "Beskonechnost' Vselennoi," *Pr*, 1940, no. 3, pp. 5–17.

Eikhenval'd, A. A. "Materiia i energiia, *Dnevnik XII S"ezda*, no. 6, pp. 331–39.

Einshteinovskii sbornik. 14 vols. Moscow, 1965–85.

Einstein, Albert. "Autobiographical Notes," in Schilpp, ed., *Einstein*, vol. 1, pp. 1–95.

————. *The Collected Papers*, vol. 2, ed. John Stachel. Princeton, N.J., 1989.

————. *Cosmic Religion with Other Opinions and Aphorisms.* New York, 1931.

———. *Essays in Science.* New York, 1934.

———. "Foreword" to Jammer, *Concepts*, pp. xi–xvi.

———. "The Foundations of the General Theory of Relativity," in Lorentz et al., *Principle of Relativity*, pp. 109–64.

———. *Ideas and Opinions.* New York, 1982.

———. "Introduction" to Rudolph Kayser, *Spinoza: Portrait of a Spiritual Leader*, tr. A. Allen and M. Newmark (New York, 1968), pp. ix–xi.

———. *Letters to Solovine*, tr. W. Baskin. New York, 1986.

———. "Max Planck als Forscher," *Naturwissenschaften*, 1913, vol. 1, pp. 1077–1079.

———. *The Meaning of Relativity.* Princeton, N.J., 1945.

———. "Neevklidova geometriia i fizika," in Grigor'ian, ed., *Einshtein*, pp. 5–9.

———. "On the Electrodynamics of Moving Bodies," in Lorentz et al., *Principle of Relativity*, pp. 35–65.

———. *Out of My Later Years.* Secaucus, N.J., 1956.

———. "Pochemu sotsializm?," *Kommunist*, 1989, no. 17, pp. 96–100.

———. "Predislovie avtora k russkom izdanii," in Einstein, *Teoriia otnositel'nosti*, p. 5.

———. "Prologue" to Planck, *Where Is Science Going?*, pp. 7–12.

———. "Quanten-Mechanik und Wirklichkeit," *Dialectica*, 1948, vol. 2, no. 4–7, pp. 320–24.

———. *Relativity: The Special and the General Theory*, tr. R. W. Lawson. New York, 1961.

———. "Remarks Concerning the Essays Brought Together in This Cooperative Volume," in Schilpp, ed., *Albert Einstein*, vol. 2, pp. 665–88.

———. *Sidelights on Relativity.* New York, 1983.

———. *Sobranie nauchnykh trudov.* 4 vols. Moscow, 1965–67.

———. *Teoriia otnositel'nosti: obshchedostupnoe izlozhenie*, tr. G. B. Itel'son. Berlin, 1921.

———. "Zum kosmologischen Problem der allgemeinen Relativitätstheorie," *Sitzungsberichte der Preussischen Akademie der Wissenschaften: Physikalisch-Mathematische Klasse*, 1913, no. 12, pp. 235–37.

Einstein, Albert, and Leopold Infeld. *The Evolution of Physics.* New York, 1966.

Einstein, Albert, and Nathan Rosen. "The Particle Problem in the General Theory of Relativity," *Physical Review*, 1935, 2d ser., vol. 48, pp. 73–77.

Einstein, Albert, Leopold Infeld, and Banesh Hoffmann. "Gravitational Equations and the Problem of Motion," *Annals of Mathematics*, 2d ser., 1938, vol. 39, pp. 65–100.

Einstein, Albert, James Murphy, and J. W. N. Sullivan. "Science and God: A German Dialogue," *The Forum*, 1930, vol. 81, pp. 373–79.

Einstein, Albert, B. Podolsky, and N. Rosen. "Can Quantum-Mechanical Description of Physical Reality Be Considered Complete?," *Physical Review*, 1935, ser. 2, vol. 47, pp. 777–80.

Einstein and the Philosophical Problems of 20th Century Physics. Moscow, 1983.

Einstein on Peace, ed. Otto Nathan and Heinz Norden. New York, 1968.

El'iashevich, M. A. "Vklad Einshteina i razvitie kvantovykh predstavlenii," in Kobzarev et al., *Einshtein*, pp. 37–78.

Engels, Friedrich. *Ludwig Feuerbach and the Outcome of Classical German Philosophy*. London, 1934.

———. *Selected Works*. 2 vols. Moscow, 1962.

Evgrafov, V. E., et al., eds. *Istoriia filosofii v SSSR*, vol. 5, part 1. Moscow, 1985.

F. G. "Bor'ba materializma i idealizma vokrug nekotorykh voprosov sovremennoi fiziki," *PZM*, 1934, no. 4, pp. 23–25.

Fataliev, Kh. M. *Dialekticheskii materializm i voprosy estestvoznaniia*. Moscow, 1958.

Fedoseev, P. N. "Filosofskie idei V. I. Lenina i metodologiia sovremennoi fiziki," in Chudinov, ed., *Leninskoe filosofskoe nasledie*, pp. 120–42.

[———]. "Zakliuchitel'noe slovo," in Fedoseev et al., eds., *Filosofskie problemy*, pp. 589–601.

Fedoseev, P. N., et al., eds. *Filosofskie problemy sovremennogo estestvoznaniia*. Moscow, 1959.

Feinberg, E. L. *Art in the Science Dominated World*, tr. J. A. Cooper. New York, 1987.

———. "Deviat' rubtsov," *Nauka i zhizn'*, 1990, no. 8, pp. 34–46.

———. "Intellektual'naia revoliutsiia," *VF*, 1986, no. 8, pp. 33–45.

———. "Vzaimosviaz' nauki i iskusstva v mirovozzrenii Einshteina," *VF*, 1979, no. 3, pp. 32–46.

Fesenkov, V. G. "Astronomicheskie dokazatel'stva printsipa otnositel'nosti," *VKA*, 1925, no. 8, pp. 200–216.

Filatov, V. P. "Ob istokakh lysenkovskoi 'agrobiologii'," *VF*, 1988, no. 8, pp. 3–22.

Filosofskie problemy sovremennoi fiziki. Kiev, 1956.

"Fizika," *Dnevnik XII S"ezda*, no. 5, pp. 170–72.

Florenskii, Pavel. *Mnimosti v geometrii: rasshirenie oblasti dvukhmernykh obrazov geometrii*. Moscow, 1922.

Fock [Fok], V. A. "Al'bert Einshtein (Po povodu 60-letiia so dnia rozhdeniia), *Pr*, 1939, no. 7, pp. 95–98.

———. "Diskussiia s Nil'som Borom," *VF*, 1964, no. 8, pp. 49–53.

———. "Fizicheskie printsipy teorii otnositel'nosti A. Einshteina," *VF*, 1966, no. 8, pp. 15–22.

———. "K diskussii po voprosam fiziki," *PZM*, 1938, no. 1, pp. 149–59.

———. "K stat'e Nikol'skogo 'Printsipy kvantovoi mekhaniki,'" *UFN*, 1937, vol. 17, no. 4, pp. 552–54.

———. "Massa i energiia," *UFN*, 1952, vol. 48, no. 2, pp. 161–65.

———. "Nekotorye primeneniia idei neevklidovoi geometrii Lobachevskogo v fizike," in Kotel'nikov and Fock, *Nekotorye primeneniia*, pp. 48–86.

———. "O dvizhenii konechnykh mass v obshchei teorii otnositel'nosti," *Zhurnal eksperimental'noi i teoreticheskoi fiziki*, 1939, vol. 9, no. 4, pp. 375–410.

———. "O roli printsipov otnositel'nosti i ekvivalentnosti v teorii tiagoteniia Einshteina," *VF*, 1961, no. 12, pp. 45–52.

[———]. "Otnositel'nosti teoriia," *BSE*, 2d ed., vol. 31 (1955), pp. 405–11.

———. "The Physical Principles of Einstein's Gravitational Theory," in *Einstein and Philosophical Problems*, pp. 211–13.

———. "Poniatiia odnorodnosti i otnositel'nosti v teorii prostranstva i vremeni," *VF*, 1955, no. 4, pp. 131–35.

———. "Problema dvizheniia mass i teoriia tiagoteniia Einshteina," in Lukirskii et al., *Sbornik*, pp. 31–43.

———. "Protiv nevezhestvennoi kritiki sovremennykh fizicheskikh teorii," *VF*, 1953, no. 1, pp. 168–74.

———. "Raboty A. A. Fridmana po teorii tiagoteniia Einshteina," in Friedmann, *Izbrannye trudy*, pp. 398–402.

———. "Sistema Kopernika i sistema Ptolomeia v svete obshchei teorii otnositel'nosti," in A. A. Mikhailov, ed., *Nikolai Kopernik* (Moscow-Leningrad, 1947), pp. 180–86.

———. "Sovremennaia teoriia prostranstva i vremeni," *Pr*, 1953, no. 12, pp. 13–26.

———. *Teoriia Einshteina i fizicheskaia otnositel'nost'*. Moscow, 1967.

———. *The Theory of Space, Time, and Gravitation*. 2d ed., tr. N. Kemmer. New York, 1964.

———. "Zamechaniia k tvorcheskoi avtobiografii Al'berta Einshteina," *UFN*, 1956, vol. 59, no. 1, pp. 107–17.

Fock, V. A., A. Einstein, B. Podol'skii, N. Rosen, and N. Bohr [Fok, Einshtein, Podol'skii, Rozen, and Bor]. "Mozhno li schitat', chto kvantovo-mekhanicheskoe opisanie fizicheskoi real'nosti iavliaetsia polnym?," *UFN*, 1936, vol. 16, pp. 436–41.

Frank, G. M. "Ob idealizme v fizike i biologii," *Sovetskaia nauka*, 1939, no. 9–10, pp. 102–9.

Frank, Joseph. *Through the Russian Prism: Essays on Literature and Culture*. Princeton, N.J., 1990.

Frankfurt, U. I. *Ocherki po istorii spetsial'noi teorii otnositel'nosti*. Moscow, 1961.

———. *Spetsial'naia i obshchaia teoriia otnositel'nosti*. Moscow, 1968.

Frank-Kamenskii, D. A., "Invariantnost' v sovremennoi fizike," *Pr*, 1968, no. 10, pp. 3–12.

Frederiks, V. K. "Nachala mekhaniki N'iutona i printsip otnositel'nosti," *UFN*, 1927, vol. 7, no. 2, pp. 75–86.

———. "Obshchii printsip otnositel'nosti Einshteina," *UFN*, 1921, vol. 2, no. 2, pp. 162–88.

Frenkel, Ia. I. "Einshtein, Al'bert," *Entsiklopedicheskii slovar'* (Granat), 1933, vol. 51, pp. 143–52.

———. *Lehrbuch der Elektrodynamik*, vol. 1. Berlin, 1926.

———. *Na zare novoi fiziki*. Leningrad, 1970.

———. *Sobranie izbrannykh trudov*, vol. 2. Moscow-Leningrad, 1958.

———. "Teoreticheskaia fizika v SSSR za 30 let," *UFN*, 1947, vol. 33, no. 3, pp. 294–317.

———. *Teoriia otnositel'nosti*. Moscow, 1923.

———. "Teoriia otnositel'nosti Einshteina," *Front nauki i tekhniki*, 1935, no. 5–6, pp. 38–48.

Frenkel, V. Ia. "Aleksandr Aleksandrovich Fridman," *UFN*, 1988, vol. 155, no. 3, pp. 481–515.

———. *Iakov Il'ich Frenkel*. Moscow-Leningrad, 1966

———. "Iurii A. Krutkov," in Grigor'ian, ed., *Issledovaniia*, pp. 210–29.

———. "Novye materialy o diskussii Einshteina i Fridmana po reliativistskoi kosmologii," *Einshteinovskii sbornik: 1973* (Moscow, 1974), pp. 5–18.

Frenkel, V. Ia., and B. F. Iavelov. *Einshtein-izobretatel'*. Moscow, 1981.

Fridlender, G. M. *Dostoevskii: materialy i issledovaniia*, vol. 6. Leningrad, 1985.

Friedmann [Fridman], A. A. *Izbrannye trudy*, ed. L. S. Polak. Moscow, 1966.

———. *Mir kak prostranstvo i vremia*. Moscow, 1985.

———. "Über die Krümmung des Raumes," *Zeitschrift für Physik*, 1922, vol. 10, pp. 377–87.

———. "Über die Möglichkeit einer Welt mit konstanter negativer Krümmung des Raumes," *Zeitschrift für Physik*, 1924, vol. 21, pp. 326–33.

Frish, S. E. *Skvoz' prizmu vremeni*. Moscow, 1992.

Frolov, I. T., and L. I. Grekov, eds. *Filosofiia, estestvoznanie, sovremennost'*. Moscow, 1981.

G. R. "Review of Ia. I. Grdina, 'K voprosu o printsipe otnositel'nosti,'" *Zhurnal Fiziko-Khimicheskogo obshchestva*, 1914, vol. 46, pp. 391–96.

Gamow, George. "The Evolutionary Universe," in Owen Gingerich, ed., *New Frontiers in Astronomy* (San Francisco, 1956), pp. 316–33.

Gavrilov, A. F., "Vospominaniia o Fridmane," in Friedmann, *Izbrannye trudy*, pp. 417–27.

German, L. "Intuitsiia," *BSE*, vol. 29 (1935), pp. 39–41.

Ginzburg, V. L. "The Heliocentric System and the General Theory of Relativity (From Copernicus to Einstein)," in Gribanov et al., *Einstein*, pp. 254–303.

———. *Kak ustroena Vselennaia i kak ona razvivaetsia vo vremeni*. Moscow, 1968.

———. *Key Problems of Physics and Astrophysics*. 2d ed., tr. O. Glebov. Moscow, 1978.

———. "Novye fizicheskie zakony," *Nauka i zhizn'*, 1988, no. 5, pp. 66–72.

———. *O teorii otnositel'nosti: sbornik statei*. Moscow, 1979.

———. "Obshchaia teoriia otnositel'nosti: posledovatel'naia li ona? Otvechaet li ona fizicheskoi real'nosti?," *Nauka i zhizn'*, 1987, no. 4, pp. 41–48.

———. "Zamechaniia o metodologii i razvitii fiziki i astrofiziki," in V. V. Kaiziutinskii et al., eds., *Dialektika-mirovozzrenie* (Moscow, 1983), pp. 71–110.

Glick, Thomas F., ed. *The Comparative Reception of Relativity*. Dordrecht, Netherlands, 1987.

Gol'dgammer, D. A. "Vremia, prostranstvo, efir," *Zhurnal Fiziko-Khimicheskogo obshchestva*, 1912, vol. 44, no. 5, physics section, pp. 180–88.

Gol'tsman, A. A. "Einshtein i materializm," *PZM*, 1924, no. 1, pp. 114–26.

———. "Nastuplenie na materializm," *PZM*, 1923, no. 1, pp. 78–112.

Gomilko, O. E. "Rol' esteticheskikh printsipov v postroenii nauchnoi teorii," *Filosofskie problemy*, 1988, vol. 77, pp. 41–46.

Gorelik, G. E. "Fizika universitetskaia i akademicheskaia," *VIET*, 1991, no. 2, pp. 31–46.

———. "Fiziki i sotsializm v arkhive KGB," *Svobodnaia mysl'*, 1992, no. 1, pp. 45–53.

———. "Moskva, fizika, 1937 god," *VIET*, 1992, no. 1, pp. 15–32.

———. "Naturfilosofskie problemy fiziki v 1937 godu," *Nauka i zhizn'*, 1990, no. 2, pp. 93–102.

———. "O chuvstve prekrasnogo ili fizikoeticheskie problemy" (2 parts), *Znanie-sila*, 1990, no. 9, pp. 38–44; no. 10, pp. 50–57.

———. "O roli gumanitarnykh predstavlenii v fizicheskom mirovozzrenii Einshteina," in *Issledovaniia: 1985*, pp. 77–94.

————. "O 'stalinizme' Einshteina i eshche koe o chem," *Kommunist*, 1990, no. 2, pp. 76–81.

————. "Obsuzhdenie 'naturfilosofskikh ustanovok sovremennoi fiziki' v Akademii nauk SSSR v 1937–1936 godakh," *VIET*, 1990, no. 4, pp. 17–31.

————. "Rasshireniiu Vselennoi—20 milliardov i 66 let," *VIET*, 1988, no. 4, pp. 95–104.

————. "U istokov novogo politicheskogo myshleniia, in *Einshteinovskii sbornik: 1986–1990* (Moscow, 1990), pp. 9–32.

————. "V. A. Fok: filosofiia tiagoteniia i tiazhest' filosofii," *Pr*, 1993, no. 10, pp. 81–93.

————. "Vikhri efirnye," *Znanie-sila*, 1992, no. 8, pp. 104–12.

————. "Zakony OTO i zakony sokhraneniia," *Znanie-sila*, 1988, no. 1, pp. 24–31.

Gorelik, G. E., and V. Ia. Frenkel. *Matvei Petrovich Bronshtein* (Moscow, 1990), pp. 245–52.

Gorelov, A. A., and R. O. Kurbanov. "Ekologiia i religiia," in Akhundov and Bazhenov, eds., *Estestvoznanie*, pp. 178–93.

"Goriachie tochki kosmologii," *Pr*, 1989, no. 7, pp. 3–18.

Gott, V. S. *Filosofskie voprosy sovremennoi fiziki*. Moscow, 1967.

————. "Lenin i Einshtein," *FN*, 1981, no. 11, pp. 97–104.

————. "Material'noe edinstvo mira i edinstvo nauchnogo znaniia," in I. T. Frolov et al., eds., *Edinstvo i mnogoobrazie mira* (Moscow, 1981), pp. 194–223.

Gott, V. S., and K. Kh. Delokarov. "Problemy vzaimosviazi filosofii i estestvennykh nauk," *FN*, 1989, no. 4, pp. 3–16.

Gott, V. S., and V. G. Sidorov. *Filosofiia i progress fiziki*. Moscow, 1986.

Graham, Loren R. "The Reception of Einstein's Ideas: Two Examples from Contrasting Political Cultures," in Gerald Holton and Yehuda Elkana, eds., *Albert Einstein: Historical and Cultural Perspectives* (Princeton, N.J., 1982), pp. 107–36.

————. *Science and Philosophy in the Soviet Union*. New York, 1972.

————. *Science, Philosophy and Human Behavior in the Soviet Union*. New York, 1987.

————. "The Socio-Political Roots of Boris Hessen: Soviet Marxism and the History of Science," *Social Studies of Science*, 1985, vol. 15, pp. 705–22.

Graves, John C. *The Conceptual Foundations of Contemporary Relativity Theory*. Cambridge, Mass., 1971.

Grdina, Ia. I. "Fizicheskii ili ogranichennyi printsip otnositel'nosti," *Zhurnal Fiziko-Khimicheskogo obshchestva*, 1916, vol. 48, pp. 1–38.

Griaznov, B. S. *Logika, ratsional'nost', tvorchestvo*. Moscow, 1982.

Gribanov, D. P. *Albert Einstein's Philosophical Views and the Theory of Relativity*, tr. H. C. Creighton. Moscow, 1987.

————. *Filosofskie osnovaniia teorii otnositel'nosti*. Moscow, 1982.

————. *Filosofskie vzgliady A. Einshteina i razvitie teorii otnositel'nosti*. Moscow, 1987.

————. "Sootnoshenie empiricheskogo i ratsional'nogo v nauchnom tvorchestve A. Einshteina," *VF*, 1980, no. 9, pp. 40–50.

Gribanov, D. P., et al. *Einstein and the Philosophical Problems of 20th Century Physics*. Moscow, 1983.

Grigor'ian, A. T. "Al'bert Einshtein kak istorik estestvoznaniia," *VIET*, 1990, no. 1, pp. 85–89.

———, ed. *Einshtein i razvitie fiziko-matematicheskoi mysli*. Moscow, 1962.

———. *Issledovaniia po istorii fiziki i mekhaniki: 1990*. Moscow, 1990.

Grigor'ian, A. T., and L. S. Polak, eds. *Ocherki razvitiia osnovnykh fizicheskikh idei*. Moscow, 1959.

Grishchuk, L. P. "Obshchaia teoriia otnositel'nosti—znakomaia i neznakomaia," *UFN*, 1990, vol. 160, no. 6, pp. 147–54.

Gulo, D. D. *Nikolai Alekseevich Umov*. Moscow, 1971.

Gunsheskaia, B. I. "Na putiakh nauchno-khudozhestvennogo dokumentalizma," in Meliakh, ed., *Khudozhestvennoe i nauchnoe tvorchestvo*, pp. 156–60.

Gurevich, G. A. "Velikaia nauchnaia revoliutsiia" (3 parts), *Russkie zapiski*, 1916, no. 5, pp. 113–34; no. 6, pp. 32–54; no. 7, pp. 92–113.

Gus, M. *Idei i obrazy F. M. Dostoevskogo*. 2d ed. Moscow, 1971.

Hadamard, Jacques. *The Psychology of Invention in the Mathematical Field*. New York, 1954.

Hawking, Stephen W. *A Brief History of Time: From the Big Bang to Black Holes*. Toronto, 1988.

Heisenberg, Werner. *Across the Frontiers*. New York, 1971.

———. *Physics and Philosophy*. New York, 1958.

——— [V. Geizenberg]. "Smysl i znachenie krasoty v tochnykh naukakh," *VF*, 1979, no. 12, pp. 49–60.

Hernek, F. "K pis'mu Al' berta Einshteina Ernstu Makhu, " *VF*, 1960, no. 6, pp. 104–8.

Hessen, B. [B. Gessen]. "Efir," *BSE*, vol. 65 (1931), pp. 10–18.

———. "Einshtein, Al'bert: filosofskie vzgliady," *BSE*, vol. 65 (1931), pp. 152–54.

———. *Osnovnye idei teorii otnositel'nosti*. Moscow, 1928.

Hoffmann, Banesh, in collaboration with Helen Dukes. *Albert Einstein, Creator and Rebel*. New York, 1973.

Holloway, David. "Physics, the State, and Civil Society in the Soviet Union," *Historical Studies in the Physical and Biological Sciences*, 1999, vol. 30, part 1, pp. 173–92.

———. "Science and Power in the Soviet Union," in N. A. Rupke, ed., *Science, Politics and the Public Good* (London, 1988), pp. 141–59.

———. *Stalin and the Bomb*. New Haven, Conn., 1994.

Holton, Gerald. *Thematic Origins in Scientific Thought: Kepler to Einstein*. Cambridge, Mass., 1975.

Idlis, G. M. *Revoliutsii v astronomii, fizike i kosmologii*. Moscow, 1985.

[Il'ichev, L. F.] "Doklad," *VAN*, 1963, no. 11, pp. 4–46.

Il'in, A. Ia., and V. V. Gapich. "Nekotorye osobennosti gnoseologicheskikh vzgliadov A. Einshteina," *FN*, 1970, no. 1, pp. 89–99.

Illarionov, S. V. "Diskussiia Einshteina i Bora," in Delokarov, ed., *Einshtein i filosofskie problemy*, pp. 465–83.

Illarionov, S. V., and S. M. Polovinkin. "Einshtein i sovremennost'," *VF*, 1979, no. 8, pp. 162–64.

Ilyin, V., and A. Kalinkin. *The Nature of Science: An Epistemological Analysis*, tr. Sergei Syrovatnik. Moscow, 1985.

In Memoriam N. I. Lobachevskii, vol. 2. Moscow, 1927.

Infeld, Leopold. "Moi vospominaniia ob Einshteine," *UFN*, 1956, vol. 59, pp. 135–84.

————. *Why I Left Canada*, tr. Helen Infeld. Montreal, 1971.

"Intuitsiia," *BSE*, 2d ed., vol. 18 (1953), pp. 319–20.

Ioffe, A. F. "Chto govoriat opyty o teorii otnositel'nosti Einshteina," *Pravda*, Jan. 1, 1927.

————. "K voprosu o filosofskikh oshibkakh moei knigi 'Osnovnye predstavleniia sovremennoi fiziki,'" *UFN*, 1954, vol. 53, no. 4, pp. 589–98.

————. *O fizike i fizikakh*. Leningrad, 1977.

————. "O polozhenii na filosofskom fronte sovetskoi fiziki," *PZM*, 1937, no. 11–12, pp. 131–43.

————. *Osnovnye predstavleniia sovremennoi fiziki*. Leningrad-Moscow, 1949.

————. "Pamiati Al'berta Einshteina," *UFN*, 1955, vol. 57, pp. 187–92.

————. "Razvitie atomisticheskikh vozzrenii v XX veke," in Bukharin and Deborin, eds., *Pamiati V. I. Lenina*, pp. 449–68.

————. "Sovetskaia fizika za 20 let," *Pr*, 1917, no. 10, pp. 43–49.

Ioffe, A. F., P. Lazarev, and V. Steklov. "Zapiska ob uchenykh trudakh Al'berta Einshteina," *Izvestiia Rossiiskoi Akademii nauk*, 6th ser., 1922, vol. 16, p. 42.

Isakova, A. V., et al., eds. *Vladimir Aleksandrovich Fok*. Moscow, 1956.

Issledovaniia po istorii fiziki i mekhaniki: 1985. Moscow, 1985.

Issledovaniia po istorii fiziki i mekhaniki: 1989. Moscow, 1989.

"Istoricheskii dokument voinstvuiushchego materializma," *VF*, 1962, no. 3, pp. 3–13.

Iurinets, V. "Filosofskii front na Ukraine," *PZM*, 1930, no. 5, pp. 129–38.

Iushkevich, P. *Materializm i kriticheskii realizm*. St. Petersburg, 1908.

————. "Printsip otnositel'nosti i novoe uchenie o vremeni" (2 parts), *Letopis'*, 1916, no. 7, pp. 204–13; no. 8, pp. 190–210.

————. "Review of 'Novye idei v fizike,'" *Vestnik Evropy*, 1912, no. 2, pp. 374–75.

Ivanenko, D. D. "Bemerkungen zur einer einheitlichen nichtlinearen Theorie der Materie," in W. Frank, ed., *Max Planck Festschrift* (Berlin, 1959), pp. 353–69.

————. "K stoletiiu so dnia rozhdeniia A. Einshteina," *VIET*, 1979, no. 67–68, pp. 3–14.

————. "The Neutron Hypothesis," *Nature*, 1932, vol. 129, no. 3265, p. 798.

————. "Ob edinoi fizicheskoi kartine mira i nekotorykh problemakh teorii elementarnykh chastits," *VF*, 1959, no. 6, pp. 74–87.

————. "Ob aktual'nosti truda Einshteina," *VIET*, 1975, n.s. no. 3 (52), pp. 13–16.

————. "Osnovnye idei obshchei teorii otnositel'nosti," in Grigor'ian and Polak, eds., *Ocherki*, pp. 238–322.

————. "Perennial Modernity of Einstein's Theory of Gravitation," in Francesco de Finis, ed., *Relativity, Quanta and Cosmology*, vol. 1 (New York, 1979), pp. 295–359.

————. "Vozmozhnosti edinoi teorii polia," in Dyshlevyi and Petrov, eds., *Filosofskie problemy*, pp. 43–56.

Ivanenko, D. D., and V. K. Frederiks. "Primechaniia," in Lorentz et al., *Printsip otnositel'nosti*, pp. 360–87.

Ivanenko, D. D., and B. G. Kuznetsov. "Pamiati Al'berta Einshteina," *VIET*, 1955, vol. 5, pp. 3–22.

Ivanenko, D. D., and G. Sardanashvili. "Novye kontseptsii edinikh teorii i model' praspinorov," in Bazhenov et al., eds., *Evoliutsiia materii*, pp. 366–70.

Ivanov, V. G. *Fizika i mirovozzrenie*. Moscow, 1975.

Jakobson, Roman. *Selected Writings*, vol. 3, ed. Stephen Rudy. The Hague, 1961.

Jammer, Max. *Concepts of Space: The History of Space in Physics*. Cambridge, Mass., 1954.

————. *Philosophy of Quantum Mechanics*. New York, 1974.

Joravsky, David. *Soviet Marxism and Natural Science*. New York, 1961.

Josephson, Paul. *Physics and Politics in Revolutionary Russia*. Berkeley, Calif., 1991.

"K itogam diskussii po teorii otnositel'nosti," *VF*, 1955, no. 1, pp. 134–38.

"K stat'e A. Einshteina 'Fizika i real'nost'," *PZM*, 1937, no. 11–12, pp. 110–13.

Kalantar, A. L. *Krasota istiny: ob esteticheskom nachale nauchnogo poznaniia*. Erevan, 1980.

Kant, Immanuel. *Philosophical Writings*, ed. E. Behler. New York, 1986.

Kapitsa, P. L. *Experiment, Theory, Practice*. Dordrecht, Netherlands, 1980.

Karasov, M., and V. Nozder. "O knige M. E. Omel'ianovskogo 'V. I. Lenin i fizika XX veka,'" *VF*, 1949, no. 1, pp. 338–42.

Karmin, A. S. "Nauchnoe myshlenie i intuitsiia: einshteinovskaia postanovka problemy," in Dyshlevyi and Luk'ianets, eds., *Nauchnaia kartina*, pp. 240–59.

Karpov, M. M. "Kritika filosofskikh vzgliadov A. Einshteina," in Maksimov et al., eds., *Filosofskie voprosy*, pp. 216–33.

————. "O filosofskikh vzgliadakh Einshteina," *VF*, 1951, no. 1, pp. 130–41.

Kas'ian, A. A. "Soderzhanie nauchnogo znaniia i vozmozhnosti aksiomaticheskogo metoda," in Aronov et al., eds., *Filosofskie voprosy*, pp. 49–65.

Kasterin, N. P. *Obobshchenie osnovnykh uravnenii aerodinamiki i elektrodinamiki*. Moscow, 1957.

————. "Sur une contradiction essentielle entre la théorie de relativité d'Einstein et l'expérience," *Izvestiia Rossiiskoi Akademii nauk*, 6th ser., 1918, no. 2–3, pp. 89–98.

Kaziutinskii, V. V. "Antropnyi printsip i problemy mirovozzreniia," in Akhundov and Bazhenov, eds., *Estestvoznanie*, pp. 151–65.

————. "Prostranstvo-vremia-Vselennaia," in Kuznetsov, ed., *Prostranstvo*, pp. 190–214.

————. "'Sistema mira' N'iutona i sovremennaia kosmologiia," in Akhundov and Illarionov, eds., *N'iuton*, pp. 125–39.

Kaziutinskii, V. V., et al., eds. *Dialektika-mirovozzrenie i metodologiia sovremennogo estestvoznaniia*. Moscow, 1983.

Kedrov, B. M. "O teorii Einshteina i vzgliadakh Engel'sa i Lenina," in Frolov and Grekov, eds., *Filosofiia*, pp. 168–80.

Kedrov, B. M., et al. "Ernest Kol'man (k 75-letiiu so dnia rozhdenii)," *VIET*, 1969, no. 2 (27), pp. 71–73.

Kedrov, B. M., and N. F. Ovchinnikov, eds. *Printsip sootvetstviia: istoriko-metodologicheskii ocherk*. Moscow, 1979.

Keldysh, M. "Estestvennye nauki i ikh znachenie dlia razvitiia mirovozzreniia i tekhnicheskogo progressa," *Kommunist*, 1966, no. 7, pp. 29–47.

Kelle, V. Zh. *Nauka kak komponent sotsial'noi sistemy*. Moscow, 1965.

Khramova, V. A. *Filosofskii analiz problemy sootnosheniia teorii i eksperimenta v reliativistskoi fizike*. Kiev, 1974.

Khvol'son [Chwolson], O. D. *Die Evolution des Geistes der Physik.* Braunschweig, 1925.
———. *Populiarnye stat'i i rechi.* Petrograd, 1923.
———. "Printsip otnositel'nosti," *Pr*, 1912, no. 11, pp. 1275–1316.
———. *Printsip otnositel'nosti.* St. Petersburg, 1914.
———. *Teoriia otnositel'nosti A. Einshteina i novoe miroponimanie.* Rev. ed. Petrograd, 1922.
Kiiko, E. I. "Vospriiatie Dostoevskim neevklidovoi geometrii," in Fridlender, ed., *Dostoevskii*, pp. 120–25.
Kisun'ko, V. G. "A. Einshtein i gumanitarnye aspekty estestvennonauchnogo znaniia," in Borisovskii et al., *Iskusstvo*, pp. 246–94.
Klein, Martin J. *The Making of a Theoretical Physicist.* Amsterdam, 1970.
Kniazeva, L. ed. *Dialekticheskii materializm i sovremennoe estestvoznanie.* Moscow, 1957.
Kobzarev, I. Iu. "A. Einshtein, M. Plank i atomnaia fizika," *Pr*, 1979, no. 3, pp. 8–26.
———. "Relativity, Theory of," *BSE*, 3d ed., vol. 18 (1974), pp. 648–54.
Kobzarev, I. Iu., et al. *Einshtein i sovremennaia fizika.* Moscow, 1979.
———. *Einshtein i teoreticheskaia fizika pervoi treti XX veka.* Moscow, 1979.
Kochergin, A. N., ed. *Nauka i tsennosti.* Novosibirsk, 1987.
Koialovich, Nikolai. "Obzor literatury po teorii otnositel'nosti na russkom iazyke" (2 parts), *Kniga i revoliutsiia*, 1922, n.s. no. 9–10 (21–22), pp. 29–34; 1923, n.s. no. 11–12 (23–24), pp. 26–32.
Kol'man, Ernest. "Boevye voprosy estestvoznaniia i tekhniki v rekonstruktivnyi period," *PZM*, 1931, no. 3, pp. 56–78.
———. *Filosofskie problemy sovremennoi fiziki.* Moscow, 1957.
———. "K sporam o teorii otnositel'nosti," *VF*, 1954, no. 5, pp. 178–89.
———. "Materializm i empiriokrititsizm V. I. Lenina i sovetskaia fizika," *Sovetskaia nauka*, 1939, no. 2, pp. 57–70.
———. "Na tekushchie temy," *PZM*, 1932, no. 9–10, pp. 163–70.
———. *Noveishie otkrytiia sovremennoi atomnoi fiziki v svete dialekticheskogo materializma.* Moscow, 1943.
———. "Pis'mo tov. Stalina i zadachi fronta estestvoznaniia i meditsiny," *PZM*, 1931, no. 9–10, pp. 163–72.
———. "Teoriia kvant i dialekticheskii materializm," *PZM*, 1939, no. 10, pp. 129–45.
———. "Teoriia otnositel'nosti i dialekticheskii materializm," *PZM*, 1939, no. 6, pp. 106–20.
———. "Uzlovye problemy sovremennoi atomnoi fiziki," *Front nauki i tekhniki*, 1936, no. 2, pp. 24–34.
———. *Velikii russkii myslitel' N. I. Lobachevskii.* Moscow, 1944.
———. "Za marksistsko-leninskuiu nauku," *Front nauki i tekhniki*, 1932, no. 1, pp. 18–25.
Kompaneets, A. I. *Leninskaia filosofiia i progress fizicheskikh nauk.* Moscow, 1967.
Konopleva, N. P. "Einshtein i sovremennye geometricheskie teorii vzaimodeistvii," in *Issledovaniia: 1985*, pp. 125–39.
Konstantinov, F. V., et al., eds. *Osnovy marksistskoi filosofii.* Moscow, 1959.
"Koordinatsiia issledovanii po kosmomikrofizike," *VAN*, 1989, no. 4, pp. 40–50.

Kopnin, P. V. *Dialektika, logika, nauka.* Moscow, 1973.

———. "Teoreticheskie postroeniia Al'berta Einshteina i sovremennaia filosofiia," in A. Petrov and Dyshlevyi, eds., *Prostranstvo*, pp. 5–12.

Korolev, F. "Tiagotenie," *BSE*, vol. 55 (1947), pp. 458–61.

"Kosmologiia," *BSE*, 2d ed., vol. 23 (1953), pp. 105–13.

Kotel'nikov, A. P. "Printsip otnositel'nosti i geometriia Lobachevskogo," in *In Memoriam*, vol. 2, pp. 37–66.

Kotel'niov, A. P., and V. A. Fock. *Nekotorye primeneniia idei Lobachevskogo v mekhanike i fizike.* Moscow-Leningrad, 1950.

Kotel'nikov, V. A., et al., eds. *Nikolai Kopernik: k 500-letiiu so dnia rozhdeniia.* Moscow, 1973.

Kotina, S. V. "Printsip krasoty v sisteme metodologicheskikh reguliatorov estestven-nonauchnogo poznaniia," *FN*, 1989, no. 11, pp. 110–17.

Kravets, T. P. "Evoliutsiia ucheniia ob energii (1847–1947)," *UFN*, 1948, vol. 36, no. 3, pp. 338–58.

Ksenofontov, V. I. *Dialekticheskii materializm i nauchnoe poznanie.* Leningrad, 1981.

———. *Leninskie idei v sovetskoi filosofskoi nauke 20-kh godov.* Leningrad, 1975.

Kudriavtsev, P. S. "Kopernik i razvitie fiziki," in V. Kotel'nikov et al., *Nikolai Kopernik*, pp. 78–83.

———. *Kurs istorii fiziki.* Moscow, 1972.

Kursanov, G. A. "K otsenke filosofskikh vzgliadov A. Einshteina," in *Filosofskie voprosy sovremennoi fiziki* (Moscow, 1959), pp. 393–410.

Kuznetsov, B. G. *Einshtein.* Moscow, 1962.

———. *Einstein*, tr. V. Talmy. Moscow, 1965.

———. "Einstein and Dostoevsky," *Diogenes*, 1966, no. 53, pp. 1–16.

———. *Einstein and Dostoevsky*, tr. V. Talmy, London, 1972.

———. "Ob esteticheskikh kriteriiakh v sovremennom fizicheskom myshlenii," in: Meilakh, ed., *Khudozhestvennoe i nauchnoe tvorchestvo*, pp. 89–101.

———. *Osnovy teorii otnositel'nosti i kvantovoi mekhaniki v ikh istoricheskom razvitii.* Moscow, 1957.

Kuznetsov, I. V. "Ob osnovnykh voprosakh teorii otnositel'nosti," in Omel'ianovskii and Kuznetsov, eds., *Filosofskie voprosy*, pp. 155–201.

———. "Printsip sootvetstviia v sovremennoi fizike i ego filosofskoe znachenie," in Kedrov and Ovchinnikov, eds., *Printsip sootvestviia*, pp. 5–95.

———. "Protiv idealisticheskikh izvrashchenii poniatii massy i energii," *UFN*, 1952, vol. 48, no. 2, pp. 221–62.

———. "Sovetskaia fizika i dialekticheskii materializm," in Maksimov et al., eds., *Filosofskie problemy*, pp. 31–86.

———, ed. *Prostranstvo, vremia, dvizhenie.* Moscow, 1971.

Kuznetsov, I. V., and N. F. Ovchinnikov. "Za posledovatel'noe dialektiko-materialist-icheskoe osveshchenie dostizhenii sovremennoi fiziki," *UFN*, 1951, vol. 45, pp. 110–40.

Kuznetsov, I. V., et al., eds. *Philosophical Problems of Elementary Particle Physics*, tr. G. Yankovsky. Moscow, 1968.

Lakatos, Imre. "Falsification and the Methodology of Scientific Research Programmes," in I. Lakatos and A. Musgrave, eds., *Criticism and the Growth of Knowledge* (Cambridge, Eng., 1970), pp. 51–196.

Landau, L. D., and E. Lifshitz. *The Classical Theory of Fields*, tr. M. Harmermest. Cambridge, Mass., 1951.

———. *Teoriia polia*. 2d rev. ed. Moscow-Leningrad, 1948.

———. "Tiagotenie," *BSE*, 2d ed., vol. 43 (1956), pp. 556–59.

Lavrova, N. B. "Trudy Einshteina na russkom iazyke," *VIET*, 1975, n.s. no. 3 (52), pp. 38–42.

Lazarev, V. V. *Stanovlenie filosofskogo soznaniia novogo vremeni*. Moscow, 1987.

Lazukin, V. N. "50-letie teorii otnositel'nosti," *VAN*, 1956, no. 2, pp. 106–10.

Lenin, V. I. *Materialism and Empirio-Criticism*. New York, 1970.

———. *Polnoe sobranie sochinenii*. 5th ed., 55 vols. Moscow, 1958–65.

Leonov, N. A. *Ocherk dialekticheskogo materializma*. Moscow, 1948.

Liubutin, K. N., and D. V. Pivovarov. "Problema nauchnosti filosofii i 'kontrfilosofiia,'" *FN*, 1989, no. 6, pp. 62–72.

Logunov, A. A. *K rabotam Anri Puankare 'O dinamike elektrona.'*" Moscow, 1984.

———. *Lektsii po teorii otnositel'nosti i gravitatsii*. Moscow, 1987.

———. "Novaia teoriia gravitatsii," *Nauka i zhizn'* (3 parts), 1988, no. 2, pp. 38–44; no. 3, pp. 60–71; no. 5, pp. 66–72.

———. "Osnovnye printsipy reliativistskoi teorii gravitatsii," *Teoreticheskaia i matematicheskaia fizika*, 1989, vol. 80, no. 2, pp. 155–72.

———. "Reliativistskaia teoriia gravitatsii," *Pr*, 1987, no. 1, pp. 36–47.

———. "Reliativistskaia teoriia gravitatsii i novye predstavleniia o prostranstve-vremeni," *Teoreticheskaia i matematicheskaia fizika*, 1987, vol. 70, no. 1, pp. 3–17.

Lorentz, H. A., et al. *The Principle of Relativity*, tr. W. Perrett and G. B. Jeffrey. New York, 1952.

———. *Printsip otnositel'nosti*. Leningrad, 1935.

Lukirskii, L. I., et al. *Sbornik posviashshennyi semidesiatiletiiu akademika A. F. Ioffe*. Moscow, 1950.

L'vov, V. E. "Voprosy ideologii v uchebnike fiziki," *Front nauki i tekhniki*, 1932, no. 9, pp. 109–17.

———. *Zhizn' Al'berta Einshteina*. Moscow, 1959.

Maikov, V. V. "Nauchnye i misticheskie aspekty golograficheskoi paradigmy," in Akhundov and Bazhenov, eds., *Estestvoznanie*, pp. 193–208.

Makedonov, A. V. "Uchenie V. I. Vernadskogo o dissimmetrii geologicheskikh ob"ektov," in Sokolov and Ianshin, *V. I. Vernadskii*, pp. 139–46.

Maksimov, A. A. "Bor'ba za materializm v sovremennoi fizike," *VF*, 1953, no. 1, pp. 175–94.

———. "Dialekticheskii i materializm," *Sovetskaia kniga*, 1939, no. 9–10, pp. 68–83.

———. "Eshche raz o populiarno-nauchnoi literature o printsipe otnositel'nosti," *PZM*, 1922, no. 11–12, pp. 123–41.

———. "Klassovaia bor'ba v sovremennom estestvoznanii," *Front nauki i tekhniki*, 1932, pp. 21–33.

———. "Marksistskii filosofskii materializm i sovremennaia fizika," *VF*, 1948, no. 3, pp. 104–24.

———. "'Materializm i empiriokrititsizm'—materialisticheskoe obobshchenie dannykh estestvoznaniia," *PZM*, 1938, no. 11, pp. 42–68.

———. "O filosofskikh vozzreniakh akad. V. F. Mitkevicha i o putiakh razvitiia sovetskoi fiziki," *PZM*, 1937, no. 7, pp. 25–55.

———. "O fizicheskom idealizme i zashchite ego akad. A. F. Ioffe," *PZM*, 1937, no. 11–12, pp. 157–91.

———. "O printsipe otnositel'nosti A. Einshteina," *PZM*, 1922, no. 9–10, pp. 180–208.

———. "O protivorechiiakh sovremennoi fiziki," *Sovetskaia kniga*, 1939, no. 3, pp. 108–20.

———. "Ob otrazhenii klassovoi bor'by v sovremennom estestvoznanii," *PZM*, 1932, no. 5–6, pp. 16–23.

———. "Obsuzhdenie knigi I. V. Kuznetsova, 'Printsip sootvetstviia v sovremennoi fizike i ego filosofskoe znachenie,'" *VF*, 1950, no. 2, pp. 378–87.

———. "Review of A. K. Timiriazev, 'Estestvoznanie i dialekticheskii materializm,'" *PZM*, 1925, no. 8–9, pp. 303–39.

———. "Sovremennoe fizicheskoe uchenie o materii i dvizhenii i dialekticheskii materializm," *PZM*, 1939, no. 10, pp. 86–111.

———. "Sovremennoe sostoianie diskussii o printsipe otnositel'nosti v Germanii," *PZM*, 1923, no. 1, pp. 101–19.

———. "Teoriia otnositel'nosti i materializm: Otvet t. Skukovu," *PZM*, 1923, no. 4–5, pp. 140–56.

Maksimov, A. A., et al., eds. *Filosofskie voprosy sovremennoi fiziki*. Moscow, 1952.

Mamchur, E. A. *Problemy sotsio-kul'turnoi determinatsiii nauchnogo znaniia*. Moscow, 1987.

Mandel'shtam, L. I. *Polnoe sobranie sochinenii*, vol. 5. Moscow, 1950

Maneev, A. K. *K kritike obosnovaniia teorii otnositel'nosti*. Minsk, 1960.

Markov, M. A. "O edinstve i mnogoobrazii form materii i fizicheskoi kartiny mira," *VF*, 1980, no. 11, pp. 60–75.

———. "O prirode fizicheskogo znaniia," *VF*, 1947, no. 2, pp. 140–76.

———. "Otvet—v beskompromissnom patsifizme," *Vestnik Rossiiskoi Akademii nauk*, 1992, no. 7, pp. 106–16.

Markov, M. A., et al. *Leninskoe filosofskoe nasledie i sovremennaia fizika*. Moscow, 1981.

Marx, K., and F. Engels. *Izbrannye proizvedeniia v trekh tomakh*, vol. 1, Moscow, 1966.

McCutcheon, Robert A. "The 1936–1937 Purge of Soviet Astronomy," *Slavic Review*, 1991, vol. 50, no. 1, pp. 100–117.

———. "The Purge of Soviet Astronomy: 1936–1937, with a Discussion of Its Background and Aftermath." M.A. thesis, Georgetown University, 1985.

Meilakh, B. S., ed. *Khudozhestvennoe i nauchnoe tvorchestvo*. Leningrad, 1972.

Merton, Robert K. *Social Theory and Social Structure*. Glencoe, Ill., 1949.

Miakishev, G. Ia. "Energiia," *BSE*, 3d ed., vol. 30 (1978), p. 191.

Migdal, A. B. "Fizika i filosofiia," *VF*, 1990, no. 1, pp. 5–32.

Migunov, A. S. "Iskusstvo i nauka: o nekotorykh tendentsiiakh k sblizheniiu i vzaimodeistviiu," *VF*, 1986, no. 7, pp. 91–99.

Mikhailov, A. A., ed. *Nikolai Kopernik*. Moscow-Leningrad, 1947.

Mikulak, Maxim W. "Soviet Philosophic-Cosmological Thought," *Philosophy of Science*, 1958, vol. 25, no. 1, pp. 35–50.

Miller, Arthur. "Visualization Lost and Regained," in Judith Wechsler, ed., *On Aesthetics and Science* (Cambridge, Mass., 1981), pp. 73–102.

Minkowski, H. "Time and Space," in Lorentz et al., *Principle of Relativity*, pp. 73–91.

Mitin, M. B. *Boevye voprosy materialisticheskoi dialektiki*. Moscow, 1936.

———. "Filosofskaia nauka v SSSR za. 25 let," in *Iubileinaia sessiia Akademii nauk-posviashchenaia 25-letiiu Velikoi Oktiabr'skoi Revoliutsii* (Moscow-Leningrad, 1943), pp. 116–39.

———. "Istoricheskoe znachenie 'Materializma i empiriokrititsizma' Lenina," *Front nauki i tekhniki*, 1934, no. 7, pp. 71–80.

———. "Mekhanisty," *Filosofskaia entsiklopediia*, vol. 3 (1964), p. 424.

———. "Ocherednye zadachi raboty na filosofskom fronte v sviazi s itogami diskussii," *PZM*, 1931, no. 3, pp. 12–35.

———. "Velikoe ideinoe oruzhie poznaniia i preobrazovaniia mira," in Fedoseev et al., eds., *Filosofskie problemy*, pp. 12–31.

———. "V. I. Lenin i bor'ba protiv sovremennogo pozitivizma," *VF*, 1961, no. 3, pp. 3–14.

Mitin, M. B., and I. Razumovskii, eds. *Dialekticheskii i istoricheskii materializm*. 2 vols. Moscow, 1932.

Mitketich, V. F. "Osnovnye vozzreniia sovremennoi fiziki," in Bukharin and Deborin, eds., *Pamiati Karla Marksa*, pp. 223–44.

Mochalov, I. "Tvorchestvo V. I. Vernadskogo i filosofiia," *Kommunist*, 1988, no. 18, pp. 65–67.

Molchanov, Iu. B. "Problema sub'ekta (nabliudatelia) v sovremennoi fizike," in Sachkov, ed., *Teoriia poznaniia*, pp. 225–41.

Morson, Gary S., and Caryl Emerson. *Mikhail Bakhtin: Creation of a Prosaics*. Stanford, Calif., 1990.

Mostepanenko, A. M. *Filosofiia i razvitie estestvonauchnoi kartiny mira*. Leningrad, 1981.

———. "Problema mnogoobraziia svoistv prostranstva i vremeni i ee metodologicheskoe znachenie," in Dyshlevyi and Petrov, eds., *Prostranstvo*, pp. 172–78.

———. "Problema sushchestvovaniia i real'nosti v fizicheskom poznanii," in Sachkov, ed., *Teoriia poznaniia*, pp. 182–96.

Mostepanenko, M. V. *Materialisticheskaia sushchnost' teorii otnositel'nosti Einshteina*. Moscow, 1962.

Moszkowski, Alexander. *Einstein the Searcher*, tr. H. L. Brose. London, 1921.

Müller-Markus, Siegfried. "Diamat and Einstein," *Survey*, 1961, no. 37, pp. 68–78.

———. *Einstein und die Sowietphilosophie*. 2 vols. Dordrecht, Netherlands, 1966.

Naan, G. I. "Gravitatsiia i beskonechnost'," in Dyshlevyi and Petrov, eds., *Filosofskie problemy*, pp. 268–85.

———. "K voprosu o printsipe otnositel'nosti v fizike," *VF*, 1951, no. 2, pp. 57–77.

———. "Kosmologiia," *BSE*, 3d ed., vol. 13 (1973), pp. 256–58.

———. "O beskonechnosti Vselennoi," *VF*, 1961, no. 6, pp. 93–105.

Narskii, I. S. *Sovremennyi pozitivizm*. Moscow, 1961.

Nevskii, V. "Marksizm i estestvoznanie," *PZM*, 1923, no. 11–12, pp. 204–13.

———. "Restavratsiia idealizma i bor'ba 'novoi' burzhuaziei," *PZM*, 1922, no. 7–8, pp. 117–31.

Nietzsche, Friedrich. *Beyond Good and Evil*. Chicago, 1955.

Nikiforov, A. L. "Iavliaetsia li filosofiia naukoi?," *FN*, 1989, no. 6, pp. 52–62.

Nikitina, A. G. "Apokalipsis i termoiadernaia voina," *FN*, 1990, no. 2, pp. 9–13.

Nikol'skii, K. V. "O putiakh razvitiia teoreticheskoi fiziki v SSSR," *PZM*, 1938, no. 1, pp. 160–72.

———. "Otvet V. A. Foku," *UFN*, 1937, vol. 17, no. 4, pp. 554–60

[Nikol'skii, K. V, and R. Shteinman]. "Otnositel'nosti teoriia," *BSE*, vol. 43 (1939), pp. 583–618.

Novikov, I. D. *Evolution of the Universe*, tr. M. M. Basko. Cambridge, Eng., 1983.

Novikova, L. I. "Esteticheskoe v strukture nauchnoi deiatel'nosti," in Meilakh, ed., *Khudozhestvennoe*, pp. 67–84.

Novinskii, I. I., ed. *Voprosy dialekticheskogo materializma*. Moscow, 1951.

Nysanbaev, A. N., and F. M. Suleimanov. *Ot 'edinomysliia' k pliuralizmu mnenii*. Moscow, 1990.

"O polozhenii na fronte estestvoznaniia," *VKA*, 1931, no. 1, pp. 23–32.

"O raznoglasiiakh na filosofskom fronte," *VKA*, 1930, no. 42, pp. 20–89.

"Ob odnom filosofskom dispute," *FN*, 1967, no. 4, pp. 133–39.

"Obsuzhdenie filosofskikh voprosov teorii otnositel'nosti," *VF*, 1959, no. 2, pp. 77–82.

"Obsuzhdenie okladov," in Fedoseev et al., eds., *Filosofskie problemy*, pp. 365–69.

"Obsuzhdeniia doklada [L. F. Il'icheva]," *VAN*, 1963, no. 11, pp. 47–67.

Ogurtsov, Aleksandr, "Podavlenie filosofii," in Senokosov, ed., *Surovaia drama*, pp. 353–74.

Omel'ianovskii, M. E. "Bor'ba materializma protiv idealizma i sovremennaia fizika," in Novinskii, ed., *Voprosy dialekticheskogo materializma*, pp. 143–70.

———. "Dialekticheskii materializm i sovremennaia fizika," in Omel'ianovskii and Kuznetsov, eds., *Filosofskie voprosy sovremennoi fiziki*, pp. 3–29.

———. *Dialektika i sovremennaia fizika*. Moscow, 1973.

———. "Falsifikatory nauki: ob idealizme v sovremennoi fizike," *VF*, 1948, no. 3, pp. 142–62.

———. "Filosofskaia bor'ba v sovremennoi fizike vokrug problemy ob"ektivnogo i sub"ektivnogo," *VF*, 1976, no. 2, pp. 126–35.

———. "O knige akademika A. F. Ioffe," *VF*, 1951, no. 2, pp. 203–9.

———. "Protiv idealizma i idealisticheskikh shatanii v kvantovoi mekhanike," *VF*, 1957, no. 4, pp. 151–66.

———. *Razvitie osnovanii fiziki XX veka i dialektika*. Moscow, 1984

———. "Teoriia otnositel'nosti Einshteina i dialektika," *FN*, 1979, no. 4, pp. 76–86.

———. "V. I. Lenin i dialektika v sovremennoi fizike," *VF*, 1967, no. 10, pp. 125–34.

———. *V. I. Lenin i fizika XX veka*. Moscow, 1947.

———, ed. *Lenin i sovremennoe estestvoznanie*. Moscow, 1969.

———. *Sovremennoe estestvoznanie i materialisticheskaia dialektika*. Moscow, 1977.

Omel'ianovskii, M. E., and I. V. Kuznetsov, eds. *Dialektika v naukakh o nezhivoi prirode*. Moscow, 1964.

————. *Filosofskie voprosy sovremennoi fiziki.* Kiev, 1956.

Orlov, I. E. "Klassicheskaia fizika i reliativizm," *PZM*, 1924, no. 3, pp. 49–76.

————. "Osnovnye formuly printsipa otnositel'nosti s tochki zreniia klassicheskoi mekhaniki," *Zhurnal Fiziko-Khimicheskogo obshchestva*, 1914, vol. 46, pp. 163–75.

————. "Review of V. K. Frederiks and A. A. Fridman, 'Osnovy teorii otnositel'nosti,' vol. 1," *PZM*, 1925, no. 7, pp. 232–34.

————. "Sushchestvuet li aktual'naia beskonechnost'," *PZM*, 1924, no. 1, pp. 136–47.

Osmanov, N. O. "Sootnoshenie printsipov krasoty i simmetrii v poznanii fizicheskoi real'nosti," in I. S. Khashimova et al., eds., *Metodologiia i metody nauchno-tekhnicheskoi revoliutsii* (Tashkent, 1986), pp. 194–205.

"Ot redaktsii," *UFN*, 1952, vol. 48, no. 2, pp. 145–46.

"Ot redaktsii," in Grigor'ian, ed., *Einshtein*, pp. 3–4.

"Ot redaktsii: k stat'e A. Einshteina 'Fizika i real'nost'," *PZM*, 1937, pp. 19–37.

"Ot redkollegii," in Vernadskii, *Filosofskie mysli*, pp. 3–18.

"Otkliki na stat'iu A. L. Nikiforova. 'Iavliaetsia li filosofiia naukoi?'" (4 parts), *FN*, 1989, no. 12, pp. 69–78; 1990, no. 1, pp. 82–87; no. 2, pp. 64–71; no. 3, pp. 102–10.

Ovchinnikov, N. F. "Tvorchestvo i traditsii v nauchnom issledovanii," in *Issledovaniia: 1985*, pp. 5–41.

Pais, Abraham. *Inward Bound.* Oxford, 1968.

————. "*Subtle Is the Lord . . .*": *The Science and the Life of Albert Einstein.* Oxford, 1982.

Panchenko, A. I. *Logiko-gnoseologicheskie problemy kvantovoi mekhaniki.* Moscow, 1981.

Panov, M. I. *Metodologicheskie problemy intuitsionistskoi matematiki.* Moscow, 1984.

Pauli, Wolfgang. *Theory of Relativity,* tr. G. Field. New York, 1958.

Pavlenko, A. N. "Dinamika razvitiia sovremennogo kosmicheskogo znaniia," *Vestnik Moskovskogo universiteta,* ser. 7, 1988, no. 3, pp. 50–58.

Perederii, V. I. "Determinizm i teoriia otnositel'nosti," in Shtokalo et al., eds., *Filosofskie voprosy,* pp. 244–47.

"Peredovaia," *PZM*, 1929, no. 5, pp. 1–5.

Perfyr'ev, V. V. "O knige M. E. Omel'ianovskogo 'V. I. Lenin i fizika XX veka,'" *VF*, 1948, no. 1, pp. 311–12.

Petrov, A. Z. *Einstein Spaces,* tr. R. F. Kollher. New York, 1969.

————. "Gravitatsiia i prostranstvo," in I. Kuznetsov, ed., *Prostranstvo,* pp. 167–89.

————. *Novye metody v obshchei teorii otnositel'nosti.* Moscow, 1966.

————. "Obshchaia teoriia otnositel'nosti," in Artsimovich et al., eds., *Razvitie fiziki,* vol. 1, pp. 58–73.

————. *Prostranstvo Einshteina.* Moscow, 1961.

————. *Prostranstvo-vremia i materiia.* Kazan, 1961.

Petrov, A. Z., and P. S. Dyshlevyi, eds. *Prostranstvo i vremiia v elementarnoi fizike.* Kiev, 1963.

Petrov, V. N. "Ob astronomicheskikh sledstviiakh obshchei teorii otnositel'nosti," *Pr,* 1940, no. 11, pp. 3–12.

Pippard, Brian. "Picturing the Atom," *Times Literary Supplement,* 1992, no. 4634, pp. 3–4.

Pisarzhevskii, L. "II. Mendeleevskii s"ezd," *PR,* 1912, no. 1, pp. 121–24.

Planck, Max. *The Universe in the Light of Modern Physics*, tr. W. H. Johnson. London, 1931.

———. *Where Is Science Going?* New York, 1932.

Planck, Max, A. Einstein, and J. Murphy. "Epilogue," in Planck, *Where Is Science Going?*, pp. 201–21.

Pogodin, Nikolai. "Al'bert Einshtein," *Teatr*, 1968, no. 9, pp. 161–89.

Poincaré, Henri. *Mathematics and Science: Last Essays.* New York, 1963.

Polak, L. S. "Zhizn' i nauchnoe tvorchestvo Aleksandra Aleksandrovicha Fridmana, 1888–1925," in Friedmann, *Izbrannye trudy*, pp. 427–47.

Polikarov, A. P. "Iz istorii ideologicheskoi bor'by vokrug teorii otnositel'nosti," in Omel'ianovskii and Kuznetsov, eds., *Filosofskie voprosy*, pp. 411–26.

Prazdnovanie Kazanskim universitetom stoletiia otkrytiia neevklidovoi geometrii. Kazan, 1927.

"Predislovie," in Dyshlevyi and Luk'ianets, eds., *Nauchnaia kartina*, pp. 3–4.

Predvoditelev, A. S., et al. eds. *Razvitie fiziki v Rossii.* 2 vols. Moscow, 1970.

Prokof'eva, I. A. "Konferentsiia po ideologicheskim voprosam astronomii," *Pr*, 1949, no. 6, pp. 71–77.

Pronin, P. I., and Yu. N. Obukhov, eds. *Modern Problems of Theoretical Physics: Festschrift for Professor D. Ivanenko.* Singapore, 1991.

"Protokol Obshchego sobraniia chlenov Kommunisticheskoi Akademii," *VKA*, 1925, no. 12, pp. 367–89.

Rainov, T. "Review of H. A. Lorentz et al., 'Printsip otnositel'nosti,'" *Sotsialisticheskaia rekonstruktsiia i nauka*, 1936, no. 3, pp. 116–18.

"Razvertyvat' kritiku i bor'bu mnenii v nauke," *Pravda*, Nov. 17, 1952.

"Reliativizm," *Filosofskii entsiklopedicheskii slovar'*, 1983, p. 578.

Rozental', M. "Marksistskaia teoriia poznaniia," *PZM*, 1938, no. 12, pp. 34–57.

Rozov, M. A. "Problema tsennostei i razvitie nauki," in Kochergin, ed. *Nauka i tsennosti*, pp. 5–27.

Runin, B. "Lichnost' i tvorchestvo: fiziki i liriki," in Agapov et al., *Khudozhnik*, pp. 94–159.

———. "Logika nauki i logika iskusstva," in B. S. Meliakh, ed., *Sodruzhestvo nauk i tainy tvorchesta* (Moscow, 1968), pp. 114–38.

———. "Puti tvorchestva," *Voprosy literatury*, 1964, no. 4, pp. 112–31.

Russell, Bertrand. *Human Knowledge, Its Scope and Limits.* New York, 1948.

———. "Introduction" to Vasil'ev, *Space, Time*, pp. xi–xxiii.

Ruzavin, G. I. "Dialektika i sovremennoe nauchnoe myshlenie," FN, 1991, no. 6, pp. 3–15.

Sachkov, Iu. V., ed. *Teoriia poznaniia i sovremennaia fizika.* Moscow, 1984.

Sachkov, Iu. V., et al., eds. *Materialisticheskaia dialektika kak obshchaia teoriia razvitiia*, vol. 3. Moscow, 1983.

Sakharov, A. D. "Kosmomikrofizika—mezhdistsiplinarnaia problema," *VAN*, 1989, no. 4, p. 39.

———. *Memoirs*, tr. Richard Lourie. New York, 1990.

Schilpp, P. A., ed. *Albert Einstein: Philosopher-Scientist.* 2 vols. New York, 1949.

Schlick, M., et al. *Teoriia otnositel'nosti i ee fizicheskoe istolkovanie.* Moscow, 1923.

Seelig, Carl. *Albert Einstein: Leben und Werk eines Genies unserer Zeit.* 2d ed. Zurich, 1960.

Selinov, P. "Fizicheskii idealizm i sovremennoe uchenie o stroenii materii," *Front nauki i tekhniki,*1934, no. 4, pp. 25–31.

Semenov, A. A. "Ob itogakh suzhdeniia filosofskikh vozzrenii akademika L. I. Mandel'shtama," *VF*, 1953, no. 3, pp. 199–206.

Semenov, N. N. *Nauka i obshchestvo.* Moscow, 1973.

Semenova, N. N. "Metodologicheskie aspekty izucheniia etiki nauchnoi deiatel'nosti," in Kochergin, ed., *Nauka*, pp. 78–111.

Semkovskii, S. Iu. *Dialekticheskii materializm i printsip otnositel'nosti.* Moscow, 1926.

———. "'Dialektika prirody' Engel'sa i teoriia otnositel'nosti," *Front nauki i tekhniki,* 1935, no. 9, pp. 8–15.

———. "K sporu v marksizme o teorii otnositel'nosti," *PZM*, 1925, no. 8–9, pp. 126–69.

———. *Teoriia otnositel'nosti i materializm.* Kharkov, 1924.

Senokosov, Iu. V., ed. *Surovaia drama naroda: uchenye i publitsisty o prirode stalinizma.* Moscow, 1989.

"Sessiia Instituta filosofii Komakademii," *VKA*, no. 4, pp. 88–96.

Shafirikin, V. "O stroenii Vselennoi i nekotorykh reaktsionnykh ideiakh burzhuaznoi kosmologii," *PZM*, 1938, no. 7, pp. 115–36.

Sheptulin, A. B., ed. *Dialekticheskii i istoricheskii materializm.* Moscow, 1985.

Shinkarenko, E. N. "Kul'turnyi reliativizm: analiz osnovnykh napravlenii," *Problemy filosofii*, 1990, vol. 86, pp. 109–16.

Shirokov, M. F. "O materialisticheskoi sushchnosti teorii otnositel'nosti," in Omel'ianovskii and Kuznetsov, eds., *Filosofskie voprosy*, pp. 324–68.

———. "O preimushchestvennykh sistemakh otcheta v n'iutonovskoi mekhanike i teorii otnositel'nosti," in Kniazeva, ed., *Dialekticheskii materializm*, pp. 59–81.

———. "Obshchaia teoriia otnositel'nosti ili teoriia tiagoteniia," *Zhurnal eksperimental'noi i teoreticheskoi fiziki*, 1956 vol. 30, no. 1, pp. 180–84.

Shishkin, A. F. "O neketorykh voprosakh issledovatel'skoi raboty v oblasti etiki," *VF*, 1973, no. 1, pp. 14–26.

Shmidt, O. Iu. "Doklad," in O. Iu. Shmidt et al., *Zadachi marksistov v oblasti estestvoznaniia* (Moscow, 1929), pp. 7–25.

Shtokalo, I. Z., et al., eds. *Filosofskie problemy sovremennoi fiziki.* Kiev, 1974.

Shpol'skii, E. V. "Al'bert Einshtein (1879–1955)," *UFN*, 1955, vol. 57, pp. 177–86.

———. "Fizika v SSSR, 1917–1937," *UFN*, 1937, vol. 18, no. 3, pp. 295–322.

Shteinman, R. Ia. "O reaktsionnoi roli idealizma v fizike," *VF*, 1948, no. 3, pp. 143–73.

———. "Za materialisticheskuiu teoriiu bystrykh dvizhenii," in Maksimov et al., eds., *Filosofskie voprosy*, pp. 234–93.

Sidorov, A. G. "O spetsifike sub"ektivno-ob"ektivnogo otnosheniia v spetsial'noi teorii otnositel'nosti," in Aronov et al., eds., *Filosofskie voprosy*, pp. 36–48.

Sitkovskii, E. "Ob antimarksistskoi sushchnosti mekhanizma i men'shevistvuiushchego idealizma," *PZM*, 1941, no. 1, pp. 30–65.

Skorobogatskii, V. V. "Razvitie filosofskogo znaniia: neobkhodimost' novogo myshleniia," *FN*, 1990, no. 2, pp. 14–23.

Smorodinskii, Ia. A. "Al'bert Einshtein i ego znachenie v razvitii fiziki," *Pr*, 1956, no. 6, pp. 23–32.

Sokolov, B. S., and A. L. Ianshin, eds. *V. I. Vernadskii i sovremennost'*. Moscow, 1986.

Solov'ev, A. A. "Metodologiia aerodinamicheskikh issledovanii N. P. Kasterina," *Istoriia i metodologiia estestvennykh nauk*, 1981, vol. 26, pp. 169–92.

Sominskii, Monus. *Akademik A. F. Ioffe*. Jerusalem, 1986.

Sonin, A. S. *"Fizicheskii idealizm": istoriia odnoi metodologicheskoi kampanii*. Moscow, 1994.

————. "Grustnaia sud'ba velikogo otkritiia," *Pr*, 1993, no. 1, pp. 94–99.

————. "Neskol'ko epizodov bor'by s 'kosmopolitizmom' v fizike," *VAN*, 1990, no. 8, pp. 122–33.

————. "Razgrom 'fizicheskogo idealizma' (ob odnoi filosofskoi diskussii)," *VAN*, 1991, no. 12, pp. 103–14.

————. "Soveshchanie, kotoroe ne sostoialos'" (3 parts), *Pr*, 1990, no. 3, pp. 97–102; no. 4, pp. 91–98; no. 5, pp. 93–99.

————. "Trevozhnye desiatiletiia sovetskoi fiziki (1920–1940)," *Znanie-sila*, 1990, no. 2, pp. 75–80.

————. "Trevozhnye desiatiletiia sovetskoi fiziki, 1947–1953," *Znanie-sila*, 1990, no. 5, pp. 80–84.

"Sovremennye zadachi marksistsko-leninskoi filosofii," *VKA*, 1931, no. 1, pp. 15–22.

Spirkin, A. G. "Intuitsiia," *BSE*, 3d ed., vol. 10 (1972), pp. 343–44.

Stadler, Friedrich. *Von Positivismus zur "Wissenschaftlichen Weltauffassung."* Vienna, 1962.

Suvorov, L. N. "Rol' filosofskikh diskussii 20–30kh godov v bor'be za leninizm protiv mekhanizma, formalisticheskikh i idealisticheskikh oshibok v filosofii," in Arkhiptsev et al., eds., *Leninskii etap*, pp. 36–45.

Suvorov, S. G. "Filosofskie vozzreniia Einshteina, ikh vzaimosviaz' s ego fizicheskimi vzgliadami," *UFN*, 1965, vol. 86, no. 3, pp. 537–84.

————. "Leninskaia teoriia poznaniia—filosofskaia osnova razvitiia fiziki," *UFN*, 1949, vol. 39, no. 1, pp. 3–50.

Sviderskii, V. I. "Bor'ba materialisticheskikh i idealisticheskikh napravlenii vokrug prostranstvenno-vremennykh predstavlenii klassicheskoi fiziki," *Vestnik Leningradskogo universiteta*, 1952, no. 6, pp. 75–102.

————. "Nekotorye filosofskie problemy teorii tiagoteniia," *Vestnik Leningradskogo universiteta*, 1956, no. 11, pp. 135–37.

————. "O nekotorykh metodologicheskikh printsipakh teorii prostranstva i vremeni," in Dyshlevyi and Petrov, eds., *Prostranstvo*, pp. 139–47.

————. *Prostranstvo i vremia*. Moscow, 1958.

Synge, J. L. *Relativity: The General Theory*. Amsterdam, 1960.

Takser, A. "Desiat' let filosofskogo IKP," *VKA*, 1932, no. 1–2, pp. 87–97.

Tamm, I. E. "A. Einshtein i sovremennaia fizika," *UFN*, 1956, vol. 59, pp. 5–10.

————. "O rabotakh N. P. Kasterina po elektrodinamike i slozhnyim voprosam," *OMEN*, 1937, no. 3, pp. 437–43.

————. "O rabote filosofov-marksistov v oblasti fiziki," *PZM*, 1933, no. 2, pp. 220–31.

Tatarinov, Iu. B. "Shest'desiat'let teorii rasshiriaiushcheisia Vselennoi A. A. Fridmana," *VIET*, 1982, no. 3, pp. 88–97.

Tepliakov, G. M. "Nikolai Petrovich Kasterin," *Istoriia i metodologiia estestvennykh nauk*, 1971, vol. 10, pp. 150–63.

Terletskii, Ia. P. "Ob odnoi iz knig akademika L. D. Landau i ego uchenikov," *VF*, 1951, no. 5, pp. 190–94.

———. "O soderzhanii sovremennoi fizicheskoi teorii prostranstva i vremeni," *VF*, 1952, no. 3, pp. 191–97.

———. "Ob izlozhenii osnov spetsial'noi teorii otnositel'nosti," *VF*, 1953, no. 4, pp. 207–12.

———. *Paradoksy teorii otnositel'nosti*. Moscow, 1966.

Timiriazev, A. K. "Eshche raz o volne idealizma v sovremennoi fizike," *PZM*, 1928, no. 4, pp. 124–52.

———. *Estestvoznanie i dialekticheskii materializm: sbornik statei*. Moscow, 1925.

———. "Novaia neudachnaia popytka primirit' teoriiu otnositel'nosti s dialekticheskim materializmom, *Voinstvuiushchii materialist*, 1925, vol. 4, pp. 243–53.

———. "Po povodu kritiki raboty N. P. Kasterina,"*OMEN*, 1938, no. 4, pp. 577–90.

———. "Printsip otnositel'nosti. (O teorii Einshteina)," *Krasnaia nov'*, 1921, no. 2, pp. 144–59.

———. "Review of A. Einshtein, 'O spetsial'noi i vseobshchei teorii otnositel'nosti,'" *PZM*, 1922, no. 1–2, pp. 70–73.

———. "Teoriia otnositel'nosti Einshteina i dialekticheskii materializm"(2 parts), *PZM*, 1924, no. 8–9, pp. 142–57; no. 10–11, pp. 92–114.

———. "Teoriia otnositel'nosti Einshteina i makhizm," *VKA*, 1924, no. 7, pp. 337–78.

———. "Volna idealizma v sovremennoi fizike na Zapade i u nas," *PZM*, no. 5, pp. 94–123.

Tiun'kin, K. I., and M. M. Stakhanova. "Opyt o cheloveke," in Dostoevsky, *Vozvrashchenie cheloveka*, pp. 5–38.

Tropp, E. A., V. Ia. Frenkel, and A. D. Cherniak. *Aleksandr Aleksandrovich Fridman: zhizn' i deiatel'nost'*. Moscow, 1988.

Tseitlin Z. A. "Neskol'ko vozrazhenii A. K. Timiriazeva," *PZM*, 1924, no. 12, pp. 159–67.

———. "O 'misticheskoi' prirode svetovykh kvant," *PZM*, 1925, no. 4, pp. 74–101.

———. "Review of A. K. Timiriazev, 'Fizika,'" *PZM*, 1926, no. 4–5, pp. 227–30.

———. "Review of S. Iu. Semkovskii, 'Dialekticheskii materializm,'" *PZM*, 1926, no. 4–5, pp. 220–28.

———. "Teoriia otnositel'nosti Einshteina i dialekticheskii materializm" (2 parts), *PZM*, 1924, no. 3, pp. 77–110; no. 4–5, pp. 115–37.

Tursunov, Akbar. *Gorizonty kosmologicheskogo zakona (istoriia i sovremennost')*. Moscow, 1969.

———. "Kosmologiia i khristianskaia teologiia: doktrina tvoreniia v svete sovremennoi nauki," in Akhundov and L. B. Bazhenov, eds., *Estestvoznanie*, pp. 136–51.

Umarov, S., and M. Mostepanenko. *Lenin i razvitie sovremennoi fiziki*. Stalinabad, 1960.

Umov, N. A. "Evoliutsiia atoma," *Nauchnoe slovo*, 1905, no. 1, pp. 5–27.

———. "Evoliutsiia fizicheskikh nauk," *Russkaia mysl'*, 1914, no. 2, section 2, pp. 1–27.

————. *Izbrannye sochineniia*. Moscow-Leningrad, 1950.

————. *Sobranie sochinenii*, vol. 3. Moscow, 1916.

————. "Znachenie Dekarta v istorii fizicheskikh nauk," *Voprosy filosofii i psikhologii*, 1898, no. 34, pp. 409–520.

Vallentin, Antonina. *The Drama of Albert Einstein*. Garden City, N.J., 1954.

Vasetskii, G. "O knige A. A. Maksimova 'Ocherki po istorii bor'by za materializm v russkom estestvoznanii,'" *Bol'shevik*, 1948, no. 1, pp. 91–96.

Vasil'ev, A. V. *Nikolai Ivanovich Lobachevskii*. Moscow, 1992.

————. "N. I. Lobachevskii i ego zavety," in *Prazdnovanie Kazanskogo universiteta*, vol. 1, pp. 21–33.

————. *Space, Time, Motion: An Historical Introduction to the General Theory of Relativity*, tr. H. M. Lucas and I. P. Samger. London, 1924.

————. "Vzgliady Ogiusta Konta na filosofiiu matematiki," *Voprosy filosofii i psikhologii*, 1899, vol. 10, no. 4, pp. 540–59.

Vavilov, S. I. "Dialektika svetovykh iavlenii," in Bukharin and Deborin, eds., *Pamiati V. I. Lenina*, pp. 469–84.

————. "Filosofskie problemy sovremennoi fiziki i zadachi sovetskikh fizikov v bor'be za peredovuiu nauku," in Maksimov et al., eds., *Filosofskie voprosy*, pp. 5–30.

————. *Lenin i fizika*. Moscow, 1966.

————. *Lenin i sovremennaia fizika*. Moscow, 1970.

————. "'Materializm i empiriokrititsizm' V. I. Lenina i filosofskie problemy sovremennoi fiziki," *VAN*, 1949, no. 6, pp. 30–39.

————. "Novaia fizika i dialekticheskii materializm," *PZM*, 1938, no. 12, pp. 27–33.

————. "Novye opytnye podverzhdeniia sledstvii obshchei teorii otnositel'nosti,' *UFN*, 1925, vol. 5, pp. 457–60.

————. "Shestoi s"ezd russkikh fizikov," in *Nauchnoe slovo*, 1928, no. 8, pp. 95–101.

————. *Sobranie sochinenii*, vols. 3, 4. Moscow, 1956.

————. "Staraia i novaia fizika," in Bukharin and Deborin, eds., *Pamiati Karla Marksa*, pp. 207–19.

————. "V. I. Lenin i fizika," *Pr*, 1934, no. 1, pp. 35–38.

Vernadskii, V. I. *Filosofskie mysli naturalista*. Moscow, 1988.

————. "Izuchenie iavlenii zhizni i novaia fizika," *OMEN*, 1931, no. 4, pp. 403–37.

————. *Khimicheskoe stroenie biosfery zemli i ee okruzheniia*. Moscow, 1965.

————. *Mysli o sovremennom znachenii istorii znaniia*. Leningrad, 1927.

————. "Po povodu kriticheskikh zamechanii akad. A. M. Deborina," *OMEN*, 1933, no. 3, pp. 395–419.

————. "Problema vremeni v sovremennoi nauke," *OMEN*, 1932, no. 4, pp. 511–41.

————. "Zapiska o vybore chlena Akademiii po otdelu filosofskikh nauk," *Kommunist*, 1988, no. 18, pp. 67–74.

Veselov, M. G. "Vladimir Aleksandrovich Fok (k shestidesiatiletiiu so dnia rozhdeniia)," *UFN*, 1958, vol. 64, no. 4, pp. 695–99.

Vizgin, V. P. "Martovskaia (1936 g.) sessiia A. N. SSSR: sovetskaia fizika v fokuse," *VIET*, 1990, no. 1, pp. 63–84.

————. "O n'iutonovskikh epigrafakh v knige S. I. Vavilova po istorii otnositel'nosti," in Akhundov and Illarionov, eds., *N'iuton*, pp. 184–206.

————. "Odin iz aspektov metodologii Einshteina," *VIET*, 1975, no. 3, pp. 16–24.

————. "On the Road to the Relativity Theory of Gravitation (1900–1911)," in *Soviet Studies in the History of Science* (Moscow, 1977), pp. 135–46.

————. *Reliativistskaia teoriia tiagoteniia: istoki i formirovanie (1900–1915)*. Moscow, 1981.

————. "Rol' idei E. Makha v genezise obshchei teorii otnositel'nosti," in *Issledovaniia: 1989*, pp. 69–83.

————. *Unified Field Theories in the First Third of the Twentieth Century*, tr. J. B. Barbour. Basel, 1994.

Vizgin, V. P., and G. L. Gorelik. "The Reception of the Theory of Relativity in Russia and the USSR," in Glick, ed., *Comparative Reception*, pp. 265–326.

Vlasenko, K. I. "O natsional'nom kharaktere russkoi filosofii," *Vestnik Moskovskogo universiteta*, 2d ser., 1990, no. 1, pp. 40–56.

Vol'kenshtein, M. V. "Krasota nauki," *Nauka i zhizn'*, no. 9, pp. 15–19.

————. "Nauka i estetika (k stoletiiu so dnia rozhdeniia V. M. Vol'kenshteina)," *VF*, 1963, no. 10, pp. 71–82.

————. *Perekrestki nauki*. Moscow, 1972.

Volkov, G. N. *Sotsiologiia nauki*. Moscow, 1968.

Volodin, E. "Lenin i filosofiia: ne postavit' li etu problemu zanovo?," *Kommunist*, 1990, no. 8, pp. 38–59.

Vorontsev-Vel'iaminov, B. A. *Ocherki po istorii astronomii v SSSR*. Moscow, 1960.

Weinberg, Steven. *The First Three Minutes*. New York, 1988.

Whitrow, G. J. *Time, Gravitation and the Universe*. London, 1973.

Yakhot, Jehoshua [I. Iakhot]. "Podavlenie filosofii v SSSR (20–30-e gody)" (3 parts), *VF*, 1991, no. 9, pp. 44–68; no. 10, pp. 72–138; 1991, no. 11, pp. 72–115.

————. "The Theory of Relativity and Soviet Philosophy (A Historical Outline)," *Crossroads*, Fall 1978, pp. 92–118.

Zabelin, I. "O kul'ture myshleniia," *Novyi mir*, 1961, no. 1, pp. 159–65.

Zahar, Elie. *Einstein's Revolution: A Study in Heuristic*. La Salle, Ill., 1989.

Zanchenko, G. A. "Sud'by neopozitivizma i postpozitivizma," *FN*, 1988, no. 2, pp. 56–67.

Zel'dovich, Ia. B. "Teoriia rasshiriaiushcheisia Vselennoi sozdana A. A. Fridmanom," in Friedmann, *Izbrannye trudy*, pp. 402–10.

————. "Tvorchestvo velikogo fizika," *VF*, 1980, no. 6, pp. 32–45.

Zel'dovich, Ia. B., and L. P. Grishchuk. "Obshchaia teoriia otnositel'nosti verna!," *UFN*, 1988, vol. 155, no. 3, pp. 515–27.

————. "Tiagotenie, obshchaia teoriia otnositel'nostii al'ternativnie teorii," *UFN*, 1986, vol. 149, no. 4, pp. 695–707.

Zeldovich, Ia. B., and I. D. Novikov. *Reliativistskaia astrofizika*. Moscow, 1967.

————. *Stroenie i evoliutsiia Vselennoi*. Moscow, 1975.

Zel'manov, A. L. "K postanovke kosmologicheskoi problemy," in *Trudy Vtoroga s"ezda Vsesoiuznogo Astronomo-geodezicheskogo obshchestva, 25 ianvaria 1955 g.* (Moscow, 1960), pp. 72–84.

————. "Kosmologiia," in Ambartsumian et al., eds., *Razvitie astronomii*, pp. 320–90.

————. "Nekotorye voprosy kosmologii i teorii gravitatsii," in *Fizicheskaia nauka i filosofiia*, pp. 275–80.

Zhdanov, A. A. *Vystuplenie na diskussii po knige G. F. Aleksandrova 'Istoriia zapadnoevropeiskoi filosofii,' 24 iiuna, 1947, g.* Moscow, 1951.

Zhdanov, G. B. "Fizika i filosofiia," *Pr*, 1963, no. 11, pp. 3–9.

Zinchenko, V. "Gumanisticheskii vektor nauki," *Kommunist*, 1990, no. 4, pp. 45–50.

Zobov, R. A. "Strukturnnyi podkhod i ego rol' v stroenii kartiny mira," in A. Mostepanenko, ed., *Filosofiia i razvitie*, pp. 198–212.

Zotov, A. F. "Fenomen filosofii: o chem govorit pliuralizm filosofskikh uchenii," *VF*, 1991, no. 12, pp. 14–21.

Index

In this index an "f" after a number indicates a separate reference on the next page, and an "ff" indicates separate references on the next two pages. A continuous discussion over two or more pages is indicated by a span of page numbers, e.g., "57–59." "Passim" is used for a cluster of references in close but not consecutive sequence.

Printed in the USA
CPSIA information can be obtained
at www.ICGtesting.com
JSHW021321221024
72173JS00012B/1633/J

9 780804 742092